Cambridge Studies in Ecology presents balanced, comprehensive, up-to-date, and critical reviews of selected topics within ecology, both botanical and zoological. The Series is aimed at advanced final-year undergraduates, graduate students, researchers, and university teachers, as well as ecologists in industry and government research.

It encompasses a wide range of approaches and spatial, temporal, and taxonomic scales in ecology, including quantitative, theoretical, population, community, ecosystem, historical, experimental, behavioural and evolutionary studies. The emphasis throughout is on ecology related to the real world of plants and animals in the field rather than on purely theoretical abstractions and mathematical models. Some books in the Series attempt to challenge existing ecological paradigms and present new concepts, empirical or theoretical models, and testable hypotheses. Others attempt to explore new approaches and present syntheses on topics of considerable importance ecologically which cut across the conventional but artificial boundaries within the science of ecology.

The ecology of recently-deglaciated terrain: a geoecological approach to glacier forelands and primary succession

CAMBRIDGE STUDIES IN ECOLOGY

Editors
R. S. K. Barnes *Department of Zoology, University of Cambridge, UK*
H. J. B. Birks *Botanical Institute, University of Bergen, Norway*
E. F. Connor *Department of Environmental Sciences, University of Virginia, USA*
R. T. Paine *Department of Zoology, University of Washington, Seattle*

The glacier foreland of Styggedalsbreen in the alpine zone of Jotunheimen, southern Norway.

The ecology of recently-deglaciated terrain

a geoecological approach to glacier forelands and primary succession

JOHN A. MATTHEWS

Department of Geology
University of Wales College of Cardiff (UWCC)

The right of the
University of Cambridge
to print and sell
all manner of books
was granted by
Henry VIII in 1534.
The University has printed
and published continuously
since 1584.

CAMBRIDGE UNIVERSITY PRESS

Cambridge

New York   Port Chester

Melbourne   Sydney

Published by the Press Syndicate of the University of Cambridge
The Pitt Building, Trumpington Street, Cambridge CB2 1RP
40 West 20th Street, New York, NY 10011-4211, USA
10 Stamford Road, Oakleigh, Victoria 3166, Australia

First published 1992

Printed in Great Britain at the University Press, Cambridge

A catalogue record for this book is available from the British Library

Library of Congress cataloguing in publication data

Matthews, John A.
The ecology of recently-deglaciated terrain : a geoecological
approach to glacier forelands and primary succession / John A. Matthews.
   p. cm. – (Cambridge studies in ecology)
Includes bibliographical references and index.
ISBN 0-521-36109-5
1. Plant succession. 2. Landscape ecology. 3. Glacial landforms.
4. Soil formation. I. Title. II. Title: Recently-deglaciated
terrain. III. Title: Glacier forelands. IV. Series.
QK910.M38 1992
581.5′264–dc20 91-25109 CIP

ISBN 0 521 36109 5 hardback

To Dr Rose Maria D'Sa–
who continues to distract and enchant,
and to maintain that there is more to life than glacier forelands

Contents

Preface

During a scientific career so far spanning 22 years, I have chosen to concentrate my research efforts on recently-deglaciated terrain (glacier forelands). The principal vehicle for this research has been 20 Jotunheimen Research Expeditions to the glacier forelands of southern Norway, which over the years have involved many staff and graduate researchers from numerous institutions and several disciplines. The expeditions have been driven by my belief in glacier forelands as field laboratories (microcosms) for a wide variety of fundamental research on the natural environment. As such, glacier forelands have a scientific importance out of all proportion to their areal extent. This book has grown out of the ecological aspects of this work, and constitutes Jotunheimen Research Expeditions Contribution No. 100.

The aim of the book is twofold: first, to review thoroughly and for the first time, available information on the ecology of recently-deglaciated terrain; second, to appraise critically the methodology currently employed in such studies and to contribute to the development of a new approach in this field (the geoecological approach). Ecological research on glacier forelands has played an interesting and important role in the history of ecology, especially in relation to theories of plant succession and soil development. The particular attraction is the existence of terrain of increasing age with increasing distance from a retreating glacier (a chronosequence). This enables the inference of ecological changes over timescales of decades and centuries: beyond the span of research grants and even research careers. Much of the book is therefore centred on an examination of theories of plant succession and related concepts in the light of the available data from glacier foreland chronosequences.

In reviewing ecological research on glacier forelands, I have tried to be as comprehensive as possible and to produce a definitive survey of the field. This includes at least a mention of all the relevant work of which I am aware. A world-wide literature is involved, ranging from the maritime regions of Europe and North America to the interior of Asia, and from Antarctica to the equatorial glaciers of Mount Kenya. There is, however, an obvious emphasis on Europe and North America, where the majority of research in this field has been carried out. There are unlikely to be major omissions, except in respect of research from the U.S.S.R. Nevertheless, I would be very interested and grateful to receive information (and hopefully copies of publications) relating to any studies that have been omitted.

I have deliberately avoided the fashionable habit of citing only recent references. The contribution and originality of the pioneers in the field are being ignored by this practice, much work of lasting value is thereby being neglected, and information of possible critical importance in the future is being lost. Where appropriate I therefore refer to older studies alongside the new. This not only gives a truer picture of the state of the science but also helps keep recent work in perspective. Throughout, taxonomic nomenclature follows that of the original author; this applies to both species' names and soil horizons. Non-standard units have, however, been converted to their SI equivalents.

Despite my wish to be comprehensive, I have had to be selective (as in any book) and have developed a particular theme. In the book I examine the thesis that the ecology of recently-deglaciated terrain in general, and of the vegetation and soils in particular, can only be fully understood when viewed as part of the developing landscape. In this geoecological approach, the spatial variation and dynamics of vegetation and soils is considered in relation to other aspects of the landscape (such as topographic variation, climate and geomorphic processes). Ecological phenomena are thereby explained in terms of physical as well as biological processes. Thus, the geoecological approach contrasts with most previous work in this field which has, in my view, overemphasized biological aspects with consequent neglect of vegetation–environment interactions and habitat change.

The development of a geoecological approach justifies the inclusion of chapters 2 and 3. After a short introductory chapter, I consider in chapter 2 the characteristics and limitations of glacier foreland timescales, and the basis for any inference of ecological change from chronosequences. Chapter 3 contains a detailed account of the physical landscape and the physical processes likely to be of ecological relevance. Further reading will show just how essential this material is to a complete understanding of glacier foreland ecology. Its relevance is perhaps most immediately apparent in relation to soil development, which is the topic of chapter 4. Those readers primarily interested in soil chronosequences will find chapter 4 fairly self-contained; those primarily interested in vegetational chronosequences will find it a useful bridge between the physical and the biological landscape.

The second half of the book deals explicitly with plant succession. Patterns and environmental relationships are considered in chapter 5, which leads on to a discussion of processes and models culminating in the geoecological model (chapter 6). In the context of glacier forelands, and in contrast to the geoecological model, existing models of succession are shown to be extremely limited in their applicability. In conclusion, chapter 7 sums up the ecological significance of recently-deglaciated terrain, emphasizing some areas for future research and pointing to some broader implications.

Current research is far from exploiting the full potential of glacier foreland landscapes for ecological purposes. It is likely, therefore, that they will remain of considerable importance on the ecological research agenda for some time to come. This book and the geoecological model will provide a useful framework and a stimulus for future research on the ecology of recently-deglaciated terrain. As a geoecological approach is likely to be appropriate in many other ecological situations, the book should also be of considerable methodological interest to a wide range of ecologists.

John A. Matthews   Storbreen

Acknowledgements

I must first thank those, many of whom have been my post-graduate students, who have been colleagues and companions on the Jotunheimen Research Expeditions, 1970–91. They have not only sustained and deepened my own interest and understanding of recently-deglaciated terrain but also have contributed greatly in their own right. This should be clear from the extensive use I have made of their work throughout this book.

Second, I thank John Birks for the invitation to write the book in the Cambridge Studies in Ecology Series and for his encouragement and support in this task. An author could not wish for a better editor or publisher.

Third, several colleagues read the whole (or substantial parts) of the manuscript and suggested many improvements, most of which have been incorporated: H. John B. Birks, Helena J. Crouch, Steven Ellis (chapter 4), John E. Etherington, Jacqui Foskett, Donald B. Lawrence, Danny McCarroll (chapters 1–3), Mark G. Noble, Lawrence R. Walker (chapters 5 & 6) and Robert J. Whittaker.

Fourth, I have made major use of previously published figures and tables, most with little or no alteration. They are acknowledged individually where they appear in the text. In addition, the following provided unpublished information or photographs: Peter J. Beckett, Richard Bickerton, H. John B. Birks, David J. Blundon, Colin J. Burrows, Reidar Elven, Knut Fægri, Bruce E. C. Fraser, Anne F. Gellatly, Donald B. Lawrence, Antony Mellor, Anne C. Messer, Mark G. Noble, Chadwick D. Oliver, Wilfred H. Theakstone, Ole Vetaas, Lawrence R. Walker, and Robert J. Whittaker.

Others, too numerous to mention, have over the years and with increasing frequency during the writing of the book, provided me with relevant references or other information. Christine Lawes, of the Interlibrary Loans Section of the Science Library (University of Wales, Cardiff) is due a special mention for obtaining numerous, often obscure, references, and Heike Neumann was most helpful in translating those written in German.

1

Introduction

*Succession – nothing in plant, community or ecosystem ecology
has been so elaborated by terminology, so much reviewed, and yet
so much the centre of controversy.*

(West, Shugart & Botkin, 1981: iv)

*Nowhere can succession be studied more profitably than in the
valley below the front of a large glacier.*

(Ellenberg, 1988: 440)

1.1   Glacier forelands and simplicity

Ecologists, like all natural environmental scientists, must simplify
in order to make progress. Simplification can take many forms. One
approach is to seek a relatively simple landscape where it is possible to
exclude, or at least significantly reduce, the effects of many of the interactions that occur in more complex landscapes. Although it is not possible to
obtain perfect experimental control in such field laboratories (or microcosms), the approach enables a reduction in levels of interference. In this
way it is possible to focus more clearly on the interactions and to progress
towards a comprehensive understanding of the whole.

Glacier forelands are considered in this book as a unique type of field
laboratory. A glacier foreland is the area of newly-formed landscape in
front of a glacier, which was recently ice covered but has since been exposed
by glacier retreat (Fig. 1.1). The term 'glacier foreland' is derived from the
German *gletschervorfeld* (Kinzl, 1929) and was introduced into the English
language by Beschel (1961). It is used here to define the land area exposed in
historical times, since the glacier maximum of the 'Little Ice Age'. This is the
glacier foreland *sensu stricto*, the glacier foreland in the narrow sense of
Holzhauser (1982). Thus, the glacier foreland often forms a distinct zone of
relatively bare terrain extending up to many kilometres from the margin of
the glacier (Fig. 1.2). This recently-deglacierized zone (deglaciated by
modern glaciers) should be distinguished from the more extensive and older
landscapes deglaciated earlier in the Quaternary.

There are several fundamental advantages of using glacier forelands as
field laboratories for ecological purposes. First, their restricted physical size

1

Fig. 1.1. Recently-deglaciated terrain, Storbreen glacier foreland, Jotunheimen, photographed in 1985. Bouldery terrain in the foreground (with moraine ridges) was deglaciated about 140 years ago.

facilitates comprehensive investigation. Second, with a relatively severe climatic environment, they support relatively simple ecosystems. Third, recently-deglaciated terrain has experienced only a short history of modification by changing natural environmental processes. Last, but by no means least, with increasing distance from a retreating glacier, a longer time period has been available for ecosystem development: hence, the pattern of ecosystems on the glacier foreland is commonly interpreted as a spatial representation of temporal change and as a vast natural experiment.

1.2   Ecology and primary succession

The first detailed study of glacier foreland ecology appears to have been made in the European Alps by Coaz (1887), who visited the Rhonegletscher in the summer of 1883. Coaz found *Saxifraga aizoides* growing on ground exposed from beneath the retreating glacier for less than three years. On progressively older terrain, more species were found, with a total of 70 species from 18 families on ground deglaciated for no more than 10 years, including 39 on the oldest terrain examined. Observations such as these, which indicate different species colonizing progressively older terrain, have inspired many ecologists with an interest in the time-dependent ecological process of primary succession. Other early studies

Fig. 1.2. Vertical aerial photograph of the Storbreen glacier foreland, Jotunhei-men, southern Norway (Widerøe's Flyveselskap A.S., 1968). The outermost of the prominent sequence of end-moraine ridges, which extends about 1.5 km from the glacier snout, defines the glacier foreland boundary and indicates the position of the glacier about A.D. 1750. See also Fig. 2.4.

include those in the European Alps (Lüdi, 1921, 1945; Frey, 1922; Braun-Blanquet & Jenny, 1926; Negri, 1934, 1936; Friedel, 1934, 1937, 1938; Oechslin, 1935), the Pyrenees (Davy de Virville, 1929), North America (Butters, 1914; Cooper, 1916, 1923a–c, 1931, 1939) and Scandinavia (Fægri, 1933). More recently, research has intensified, and other aspects of ecosystem development (particularly soils) have been investigated.

Succession, the complex of processes producing gradual, directional changes in the species composition and structure of ecosystems, is commonly encountered, yet it is far from fully understood. Indeed, Cooper (1926), a pioneer in the study of the ecology of recently-deglaciated terrain in North America, proposed that the universality of change should be regarded as a fundamental fact of ecology. Today, a variety of largely unsubstantiated succession theories remain one of the basic conceptual underpinnings of the subject (McIntosh, 1981, 1985; Miles, 1987). A recent survey of opinion amongst members of the British Ecological Society has revealed that succession is regarded as the most important ecological concept after that of the ecosystem itself (Cherrett, 1989).

There is also a continuing and increasing practical interest in succession and related concepts of stability, especially in the context of human-induced primary and secondary succession, land restoration and conservation (Bradshaw & Chadwick, 1980; Cairns, 1980; Jordan, Gilpin & Aber, 1987; Salzberg, Fredriksson & Webber, 1987; Luken, 1990). Even though succession is of fundamental importance, it is only rarely amenable to conventional observation and experimentation because its timescale ($10–10^3$ years) is too long. Primary succession, whereby a newly-formed land surface devoid of life is colonized and develops a new ecosystem, generally operates over a longer time period than the related process of secondary succession. The latter is initiated from at least a vestige of an existing ecosystem, tends to proceed at a faster rate, and may differ qualitatively from primary succession. Thus sites and methodologies which permit the detailed investigation of succession in general and of primary succession in particular are essential to the advancement of both pure and applied ecology.

The potential of glacier forelands for the study of primary succession and related aspects of ecosystem development was appreciated only when observations of these sites were made with the aid of an appropriate methodology. By substituting space for time, Coaz (1887) and later workers realized that increasing distance from the margin of a retreating glacier could be interpreted as representing a temporal sequence (chronosequence) in ecosystem development. Combined with absolute dating of terrain age,

quantitative rates of change can also be inferred. However, this methodology of space-for-time substitution is based on some largely untested
assumptions and is not without limitations (Pickett, 1988). The aim of this
book is therefore to review not only the available information on the
ecology of recently-deglaciated terrain and its relevance to concepts and
theories of succession but also the methodology used. The remainder of this
chapter provides an introduction to this methodology and to the distinctive
approach adopted in this book.

1.3   Space-for-time substitution (chronosequences)

The most influential research on the ecology of recently-deglaciated terrain, at least in the Anglo-American literature, has undoubtedly
been that from Glacier Bay, Alaska, where a series of studies on plant
succession was pioneered by Cooper (Cooper, 1923a–c, 1931, 1939).
Utilising a well-documented record of glacier retreat, three major vegetation types (pioneer community, willow–alder thicket and spruce forest)
were recognized, and a successional scheme has been elaborated to include
up to eight intergrading successional stages (Lawrence, 1958; Decker, 1966;
Reiners, Worley & Lawrence, 1971; see section 5.1.1). The research of
Crocker & Major (1955) in particular is widely cited, even in textbooks (e.g.
Krebs, 1985; Kershaw & Looney, 1985; Colinvaux, 1986; Begon, Harper &
Townsend, 1990), as they link vegetation changes to soil development and
thereby provide evidence of a mechanism for succession. For example, they
suggest that soil acidification and nitrogen accumulation under Sitka alder
(*Alnus crispa*) precede the invasion of Sitka spruce (*Picea sitchensis*). The
reaction of alder on the soil, which facilitates invasion by spruce (cf. the
facilitation model of Connell & Slatyer (1977), is invoked as the mechanism
of change. Thus the research at Glacier Bay has been interpreted as
supporting the classical view of vegetation-controlled or autogenic succession (Tansley, 1935), driven by reaction mechanisms (Clements, 1928) (but
see section 6.1.4).
Most ecologists have interpreted the vegetation patterns on recently-
deglaciated terrain in much the same way, though few have provided good
evidence for mechanisms of change. Nevertheless, valuable data have been
collected during studies in many different parts of the world. These data
together constitute one of the main sources of empirical information on
primary succession and soil development. However, the chronosequence
methodology employed in most of this research has been rather restrictive.
The concept of a chronosequence as a spatial representation of a

temporal sequence has been systematized by Jenny (1941, 1946) in the context of soil development. A similar concept and systematization in the context of vegetation succession forms part of Major's (1955) functional factorial approach to plant ecology (see also Perring, 1958; Jenik, 1986), and the concept has since been extended to ecosystem development (Olsson, 1958; Jenny, 1961, 1980). Chronosequences are spatial sequences in which environmental factors other than time are assumed to be unimportant, either because they are invariant or because they are relatively ineffective.

In front of a retreating glacier, the time factor is obviously a major control on spatial patterns in the vegetation and soils. Immediately after deglaciation the terrain is normally viewed as being devoid of living organisms (but see section 6.1.4). The ecosystems developed on older ground probably began from a broadly similar state. Selection of comparable sites (in terms of climate, parent material, relief, etc.) for the description and analysis of chronosequences reduces the effect of environmental factors other than time. In addition, the time factor can often be quantified quite precisely in terms of terrain age. A chronosequence, however, is not identical to a successional sequence at a single site. Environmental factors vary to some extent between sites in a chronosequence (and through time at particular sites) and terrain age does not always correspond precisely with the time elapsed since deglaciation.

1.4   Geoecology (landscape ecology)

Although the objective of most research has been to describe chronosequences and, more recently, to define chronofunctions quantitatively, many observations and results suggest that an interpretation of the ecology of recently-deglaciated terrain merely in terms of chronosequences is an oversimplification. Even Coaz (1887: 11) recognizes the importance of such factors as snow and avalanches, cold winds, shifting glacial meltwater streams and torrential rains as effective influences on the vegetation. Similarly, later workers have recognized this problem. For example, where vegetation maps have been produced they do not show a simple vegetation zonation corresponding to the pattern of deglaciation (Friedel, 1938; Richard, 1968; Jochimsen, 1970, Matthews, 1979a). Such maps demonstrate a complex mosaic of communities (see section 5.3.1). They also suggest the need for a broader perspective for studies of the ecology of recently-deglaciated terrain. Such a perspective is provided by geoecology (landscape ecology).

The term 'landscape ecology' was first used by the German geographer C. Troll, who later coined the term 'geoecology' to include not only the study of ecosystems but also all the phenomena and interacting processes of the natural (and cultural) landscape (Troll, 1939a, b, 1971, 1972). Although landscape ecology has been recently rediscovered in America (e.g. Risser, Karr & Forman, 1984; Forman & Godron, 1986; Urban, O'Neill & Shugart, 1987), the concept of landscape (German: *landschaft*) and the principles of geoecology (landscape ecology) have a long tradition in Europe (Naveh & Lieberman, 1984; Schreiber, 1990). Geoecology shares common ground with the study of ecosystems and has an even closer relationship with the Soviet field of biogeocoenology (Sukachev & Dylis, 1964). However, the concept of landscape differs from that of ecosystem in giving greater emphasis to spatial organization on the Earth's surface and to the interaction of ecological processes within a broader framework of environmental processes. The emphasis given to spatial organization of the landscape has been termed the horizontal (geographical) aspect of geoecology; the vertical (biological–ecological) aspect emphasizes functional interrelationships at particular sites (Troll, 1971). Both aspects are necessary to obtain a comprehensive understanding of any landscape.

From the viewpoint of geoecology, ecosystem succession in front of a retreating glacier is clearly part of the developing landscape – or landscape succession (Troll, 1963) – in which plant life changes together with the animals, the soil and the abiotic environment. The thesis that newly-developing ecosystems in front of retreating glaciers can be fully understood only when viewed as part of the developing landscape is examined further in this book. The geoecological approach that has been developed aims to provide a more comprehensive, integrated and interdisciplinary treatment of the ecology of recently-deglaciated terrain than has been attempted before. The areas in front of retreating glaciers are particularly appropriate for such a holistic approach because of the almost unique opportunity to define an accurate timescale.

2

The nature of the timescale

The ecological significance of recently-deglaciated terrain depends in large measure on the existence of a timescale. It is appropriate, therefore, that the nature and limitations of the timescale be fully assessed.

Few landscapes exist where even the relative age of the land surface exhibits a simple pattern or can be established with any certainty. Distance from the margin of a retreating glacier is exceptional in that it is directly related to terrain age and therefore represents an index of relative age. Because glacier retreat rates are rarely constant, however, the relationship between distance and time is non-linear and more information is necessary to obtain absolute (numerical) age estimates. Derivation of an absolute timescale requires an observed history of glacier variations or the application of dating techniques. A full understanding of the nature and limitations of the timescale therefore demands an appreciation of the extent and timing of glacier variations, the range and accuracy of available dating techniques, and their interaction in the context of specific glacier forelands.

2.1   Glacier variations

The size of a glacier depends essentially on the balance between winter accumulation and summer ablation and is therefore determined by climate (Meier, 1965; Andrews, 1975; Porter, 1981a). For a given climatic environment an equilibrium size and an equilibrium profile may be considered (Fig. 2.1a). The equilibrium profile is maintained by the flow of ice from areas of net accumulation to areas of net ablation. This condition theoretically pertains after a number of years with a similar climate and is characterized by a stationary ice margin. However, because of climatic variability and change the ice margin, particularly at the glacier terminus, is rarely stationary.

Although the relationship between climatic change and glacier variations is not simple (Meier, 1965; Paterson, 1981; Sutherland, 1984; Oerle-

(a)

Net mass balance
Flow line

Stationary
terminus

(b)

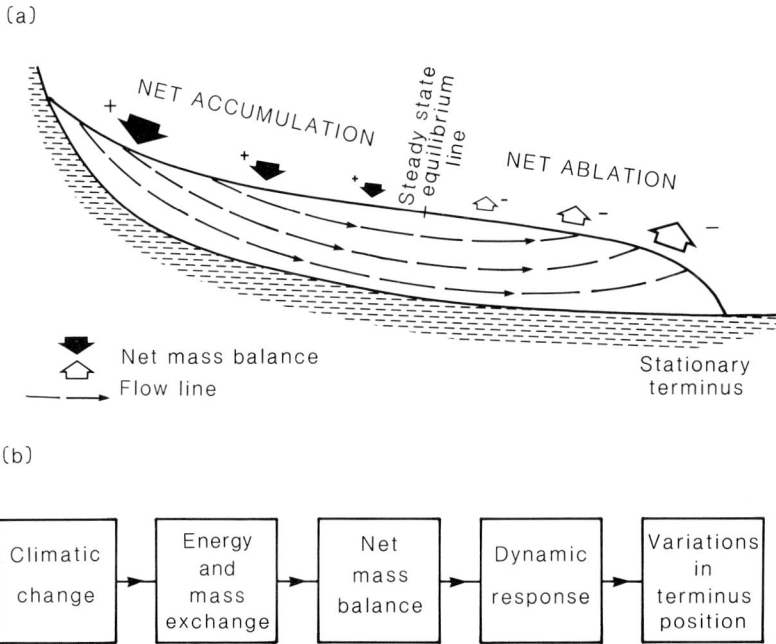

Fig. 2.1. Some theoretical aspects of the relationship of a glacier to climate: (a) mass balance and ice flow of a glacier in steady-state equilibrium with climate (from Sugden & John, 1976); (b) the dynamic response of a glacier to climatic variation (from Meier, 1965).

mans, 1989), it may be envisaged as a chain of distinct processes (Fig. 2.1b). A change in climate alters the local mass and energy exchange processes at the surface of the glacier, which produce changes in accumulation and/or ablation and hence net mass balance. The gain or loss of mass determines the dynamic response of the glacier, generally resulting in a change in glacier thickness and the rate of flow, an adjusted surface profile and an advance or retreat of the terminus. Complications include the existence of negative (damping) and positive (amplifying) feedback relationships, varying response times (the period during which adjustment to a mass balance change occurs), and varying lag times (the interval between maximum or minimum mass balance and maximum or minimum terminal position). Factors such as the size, surface slope and thermal characteristics of the glacier are effective influences. For example, response times for small glaciers generally vary from several years to decades; for large glaciers and ice sheets, values of hundreds or even thousands of years have been calculated (Paterson, 1981).

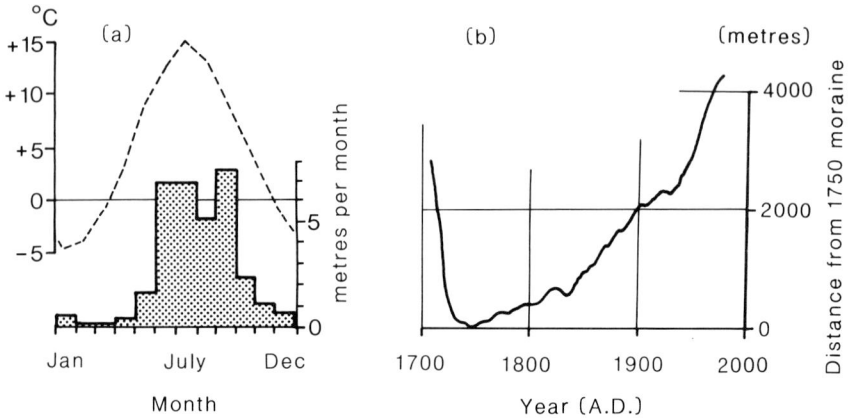

Fig. 2.2. Short- and long-term advance and retreat of the terminus of Nigards-breen, southern Norway: (a) mean monthly variations, October 1935 to July 1947 (during a period of rapid glacier retreat) compared with mean monthly temperatures at Fjærland (from Fægri, 1950); (b) longer-term variations since A.D. 1710, based on annual measurements and documentary sources, and including a major advance followed by a major recession (from Østrem, Liestøl & Wold, 1976).

Variations in the position of the glacier terminus are only partly determined by the dynamic response of the glacier to changes in mass balance described above. Major advances and retreats can be explained in such terms but minor glacier fluctuations often demonstrate a more direct and immediate response to seasonal and annual climatic fluctuations. Consider the equilibrium profile in Fig. 2.1a. The position of the glacier terminus depends on the ice velocity and the rate of ablation. The glacier advances when the former exceeds the latter and retreats when the reverse case is true. It would be expected, therefore, that a glacier in equilibrium with climate might advance in winter, when the terminus is snow covered, and retreat in summer, when ablation is at a maximum. In a year with above average ablation (as would occur, for example, in an abnormally warm summer or in a year with reduced winter snowfall) the summer retreat might be accentuated (Fig. 2.2a) and in a year with below average ablation the summer retreat would be less. Similarly, a run of years with reduced ablation might produce a short-term glacier advance during a period when the long-term dynamic response of the glacier is to retreat (Fig. 2.2b). Thus, for glaciers that are responsive to these direct seasonal and annual climatic fluctuations, ratchet-like advances and retreats can be envisaged.

The seasonal, annual and longer-term advance or retreat of a glacier does not depend wholly on climate. Large glaciers not only react more slowly to a change in climate but also they tend to be less responsive in the shorter-term (Karlén, 1973). The morphology of the glacier and the topography of the surrounding terrain can also be of significance. For example, more rapid retreat will take place across a land surface sloping away from the glacier terminus than across a horizontal surface or a surface sloping towards the glacier (Price, 1980).

In some glaciers, dramatic increases in ice-flow velocity (surges) are known to occur, producing abnormally great glacier advances within a few months or years. Normally such surges are periodic with intervening quiescent or stagnant phases of several decades (Meier & Post, 1969). Other destabilizing factors that have been recorded include landslides, volcanic activity, earthquakes and calving into fjords and lakes (Reid, 1969; Marangunic, 1972; Brugman & Post, 1981; Mann, 1986; Porter, 1989). Calving of icebergs at the terminus of tidewater glaciers produces particularly rapid glacier retreat rates (as, for example, at Glacier Bay, Alaska).

A combination of such local climatic and non-climatic factors can produce a very different pattern and timing of glacier variations in glaciers that are experiencing a similar regional climate. Variations in neighbouring glaciers may be out of phase for considerable periods (Nichols & Miller, 1952; Gellatly, Whalley & Gordon, 1986; Dugmore, 1989). Within the same region, however, glaciers tend to respond in a broadly similar way to regional climatic trends.

Fortunately for ecological studies, the dominant trend in glacier variations over at least the last century has been one of glacier retreat (Ahlmann, 1953a; Grove, 1988). The major 'Little Ice Age' glacier expansion and contraction episode of the last few centuries was a world-wide phenomenon, even though there were considerable regional differences in the precise timing of the event (Grove, 1979, 1988). Shorter-term advances superimposed on the general glacier retreat from 'Little Ice Age' maximum positions appear to have been broadly synchronous within hemispheres and generally synchronous within regions (Porter, 1981b, 1986). However, the recognition of world-wide synchroneity in major glacier expansion episodes earlier in the Holocene (e.g. Denton & Karlén, 1973; Begét, 1983; Röthlisberger, 1986) would seem to be premature.

2.2 Dating techniques

The nature of glacier variations implies that distance from a retreating glacier is very unlikely to provide more than a crude index of relative terrain age. Many dating techniques are potentially of value in improving such timescales and especially by providing absolute (numerical) age estimates. These have been classified in a variety of ways (Colman, Pierce & Birkeland, 1987). In this section, techniques are included only if they have been found useful in dating land surfaces in front of retreating glaciers and hence are of relevance to the ecology of glacier forelands. These are considered within three categories: (1) dating based on historical sources, including observation and measurement; (2) dating based on biological processes; and (3) physico-chemical dating techniques.

2.2.1   Historical sources

Where available, historical sources must be regarded not only as the most accurate means of reconstructing very recent glacier variations and terrain ages but also as the basis for calibration of many other techniques that extend the range of the timescale back in time. This kind of evidence cannot be accepted uncritically, however, a point that is particularly pertinent where some of the older documentary sources are concerned.

The oldest historical documents relating to glacier variations appear to be those dating from the time of the Icelandic Sagas (A.D. 870 to A.D. 1264), which have been interpreted as indicating that the glaciers of Iceland were considerably smaller at that time than today (Ahlmann, 1953a, b). Most of this material, as in the case of legends and traditions in the Alps and Norway, does not stand up to modern standards of historical analysis and data verification (Ingram, Underhill & Farmer, 1981; Hooke & Kain, 1982). More reliable historical sources are available from more recent times, often a fortuitous outcome of glaciers impinging on a wide variety of human activities.

Evidence of glacier variations has resulted from activities as diverse as farming, science, art and mountaineering. Agricultural land has been abandoned as a result of glacier advances in Iceland, Norway and the Alps. In the Alps, archaeological evidence in the form of aqueducts has been recently exposed by retreat of the Aletschgletscher (Holzhauser, 1984a). In Norway, the 'Little Ice Age' advance of glaciers such as Nigardsbreen (Fig. 2.2b) has been graphically described and dated in testimonies made to local courts of enquiry (Grove, 1985, 1988). These were set up in the eighteenth century shortly after the event to determine tax concessions for farmers whose meadows and, in some cases, farm houses had been overridden by advancing glaciers.

From the seventeenth century onwards, a succession of visitors to the Alps left an accumulating legacy of literary and pictorial evidence, including paintings, drawings, prints and photographs. This legacy provides an unparalleled basis for the reconstruction of eighteenth and nineteenth century glacier variations (Le Roy Ladurie, 1972). A particularly good example of the critical analysis necessary to establish a detailed record is provided by investigations of the Grindelwaldgletscher in the Swiss Alps (Zumbühl, 1980; Zumbühl, Budmiger & Haeberli, 1981). For the Lower Grindelwald Glacier alone, 323 illustrations (pictorial views) dating from the nineteenth century or earlier were consulted. Together with literary evidence, these have enabled the reconstruction of a more-or-less complete record of variations in the position of the glacier terminus since A.D. 1590. A limited quantity of high-quality documentary evidence of this kind is available from other glacierized regions, such as southern Norway. Nowhere other than the Alps, however, has it proved possible to determine the historical variations of so many glaciers, in so much detail, over so long a timescale and with such continuity.

Observation and measurement of glaciers specifically for scientific purposes also began in the Alps where, under the direction of Louis Agassiz, M. J. Wild constructed in A.D. 1842 the first precise topographic map of a glacier to a large scale (the Unteraar glacier, 1:10,000) (Zumbühl, Budmiger & Haeberli, 1981). Repeated accurate mapping of glacier termini may be considered as the ultimate approach to the establishment of glacier variations. However, the effort and cost expended to produce such maps has, with notable exceptions (Brunner, 1987), led to their infrequent production in practice. The most common approach to this problem has involved the measurement of distances to the glacier terminus along one or more sighting lines from a fixed mark or marks in front of the glacier. Whilst this has the disadvantage that a sighting line may not yield variations representative of all points along the glacier terminus, it has the advantage of providing a rapid and cost-effective way of obtaining quantitative data at frequent intervals.

Annual measurements of glacier variations are carried out in this way in many countries. In Switzerland, some 50 to 100 glaciers have been measured since A.D. 1890 and the Swiss Glacier Commission's observation network presently comprises 116 glaciers (Aellen, 1981). Although fewer glaciers are monitored in this way outside the Alps, most detailed records of twentieth century glacier variations have been obtained by variants of this technique (see, for example, Figs 2.2a & 2.2b) often supplemented by terrestrial photographs, occasional maps and, more recently, vertical aerial photographs. Most recent of all is the use of satellite imagery (e.g. Williams, 1986), which should permit a much larger number of glaciers to be monitored in the future and may prove particularly useful for glaciers in inaccessible regions.

2.2.2   Biological dating

In the absence of accurate historical information, other methods are necessary for determining terrain age. Two methods based on time-dependent biological processes have been found particularly useful for dating land surfaces deglaciated over the last few centuries. These are dendrochronology and lichenometry.

Glaciers commonly descend below the tree line in regions with a relatively maritime temperate climate, where the basis for the successful application of dendrochronology is the annual pattern of growth increments exhibited by many tree species. In the context of recently-deglaciated terrain, dating relies heavily on determining: (1) the age of living trees; (2) the date of interruptions to the normal growth pattern of living trees; and (3) dates of establishment, death or growth-rate variations from tree remains (relics) (Lawrence, 1950; Alestalo, 1971; Shroder, 1980; Schweingruber, 1988).

As trees are normally younger than the land surface on which they are growing, the age of the oldest living tree provides a minimum estimate of the time elapsed since deglaciation of the surface and hence a minimum estimate of terrain age. This technique has been widely used on glacier forelands in western North America (e.g. Heusser, 1966; Sigafoos & Hendricks, 1961, 1972; Bray & Struik, 1963; Heikkinen, 1984a). In front of Nisqually Glacier, for example (Fig. 2.3), the oldest tree inside the outermost moraine provides a minimum date for the moraine of A.D. 1842. Older trees immediately outside the moraine indicate that the glacier was no larger than in 1842 for at least the preceding 186 years, whereas the oldest trees on the younger terrain indicate the pattern and timing of deglaciation. The main source of dating inaccuracy seems to be in the time taken for the trees to establish (ecesis), which varies according to such factors as the availability of seeds and the presence of a suitable seed bed (Lutz, 1930; Sigafoos & Hendricks, 1969). On recently-deglaciated terrain in the Canadian Rockies, for example, ecesis has been estimated to require from 10 to 15 years at favourable sites on Mount Edith Cavell to about 80 years on a blocky moraine at the Bennington Glacier (Luckman, 1986).

Additional problems include those of coring (such as coring above the tree bole or missing the pith), of locating the oldest tree, and of the possible survival of trees deposited from supraglacial positions. Luckman (1986) gives an extreme example of the difficulty experienced in finding the oldest

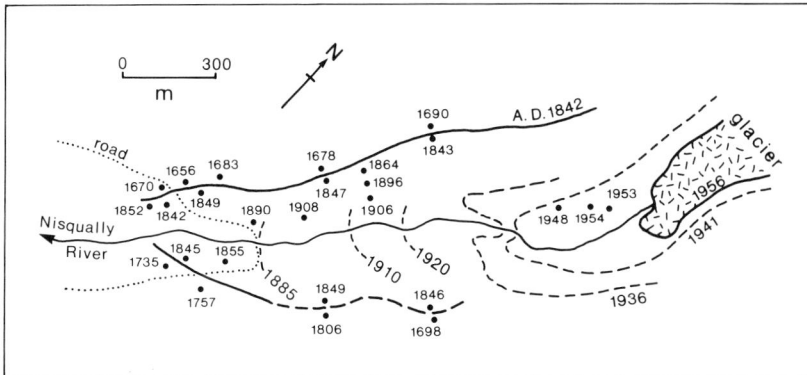

Fig. 2.3. The date of establishment of the oldest living trees in front of Nisqually Glacier, Mount Rainier, Washington, U.S.A. The outermost moraine (A.D. 1842) and younger ice-margin positions (dated) are also shown (from Sigafoos & Hendricks, 1961)

tree on ground deglaciated by the Penny Glacier in the Premier Range of British Columbia. At xeric, exposed sites on moraines in the centre of the valley, trees aged 77 years were 0.92 m tall with a basal diameter of 3.8 cm; in a sheltered, moist site on the distal slope of a lateral moraine of the same age, trees in excess of 40 m high and 90 cm basal diameter were only 84 years old. In Alaska (Stephens, 1969; Post & Streveler, 1976, Noble, 1978), Yukon Territory (Birks, 1980a) and Chile (Rabassa, Rubulis & Suarez, 1981; Veblen *et al.*, 1989) trees have been observed growing on the surface of glaciers with a thick cover of supraglacial debris. Clearly, the survival of such trees on deposition of the moraine could lead to the unusual case of an overestimate of terrain age from tree age.

A second group of dendrochronological techniques seeks to date glacially-induced variations in tree growth rate. This approach is particularly useful for dating the maximum of a glacier advance when the glacier may disturb a tree but fail to kill it. Such trees are readily recognizable where they have been tilted by the advance. Good examples have been described by Heikkinen (1984b) and Luckman (1986). Eccentric growth rings tend to produce trunk curvature following tilting and permit precise dating of the disturbance event. Other possible datable responses to physical disturbance or damage include abrupt reductions in growth and hence the thickness of growth rings (Heikkinen, 1984a), the production of ice-pressure scars (Luckman, 1988) and layering (Bray & Struik, 1963).

Growth suppression has also been observed following the approach of the ice during a glacier advance (Mathews, 1951; Holzhauser, 1985;

Villalba *et al.*, 1990). Bray & Struik (1963) cite examples of both suppression and stimulation of growth. Growth suppression appears to have been caused by the effect of the proximity of the glacier on the local climate of the tree. Growth stimulation, on the other hand, was probably produced by reduced competition following the destruction (thinning out) of surrounding trees. The direct effect of a glacier on the local climate should be distinguished from regional climatic changes to which both glaciers and trees may respond independently (cf. LaMarche & Fritts, 1971; Matthews, 1976a, 1977a; Tessier, Coûteaux & Guiot, 1986).

Trees killed by a glacier advance may be cross-dated with living trees or may be dated independently by ^{14}C dating. This third group of dendrochronological techniques has been intensively applied in the Alps to date relict tree trunks and *in situ* stumps exposed by retreating glaciers, and to date earlier glacier and climatic variations inferred from tree-growth variations (Röthlisberger, 1976; Röthlisberger *et al.*, 1980; Furrer & Holzhauser, 1984; Holzhauser, 1984b). The chronology of Alpine glacier and climatic variations has been extended by this research to cover most of the Holocene. In many other regions these techniques have contributed to the reconstruction of glacier variations beyond the life-span of living trees (e.g. Goldthwait, 1963; McKenzie & Goldthwait, 1971; Worsley, 1974a; Worsley & Alexander, 1976a; Ryder & Thomson, 1986; Porter, 1989).

Lichenometric dating has proved at least as useful as dendrochronology for determining glacier foreland timescales. Indeed, within arctic–alpine environments (often with a dearth of historical evidence and in the absence of trees) it is often the only feasible method for the determination of terrain age at the levels of accuracy required for detailed ecological studies. Under cold, dry continental conditions in West Greenland some crustose lichens reputedly live for 4500 years or more and may therefore be amongst the longest-living organisms (Beschel, 1958). This is an extreme example, however; in the majority of cases, lichenometry as an absolute dating technique has a useful range of less than 500 years (Innes, 1985a).

Lichenometric dating was developed in the Alps by Beschel (1950, 1957, 1961) specifically in the context of glacier forelands. It has since been widely applied on recently-deglaciated terrain using, in particular, the yellow-green species group of the genus *Rhizocarpon*. This has commonly been described as *Rhizocarpon geographicum* in the lichenometric literature (Innes, 1985b). There are two basic variants of the technique, both of which depend on measuring the size of lichen thalli on boulders or bedrock surfaces.

The 'indirect' variant proceeds on the assumption that a numerical relationship can be established between lichen size and substrate age. It does not require the estimation of lichen age but it does require the availability of at least one, and preferably many more, lichen-supporting substrates of known age. A lichenometric dating curve established in this way can then be used to interpolate the age of other surfaces using only the size of the lichens growing on them. Where no other dating evidence is available, extrapolation may provide the only solution, although predicted ages obtained in this way are much less reliable (Matthews, 1977b).

Practicalities of lichen sampling and lichen size measurement can have an important influence on results, as shown in the reviews of Locke, Andrews & Webber (1979), Mottershead (1980), and Innes (1985a). A major source of potential error is environmental controls on lichen growth (Jochimsen, 1973), which should be taken into account in sampling, by selecting comparable environments (such as moraine crests or moraine bases) and/or by searching a large enough area to include the optimum environment for lichen growth. Although some workers advocate the use of only the single largest lichen on a surface, on the basis that the largest lichen is likely to be the oldest and also growing in optimum conditions, indirect lichenometric dating does not require this (Matthews, 1974). Use of a mean largest lichen size is less dependent on the influence of anomalously large lichens as might result from the dumping of lichens from supraglacial debris or landslides (Matthews, 1973; Griffey, 1978). Use of percentage lichen cover (Innes, 1986a) or the whole population of lichen sizes by size–frequency methods (Innes, 1986b) are also possible, though these are more time-consuming and do not appear to produce more accurate dating results.

The greatest limitation of indirect lichenometric dating is the necessity for independently-dated surfaces of known age, which provide the so-called 'fixed points' or 'control points' for the construction of the dating curve. The limited number of such surfaces in front of glaciers often leads to the use of data from farther afield, such as mine spoil or gravestones. This introduces uncertainties relating to data comparability as well as quantity. Nevertheless, even with a small number of representative control points, useful estimates of terrain age can be made. For surfaces deglaciated within the last 250 years, age estimates with an accuracy of ± 5–10 years may be approached (e.g. Matthews, 1975a; Porter, 1981c). Errors are likely to increase in direct proportion to terrain age and in inverse proportion to the number of control points.

Justification for lichenometric dating remains its proven usefulness rather than its theoretical basis, which has been persistently criticised (e.g.

Jochimsen, 1973; Webber & Andrews, 1973; Worsley, 1981). The 'direct' variant of the technique has been developed in part as a response to this criticism. Direct lichenometric dating seeks to establish a growth curve from the direct measurement of lichen growth rates. Despite the existence of a number of methods for constructing growth curves from short-term measured growth increments, relatively little progress has been made with this approach in practice (Miller, 1973; Proctor, 1983; Schroeder-Lanz, 1983; Haworth, Calkin & Ellis, 1986). This lack of success is due to the slow growth rates of the relevant species, the inherent variability of lichen growth, and the problems of linking lichen growth curves to terrain age.

Several biological dating techniques other than dendrochronology and lichenometry have been applied to recently-deglaciated terrain. In general, however, these have proved of only limited usefulness. Beschel (1963) examined the growth rings of several species of dwarf shrub, including *Salix arctica* on Axel Heiberg Island and *Salix glauca* and *Juniperus communis* in West Greenland, and found them reliable indicators of terrain ages up to 200 years. Dwarf willows and other shrubs have also been used in this way by Matthews (1974) and by Birks (1980a), whilst Kolishchuck (1990) has suggested the possible use of a wide variety of prostrate woody plants. Another technique that has been used to date various dwarf species of *Salix* utilizes the number of bud-scale scars along the stem (Palmer & Miller, 1961). Based on investigations of *Silene acaulis* and mosses of the genus *Grimmia* on recently-deglaciated terrain in front of the glacier Hintereisferner in the Alps, Beschel (1963) suggested the use of the size of plant cushions for dating purposes. The most advanced study of this type is probably that of Benedict (1989) who has made a thorough analysis of the growth rate of *Silene acaulis* on moraines in front of Arapaho Glacier, Colorado. He concludes that the maximum cushion diameter is useful for minimum age estimates on surfaces younger than about 75 years.

Relative dating techniques based on vegetation or soil characteristics have also been proposed. These include the recognition and use of lichen zones (Bergström, 1955; Corner & Smith, 1973; Mahaney, 1987), vegetation differences and trim lines (Heusser, Shuster & Gilkey, 1954; Goldthwait, 1960; Mahaney & Spence, 1990), plant species distribution patterns (Matthews, 1978a; Caseldine & Cullingford, 1981; Solomina, 1989), soil profile development (Birkeland, 1978) and specific soil properties (Kienholz, 1975; Fitze, 1982; Gellatly, 1985; Alexandrovsky, 1989). Although these methods are capable of yielding a crude indication of terrain age, they do not provide absolute ages and would involve circular argument if applied in the context of studies of vegetation succession and soil development.

2.2.3 Physico-chemical dating

The term 'physico-chemical dating' is used here to encompass a variety of non-biological techniques based on isotopic or radiogenic processes, rock or mineral weathering rates, erosional or depositional processes, tephrochronology, varve chronology and palaeomagnetism. Several of these techniques have been applied on glacier forelands, particularly on relatively old terrain and for relative-age determination. As radiocarbon (^{14}C) dating is the only absolute physico-chemical dating technique that has been widely applied in this context, it is given most attention here.

The status of ^{14}C dates associated with recently-deglaciated terrain varies according to the stratigraphic position of the dated samples: dates may be contemporary, bracketing or limiting (Porter, 1981a). In certain circumstances the time of sediment deposition can be specified from the date of a contemporary sample. Most usually, however, a sample provides either a minimum or a maximum limiting date for deposits lying, respectively, above or below. A pair of dates, from above and below, may bracket the age of a deposit. Wood, peat and soil samples have all been widely used to produce dates with one or other of these properties. Vegetation other than wood has also been used, where the environment has been conducive to its preservation, and this has provided valuable information about glacier variations in the Arctic (Baranowski, 1977; Bergsma, Svoboda & Freedman, 1984) and in the Antarctic (Fenton, 1982; Smith, 1982).

Although plant remains have some advantages as materials for ^{14}C dating, these are counterbalanced by the ubiquity of soils, even in relatively severe arctic-alpine environments. The detailed chronology of Holocene glacier variations in the Alps, referred to in the previous section, is as much based on buried soils as it is on wood and peat (Patzelt & Bortenschlager, 1973; Patzelt, 1974; Röthlisberger, 1976; Röthlisberger *et al.*, 1980). Buried soils have been quite commonly found in the Alps, the Himalaya, New Zealand and elsewhere, sometimes in association with the remains of trees or other vegetation, within lateral moraine complexes built up by the accretion or the superimposition of individual moraines.

Radiocarbon dating of different soil organic fractions from such buried soils has, in favourable circumstances, enabled the duration of advance and retreat phases to be determined as well as the time elapsed since burial. At sites in front of the Tasman Glacier, New Zealand, dates from wood and humic acids were in close agreement and reflected the date of soil burial; dates from the total organic content of the same soils were older and reflected the time elapsed since formation of the original moraine surface in which the soil had developed. In a stacked sequence of soils, a detailed pattern of advance and retreat phases could thus be reconstructed (Geyh, Röthlisberger & Gellatly, 1985). The total organic content appears to have been dominated in these cases by the resistant micro-remains of lichens, which were early colonizers of the original moraine surfaces. According to Geyh *et al.*, this approach is capable of a maximum time resolution of ± 200 years.

The dating of thicker soils that developed for a much longer time prior to burial is subject to further complications, including problems in isolating age-differentiated organic fractions and in establishing the greater apparent mean residence times characteristic of at least some fractions (Matthews, 1985). In particular, very steep age/depth gradients have been found for comparable organic fractions within well-developed soils buried beneath 'Little Ice Age' moraines in southern Norway. At Haugabreen and Vestre Memurubreen, the 'Little Ice Age' glacier advance appears not to have been exceeded during at least the last half of the Holocene. Radiocarbon dates from these sites range from less than 1000 ^{14}C years near the buried soil surface to about 4000 ^{14}C years at depth (Matthews & Dresser, 1983; Matthews & Caseldine, 1987). In well-developed soils such as these, predicted ages based on the age/depth gradient would appear to provide the best estimates of both the maximum time elapsed since soil burial and the minimum time elapsed since the initiation of soil formation.

Where plant remains are available for ^{14}C dating these are clearly more likely than soil organic matter to provide useful estimates of terrain age. Problems remain, however, with the imprecision of the dating. Statistical limits to ^{14}C dates quoted with 95% certainty are at best about ± 100 years. There is also the ambiguity involved in calibrating ^{14}C years in terms of calendar years (Stuiver, 1978, 1982). The latter problem is of major importance in the present context because ^{14}C ages of less than 400 years are unlikely to yield unique calendar dates. Porter (1981a) cites the example of a ^{14}C age of 220 ± 50 years, which is equivalent to all calendar ages within the ranges 150–210, 280–320 and 410–420 years ($\pm 1\sigma$). Nevertheless, Matthews, Innes & Caseldine (1986) used ^{14}C dating of moss and grass remains to provide a close maximum date for the 'Little Ice Age' maximum advance of Nigardsbreen, southern Norway. Based on calibration of the extreme upper confidence limit ($\pm 2\sigma$), they obtained a calendar date of younger than A.D. 1670 for the moraine marking the limit of the advance dated historically to A.D. 1750.

The use of tephra layers for dating Holocene moraines and related deposits has been developed in several regions subject to volcanic eruptions, most notably Iceland, the home of tephrochronology (Thorarinsson, 1956, 1964, 1966, 1981; Maizels & Dugmore, 1985; Dugmore, 1989), and North America (Crandell & Miller, 1974; Black, 1981; Porter, 1981d). In principle, a tephra layer is a stratigraphic marker horizon that can indicate the relative age of the overlying or underlying deposits by correlation. Once the age of the tephra is determined from historical sources (as in the case of some Icelandic eruptions), ^{14}C dating, or other techniques, it provides a means of absolute dating at a regional scale. With frequent volcanic eruptions interspersed with glacier variations and moraine formation, tephra layers can form the basis of a detailed chronology.

Varves and other rhythmites in lake sediments may be indicative of upstream glacier variations (Karlén, 1976, 1981; Leonard, 1986a,b; Østrem & Olsen, 1987), and ice-dammed lake sediments have been used for dating outermost moraines and glacier limits on glacier forelands (Matthews, Dawson & Shakesby, 1986; Goodwin, 1988). Most other absolute dating techniques are more appropriate for much longer timescales than those involved here, although some, such as ^{137}Cs and ^{210}Pb dating (Wise, 1980) or dating from known secular palaeomagnetic variations (Thompson, 1984; Thompson & Oldfield, 1986), may find applications in the future. Relative-dating methods, based on weathering, erosion or deposition rates, have similarly found few applications on the timescale of interest in this book. Of the relative-dating methods, those based on the surface weathering of boulders show most promise.

Indices of weathering have been based on such attributes as rock colour (Mahaney, 1987), degree of disintegration (Innes, 1984), hardness (Matthews & Shakesby, 1984; McCarroll, 1989a,b) and weathering–rind thickness (Chinn, 1981; Gellatly, 1984). Multi-parameter methods have also been advocated (Carroll, 1974; Burke & Birkeland, 1979; Dowdeswell & Morris, 1983). In theory, weathering rates can be calibrated with independent dates, thus rendering the dating absolute (Colman, 1981; Colman & Dethier, 1986). The most detailed application on a Holocene timescale has been made in New Zealand, where modal weathering–rind thicknesses on sandstone boulders have been calibrated with ^{14}C dates (Chinn, 1981; Whitehouse *et al.*, 1986) and used to date moraines to a suggested accuracy of 28% (Gellatly, 1984).

In summary, a wide variety of techniques are relevant to the dating of glacier variations and terrain age on glacier forelands. Of these, a small number of relatively accurate absolute dating techniques – historical observation and measurement, dendrochronology, lichenometry, ^{14}C dating, tephrochronology and weathering rinds – are considered to be the most important. As these techniques are often applicable in different environments, for different parts of the timescale, and vary in their inherent accuracy, they are complementary. More than one dating technique usually contributes to the definition and quantification of terrain-age in front of a particular glacier; and terrain-age sequences at different glaciers may be based on different dating techniques. It is likely, therefore, that the accuracy of terrain-age estimates will vary both within and between glacier forelands, according to the dating techniques employed.

2.3   Terrain-age sequences and areal chronologies

The nature of the timescale on a glacier foreland depends not only on the availability and accuracy of dating techniques, but also on the extent and timing of glacier variations. These determine the precise terrain-age sequence with distance from the glacier and the spatial pattern of terrain ages in the landscape as a whole. Thus, the timescale may be represented as an areal chronology (e.g. Fig. 2.4b).

An areal chronology is a representation of the timescale in terms of spatial co-ordinates as well as terrain-age. In Fig. 2.4b this has been achieved by mapping isochrones (lines of equal terrain-age) in front of Storbreen, a southern Norwegian glacier. The isochrones are based on historical sources and lichenometric dating. In A.D. 1902, P. A. Øyen began annual measurements of the position of the glacier terminus along siting lines. Apart from the period between 1920 and 1932, the measurements have been maintained, with very few missing years, to the present day (Liestøl, 1967; Matthews, 1976b). Together with maps, aerial photographs and terrestrial photographs, they account for all the isochrones of 100 years or less. In general, these relatively young isochrones are based on precise dates but in some cases the spatial co-ordinates are imperfectly known. The remaining isochrones are based on lichenometric dating of a series of arcuate moraine ridges (Fig. 2.4a). Whilst the spatial co-ordinates of these older isochrones are precisely known, the accuracy of the dating probably lies within ± 10 years.

Besides dating and mapping errors, which in the case of Storbreen glacier foreland are comparatively small, other characteristics of areal chronologies that limit their usefulness should be recognized. These

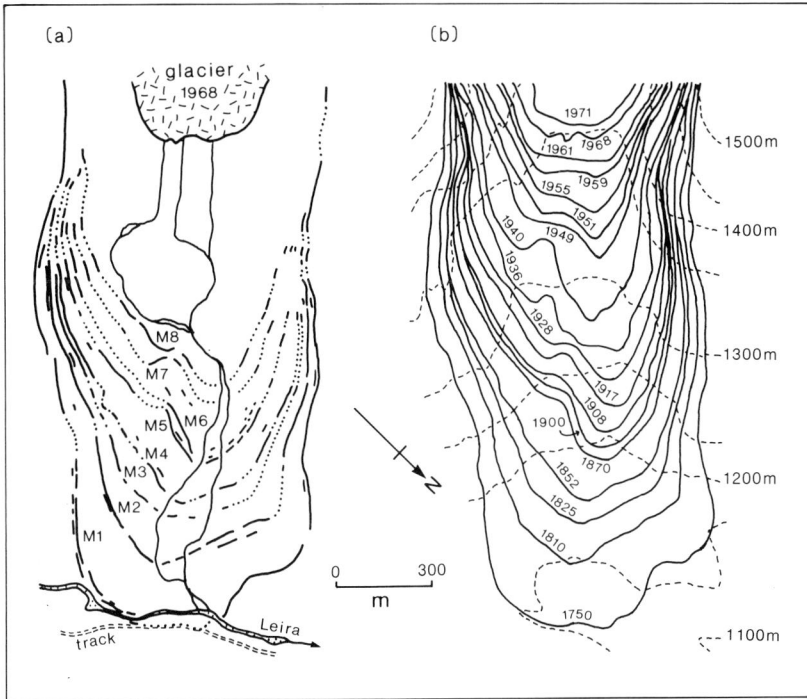

Fig. 2.4. Storbreen glacier foreland, Jotunheimen, southern Norway: (a) the end moraine sequence (M1–M8) (from Matthews, 1974); (b) the areal chronology depicted by isochrones (dated). Contours are also shown in (b) (from Matthews, 1978a).

include: (1) spatial variations and temporal discontinuities in the gradient of terrain-age; (2) the range of the timescale; and (3) the fact that terrain-age need not be equivalent to time since deglaciation (deglaciation age).

Given the dynamics of glacier variations, a constant rate of glacier retreat for any prolonged period of time is unlikely. Consequently, the spatial gradient of terrain-age is likely to vary over the glacier foreland and, as a result of glacier advances, temporal discontinuities will occur. Whereas seasonal advances (see section 2.1) present few problems and may, in rather exceptional circumstances, enhance the resolution of the timescale by the formation of annual moraine ridges (Worsley, 1974b; Worsley & Ward, 1974; Sharp, 1984), longer-term advances or still-stands of the glacier terminus produce longer 'gaps' in the timescale.

At Storbreen, the presence of moraine ridges on the glacier foreland bears witness to the occurrence of at least seven such discontinuities (Fig.

2.4b). At Nigardsbreen, also in southern Norway, many more moraine ridges, each associated with a glacier advance or still-stand, have been mapped by Andersen & Sollid (1971) and by Erikstad & Sollid (1986). Most of these moraine ridges appear to have been associated with glacier advances of the order of 5–10 years (see, for example, Fig. 2.2b). The corresponding discontinuity in the areal chronology may be substantially greater, however, as this is determined by the rate of glacier retreat prior to the advance, in addition to the duration of the advance itself. A well-documented example is provided by the position of the terminus of the Rhonegletscher before, during and after the advance from A.D. 1912 to 1922 (Aellen, 1981). The advance overrode terrain that had been deglaciated between about A.D. 1906 and 1912. Thus the temporal discontinuity in the areal chronology caused by the 10-year advance is 16 years.

Much larger temporal discontinuities are not uncommon. The advance of the Lower Grindelwaldgletscher, for example, which culminated in A.D. 1856, advanced over terrain deglaciated during the previous 215 years. In the special case of a glacier surge, relatively large areas may be over-ridden in a relatively short time. This was demonstrated in Svalbard between A.D. 1936–8 when Bråsvellbreen advanced over 15 km (Drewry & Liestøl, 1985).

Temporal discontinuities tend to be greater and/or more frequent on older terrain. Although it is possible to conceptualize this general pattern as the product of a random series (cf. Gibbons, Megeath & Pierce, 1984), the existence of regional, hemispherical and, in some cases, world-wide patterns in the timing of glacier variations indicate that climatic forcing is the dominant process. There has been a strong tendency towards climatic amelioration and glacier retreat since the mid-nineteenth century, which has accelerated (especially at high latitudes) in the twentieth century (Lamb, 1977). This has produced extensive areas of very recently degla-ciated terrain with relatively complete areal chronologies.

The greatest temporal discontinuity often occurs at the glacier foreland boundary. At Storbreen, the outermost moraine dates from the 'Little Ice Age' glacier maximum (*ca* A.D. 1750), which has overridden terrain which is at least an order of magnitude older and was probably deglaciated in the early Holocene, *ca* 9000 ^{14}C years B.P. (Matthews, 1976b). For ecological purposes, therefore, terrain beyond the glacier foreland boundary can be regarded as infinitely old. Whilst this terrain may provide a useful 'control', the effective range of the timescale on the glacier foreland itself is restricted in this case to about 240 years. A similar situation arises at most other glacier forelands in southern Norway (Matthews & Dresser, 1983; Matthews & Shakesby, 1984; Caseldine & Matthews, 1987; Matthews & Caseldine, 1987), at Glacier Bay, Alaska (Goldthwait, 1966), in some other regions of North America (Luckman, 1986; Ryder & Thomson, 1986), and in parts of Iceland (Caseldine, 1985, 1987; Sharp & Dugmore, 1985) and Greenland (Kelly, 1980; Weidick, 1985).

In many other regions, however, 'Little Ice Age' glacier advances were exceeded by earlier Neoglacial advances, as evidenced by older moraines located just beyond 'Little Ice Age' glacier limits. The term Neoglacial was defined by Porter & Denton (1967) as the period of glacier expansion subsequent to maximum Hypsithermal ('climatic optimum') shrinkage. Pre-'Little Ice Age' Neoglacial moraines have been investigated in the Alps (Patzelt & Bortenschlager, 1973; Patzelt, 1974; Röthlisberger, 1976; Röthlisberger *et al.*, 1980), northern Scandinavia (Karlén, 1973, 1979, 1982; Karlén & Denton, 1976; Griffey, 1976; Griffey & Worsley, 1978), Iceland (Dugmore, 1989), North America (Porter & Denton, 1967; Denton & Karlén, 1977; Burke & Birkeland, 1984; Ellis & Calkin, 1984), New Zealand (Burrows & Gellatly, 1982; Gellatly, 1984), and in the Himalaya (Röthlisberger & Geyh, 1985; Röthlisberger, 1986). Most of these pre-'Little Ice Age' Neoglacial moraines date from the last 5000 years, especially the last 3000 years. Although there may be considerable errors in dating these older moraines, which often occupy very limited areas, they nevertheless present the possibility of an extended timescale for ecological studies.

Two other limitations of areal chronologies should be emphasised here. These relate to the possibility of terrain age differing from the age of deglaciation. First, some areas of the glacier foreland may contain elements that predate deglaciation. Second, it is likely that many areas postdate deglaciation, due to the activity of geomorphological processes.

Although the former case is relatively uncommon, diverse examples are to be found in the literature. 'Old' land surfaces have been exhumed by retreating glaciers, sometimes associated with an *in situ* former vegetation and soil cover. Such organic remains provide nutrients for colonizing plants and it is possible that they may contain viable seeds. Well-preserved sorted patterned ground has been exposed by glacier retreat in northern Baffin Island (Falconer, 1966) and in the Lyngen Peninsula, northern Norway (Whalley, Gordon & Thompson, 1981). An area of about 800 m² bearing undisturbed, non-sorted patterned ground and *in situ* patches of unidentified mosses partly enveloping lichen-covered boulders and cobbles was found in front of Golden Eagle Glacier, Brooks Range, Alaska (Ellis & Calkin, 1984). Karlén (1973) has described moraine ridges that survived glacier overriding in Swedish Lappland. Baranowski & Karlén (1976) found 'fossil' Viking-age tundra fragments in front of retreating glaciers in both Svalbard and northern Sweden. These fragments yielded ^{14}C dates ranging from 760 ± 145 to 1565 ± 235 ^{14}C years B.P.. Remains of *Salix herbacea, Salix* spp., *Silene acaulis* and mosses exposed on ground recently deglaciated by Omnsbreen, southern Norway, were found by ^{14}C dating to be approximately 500 years old (Elven, 1978). Similarly, a *Cassiope tetragona–Dryas integrifolia*-dominated community was 'entombed' and released after about 400 ^{14}C years beneath Twin Glacier, Ellesmere Island, Canada (Bergsma, Svoboda & Freedman, 1984). Examples from Antarctica are provided by Collins (1976), Fenton (1982) and Smith (1982).

All of these examples relate to thin ice masses or relatively inactive polar or sub-polar glaciers with little erosive power. In temperate latitudes, glacier overriding tends to be more destructive and, with the exception of relatively robust tree trunks, relics of overridden landscapes are scarcely recognizable. In a few instances, however, the remains of forests have survived with *in situ* stumps (see section 2.2.2; Cooper, 1923; Goldthwait, 1966; Delibrias *et al.*, 1975; Bezinge, 1976).

Less obvious but more widespread remnants of previous landscapes exist within moraines. Folded, inverted and variously disrupted soils and vegetation layers have been found within the outermost 'Little Ice Age' moraine ridges at several southern Norwegian glaciers (Griffey & Matthews, 1978; Matthews, 1980; Matthews, Innes & Caseldine, 1986). These inclusions result from the glacier ploughing into the land surface during glacier advance and moraine formation. A similar mechanism accounts for the occurrence of highly-weathered boulders on otherwise 'fresh' surfaces (Matthews & Shakesby, 1984). Reworking of sediments by repeated glacier advance and retreat over the same terrain is an added complication, which may have occurred extensively during Neoglaciation (Gillberg, 1977; Matthews & Petch, 1982).

The possibility of living lichens and trees being transported on the surface of a glacier and surviving deposition has already been mentioned in the context of biological dating (section 2.2.2). Thus another mechanism is provided whereby deglaciation age may underestimate terrain age. The growth of lichens, mosses and higher plants on a medial moraine of the Glacier Inférieur de l'Aar, Swiss Alps, was noted by Nicolet in 1840 (Nicolet, 1841). More complex ecosystems have been reported on many glaciers in coastal Alaska (Tarr & Martin, 1914; Post & Streveler, 1976). These may undergo succession and survive deposition. Sharp (1958) found Sitka spruce up to 160 years old on Malaspina Glacier and younger plant assemblages inwards from the glacier margin. Stephens (1969) describes a forest of Sitka spruce (*Picea sitchensis*) on the Kushtaka Glacier in which alder (*Alnus crispa*) had been almost eliminated from the site. Evergreen beech forest, dominated by *Nothofagus dombeyi*, has been found growing on the Casa Panque Glacier, Mount Tronador, Chili, where a 1–3 m thick cover of supraglacial debris is derived from avalanches that fall on to the ice from steep valley sides (Rabassa, Rubulis & Suarez, 1981; Veblen *et al.*, 1989). Despite considerable instability of the debris, forest stands are characterized by trees up to 4.0 m in diameter and about 70 years in age.

The survival of such ecosystems upon deposition would hasten succession on the glacier foreland. However, survival is unlikely in the absence of a thick and relatively stable supraglacial debris cover. This is illustrated by the distribution of higher plant species on a medial moraine on the surface of Storbreen, southern Norway. Numerous pioneer species, including *Poa alpina*, *Trisetum spicatum* and *Saxifraga groenlandica*, have been observed by the writer on the most stable up-glacier portion of the medial moraine (beneath a large nunatak). Although crustose lichens grow on boulders along the length of the moraine (Matthews, 1973), the higher plants do not. Clearly, the vascular plants are unable to survive the instabilities of supraglacial transport and hence are not deposited on the glacier foreland.

Perhaps a more likely occurrence is the survival of disturbed soil, seeds and growing plants as a glacier ploughs into the land surface during minor glacier advances. Good examples of this are given by Sharp (1958), who describes an advance into a forest of Malaspina Glacier, Alaska, and by Haeberli *et al.* (1989: 93) who present a photograph of the advancing Ghiacciaio del Belvedere in the Italian Alps. In this photograph, the margin of the glacier snout is completely covered with debris and vegetation, including small trees, that were previously growing in front of the glacier. Future glacier retreat will be accompanied by the continued growth of at least some of the plants. On such moraines, new colonizers are most likely to establish on surviving soil patches with higher organic contents (Brandani, 1983).

Whereas the above discussion suggests that terrain age may rarely exceed the age of deglaciation, many parts of the glacier foreland are subject to erosion, deposition or less drastic disturbance after deglaciation. Hence much of the glacier foreland must be regarded as being considerably younger than the age of deglaciation. This may well be the single greatest limitation of areal chronologies, such as that shown in Fig. 2.4b, which imply that terrain age is equivalent to deglaciation age. A full discussion of the geomorphological processes responsible for such disturbance forms an integral part of the next chapter.

3

The physical landscape

Although the physical landscape in areas of recently-deglaciated terrain has some unifying characteristics, such as a relatively severe climate and a limited range of landforms and sediments, the landscape is far from uniform. It would therefore be an oversimplification to regard glacier foreland ecosystems as explicable solely as a function of terrain age. Heterogeneity, systematic spatial patterns in environmental factors, and environmental gradients are characteristic.

Neither is the physical landscape static. Physical environmental change is widespread in recently-deglaciated terrain, providing a dynamic framework within which ecological changes occur. Indeed, the glacier foreland landscape may be envisaged as the product of complex physical and biological interactions, which in turn change through time. A full appreciation of the physical environmental patterns and processes within the evolving landscape is as essential to the understanding of its ecology as is the nature of the timescale itself.

3.1   The legacy of glaciation

The most important visible features of a landscape emerging from beneath a retreating glacier are the landforms, with their associated characteristic soil parent materials and potential plant substrates. These influence the initial conditions and provide constraints on the later phases of soil development and vegetation succession. Thus, in many ways, the landforms and sediments of a newly-deglaciated landscape may be viewed as the stage for the enactment of a successional play. In this section the characteristics of the stage are described, largely as the product of processes of glacial erosion and deposition.

3.1.1 Glacial erosion

Glacial erosion is highly complex. It involves a number of imperfectly understood subglacial processes including abrasion, plucking and subglacial fluvial erosion (Drewry, 1986). Although there is some experimental evidence that clean ice can abrade (Budd, Keage & Blundy, 1979), abrasion is usually defined as the process whereby bedrock is scored by debris carried in the basal layers of a glacier (Sugden & John, 1976). Plucking is a more complex process that may involve several different mechanisms for the loosening and removal of rock fragments or larger blocks, such as crushing and fracturing, freezing-on, ice movement and variations in water pressure (Röthlisberger & Iken, 1981; Drewry, 1986). Erosion by subglacial meltwater occurs due to both the mechanical action of high-velocity streams and the removal of the products of chemical weathering (Röthlisberger & Lang, 1987; Souchez & Lemmens, 1987; Souchez & Lorrain, 1987). By a combination of these processes glaciers effect major landscape modifications (including areal scouring, valley widening and deepening, and the excavation of rock basins) as well as produce a variety of medium- and small-scale landforms (including roches moutonnées, grooves, potholes and other p-forms, and friction cracks) (Embleton & King, 1975a; Sugden & John, 1976).

Once exposed by glacier retreat, former areas of glacial erosion tend to be characterized by bedrock outcrops. These provide a completely different environment for ecosystem development than areas dominated by the sedimentary products of erosion (see section 3.1.2) and depositional landforms (see section 3.1.3). Most notably, the rate of soil development and vegetation succession on bedrock is at least an order of magnitude slower than on unconsolidated sediments. For example, 250 years after deglaciation in southern Norway, bedrock and large boulder surfaces in otherwise favourable environmental conditions are characterized by a general cover of crustose lichens (Haines-Young, 1983; Innes, 1986a). Corresponding areas of unconsolidated sediments show considerable evidence of soil development (Mellor, 1985; Messer, 1988) beneath mid- and low-alpine dwarf shrub heaths (Elven, 1978; Matthews & Whittaker, 1987) or sub-alpine *Betula pubescens* woodland (Fægri, 1933). McCarroll (1990) has suggested that in arctic-alpine areas there may be fundamental differences in processes and rates of chemical weathering associated with exposed and sediment-covered rock surfaces.

The extent of bedrock outcrops and the proportion of surfaces formed by the different erosion processes may vary greatly between and within glacier forelands depending on a variety of factors, including the thermal regime of the glacier, the topography of the glacier bed and the nature of the bedrock. The thermal regime is a key factor in determining erosional activity (Drewry, 1986). In polar and high-altitude regions, particularly where glaciers are thin, ice near the glacier bed is likely to be below the pressure melting point ('cold ice'). Under such circumstances the basal ice is frozen to the bedrock, and sliding of the glacier over its bed is inhibited, together with abrasion and meltwater production. Basal ice at lower latitudes and lower altitudes, on the other hand, tends to exist at temperatures above the pressure melting point ('temperate ice') with consequent enhancement of erosion. Both types of thermal regime may exist under different parts of the same glacier. Thus, an otherwise 'temperate' glacier may be characterized by 'cold' ice in its upper parts or beneath thin ice at the glacier margin in winter.

The existence of topographic obstacles to glacier flow and structural weaknesses in the bedrock appear to promote erosion by plucking. Lithological variations may also be critical, granite and some metamorphic rocks being particularly susceptible to crushing and fracture (Boulton, 1979). Less brittle and softer rocks, such as shales or clays, are more prone to abrasion.

Glacially-abraded rock surfaces, which tend to be smooth, striated and sometimes polished, and water-eroded bedrock (which also tends to be smooth and is usually washed clean of loose debris) are the least favourable surfaces for plant colonization. Plucked surfaces, with irregularities, ledges and open joints are more favourable, particularly where pockets of debris and fissures enable the respective establishment of chomophytes and chasmophytes (cf. Tansley, 1949; Ellenberg, 1988).

3.1.2   Glacial sediments

Glacial sediments include not only those deposited directly from glacier ice (tills) but also disaggregated and redeposited till (e.g. so-called 'flow-tills'), older deposits modified by glacier overriding (so-called 'deformation tills'), glacio-fluvial sediments, glacio-lacustrine sediments and glacio-marine sediments. All of these sediments are highly variable and are often modified considerably during as well as after deposition. The variety of sediments is best understood by considering their origins and the ways in which they are subsequently modified prior to and during deposition (Goldthwait & Matsch, 1988; Lundqvist, 1988).

Debris in glaciers originates either from subglacial erosion (section

3.1.1) or from subaerial processes on the slopes surrounding the glacier. If it falls on to the glacier surface above the equilibrium line (in the accumulation area), debris from the glacier headwalls, nunataks and valley sides may be buried by snow and be transported along an englacial pathway. In the ablation area, however, this debris remains in supraglacial transport. Unless such debris reaches the glacier bed (by, for example, basal melting in a temperate glacier), it tends to be transported passively, that is it remains relatively coarse in texture with angular clasts (Reheis, 1975; Boulton, 1978). In contrast, the products of subglacial erosion originate in the basal transport zone where they are modified by abrasion and crushing (Fig. 3.1).

Abrasion produces fine-grained 'rock flour', usually rich in silt (< 0.06 mm) but almost devoid of clay (< 0.002 mm) (Boulton, 1978; Haldorsen, 1981). Somewhat coarser particles, usually medium sand (< 0.6 mm) or finer, derive from the crushing of rock fragments. The rock fragments are in turn abraded, become striated and sub-angular or sub-rounded in shape. Comminution tends rapidly to produce bimodal or multimodal grain-size distributions (Dreimanis & Vagners, 1971). These tend to be characterized by a 'rock fragment' mode and one or more 'mineral grain' modes (the latter depending on lithology), whilst prolonged transport may lead to the progressive elimination of the coarser material (Lowrison, 1974).

Several other factors may be influential in creating or modifying the bimodal grain-size distributions that have been found in tills and have been attributed to comminution in the basal transport zone (Drewry, 1986). These include: (1) the addition of new rock fragments during glacial transport; (2) meltwater flow, particularly in water films at the glacier bed, which may remove particles of fine sand size (< 0.2 mm) or smaller; (3) the presence of a mixture of bedrock lithologies, which is likely to create additional modes from the superimposition of different characteristic grain-size distributions; (4) the possibility of the glacier overriding and incorporating older deposits, including finer grain sizes from weathered material; (5) the existence of debris from different ice facies which, on deposition, may become mixed to produce apparent bimodal or multimodal distributions.

Deposition of till is believed to involve two primary mechanisms: lodgement and melt-out (Shaw, 1985; Dreimanis, 1988). Both processes occur subglacially, whereas only melt-out can occur in supraglacial locations. This gives three types of primary till: lodgement till, subglacial melt-out till and supraglacial melt-out till. Lodgement till is plastered on subglacial surfaces as the result of the drag of glacially-transported debris against the bed (Boulton, 1976). As well as possessing the textural

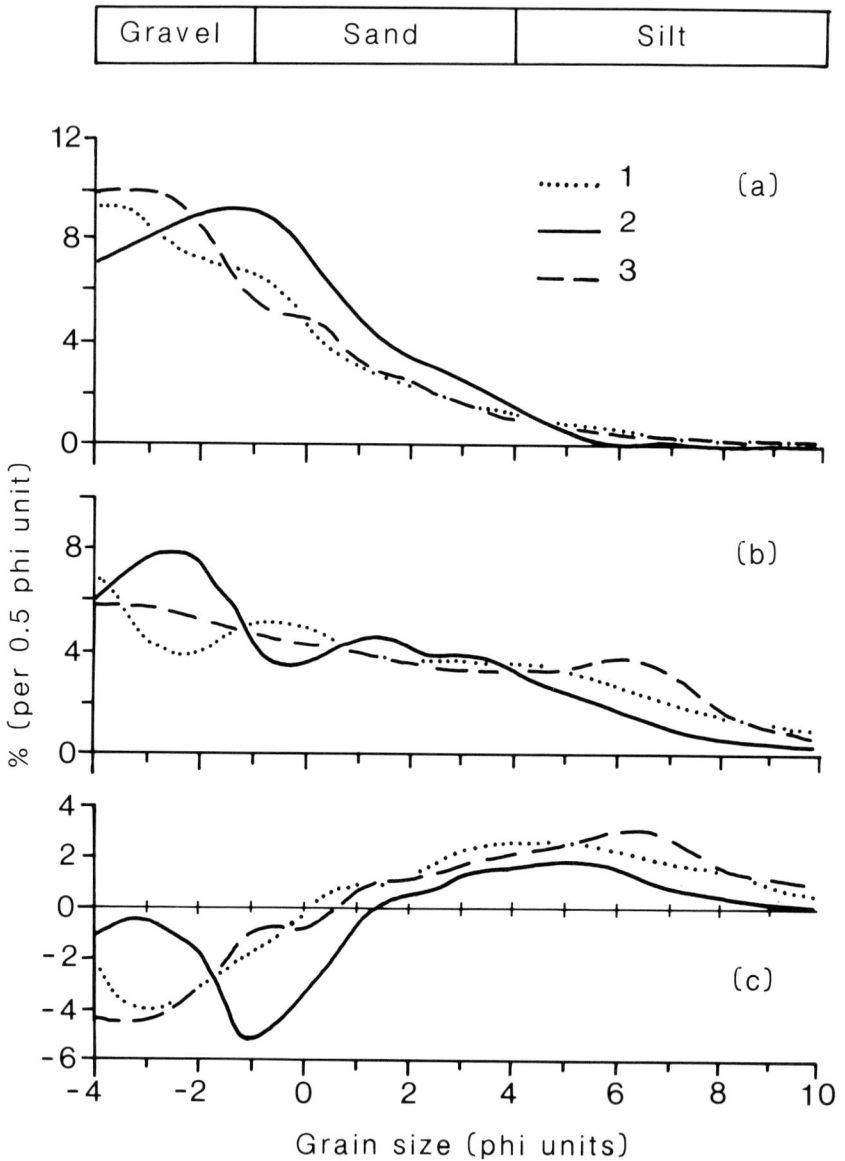

Fig. 3.1. Mean grain-size curves for sediments associated with glaciers in (1) Iceland (Breidamerkurjökull, n = 23 and 27 samples), (2) the Alps (Glacier d'Argentière, n = 24, 18), and (3) Svalbard (Søre Buchananisen, n = 11 and 10): (a) supraglacial samples; (b) subglacial samples (from the basal transport zone); (c) difference curves (from Boulton, 1978).

properties of material from the basal transport zone of the glacier, it is often compacted with a weak platy structure, and may contain small lenses of sorted material, smudges, smudged clasts, consistently striated clasts, clast clusters and features diagnostic of consolidation and deformation (Krüger, 1979; Lawson, 1979). Lodgement till can be deposited almost anywhere by flowing ice that is not eroding, and it may occur in discontinuous pockets or in sheets of variable thickness and extent.

Subglacial melt-out till is deposited by the slow *in situ* melting of debris-rich basal ice such that the deposit retains some of the structural integrity of debris in the parent ice (Boulton, 1976). It often overlies lodgement till. The most favourable conditions for formation are associated with stagnant or slow-moving active ice with minimal resedimentation. Whereas the texture and structure of melt-out tills resemble those of debris in the parent ice, some changes on deposition are inevitable, particularly in connection with melting, compaction and drainage under cryostatic pressure (Boulton & Paul, 1976) which may reduce the clarity of structures, induce crude stratification from the melting of alternate ice-rich and debris-rich layers, reduce the angle of dip of clasts, increase packing, and produce some draping of laminae over underlying irregularities (Lawson, 1981a,b).

The operation of a similar *in situ* melt-out process at the upper surface of a glacier produces supraglacial melt-out till. The preservation of structure may be enhanced as sediments accumulate and protect the underlying ice from ablation. However, as disturbance and resedimentation are likely, due to ice-surface lowering and meltwater release, supraglacial melt-out tills rarely retain exactly the structural characteristics of the parent ice (see also, Paul & Eyles, 1990). Thick sequences of supraglacial melt-out till may accumulate on the surface of high-altitude or high-latitude glaciers where sublimation lowers the ice surface very slowly without meltwater production (Shaw, 1977). Their formation is also favoured beneath supraglacial lakes, where the insulating water body slows the rate of ice melting (Boulton, 1970), and on small cirque and valley glaciers, where relatively thick supraglacial debris layers may accumulate from the surrounding slopes (Boulton & Eyles, 1979; Small, 1987a,b). These deposits may be let down and draped over subglacial tills during deglaciation.

Because supraglacial melt-out tills often originate from debris transported at high levels within the glacier, without entering the basal transport zone, this type of till can often be recognized on the basis of a coarser texture and the predominance of angular clasts (see above). It should be pointed out, however, that subglacial debris may be raised to a supraglacial position at a glacier confluence, in the lee of obstructions to glacier flow, or at the glacier margin.

Till-like deposits result from several different but related resedimentation processes (Lawson, 1979, 1982, 1988; Shaw, 1985). Working on glacial deposition at the Matanuska Glacier, Alaska, Lawson found that only about 5% of the deposits at the glacier terminus resulted from till-forming processes. Sediment flow, the downslope flow of sediment under the action of gravity, was the primary depositional process. The properties of active sediment flows and their deposits are largely determined by water content, with increasing liquifaction, flow rate and channelization, and decreasing sediment density and grain size tending to result from an increase in water content. The flows originate where sediments overlie glacier ice. Ablation of ice disaggregates the overlying sediment, which is simultaneously mixed with released meltwater further reducing its strength. The flows cease when there is a sufficient decrease in the slope, thickness or water content of the flow. Other processes intimately associated with such flows include meltwater sheet and rill flow, slope slump and slip, and the simple accumulation at the base of slopes of debris released from the ice by ablation.

Deposition by sediment flow is in some respects transitional between till deposition and the deposition of glacio-fluvial sediments. Glacio-fluvial sediments originate from the full range of deposits described above, directly from the ice itself, and from the glacio-fluvial erosion of bedrock. As a result of transport by water, inter-granular bonds are removed, the particles are reduced in size and rounded by attrition, and size-sorting leads to stratified deposits. These materials may, however, follow very complicated transport paths involving temporary deposition, resedimentation, mixing and interstratification with other deposits, particularly in the ice-marginal zone.

Glacio-fluvial deposits, ranging from lags of boulders and cobbles to pockets or more extensive linear sequences of stratified gravels and sands, rarely survive for long in unstable subglacial, englacial or supraglacial environments (Drewry, 1986). Nevertheless, at Glacier Bay, considerable thicknesses have been uncovered since the 'Little Ice Age' glacier maximum and it has been estimated that the modern sandar (see section 3.2.1) there are 30–90% complete before they appear out from under the thinning ice (Goldthwait, 1974). Such deposits are unlikely to maintain their structural integrity on deglaciation, unless they are large or associated with slowly ablating ice. Large volumes of sediment transported by meltwater streams on to the glacier foreland are almost certain to modify and extensively rework the surface layers (see section 3.2.1). Indeed, observations and measurements carried out beneath Bondhusbreen, southern Norway, show that the meltwater streams may in some instances transport more debris than the ice itself (Hagen *et al.*, 1983).

The finest particles transported by meltwater remain in suspension until they settle out in standing bodies of water. However, coarser material from both the bedload and the suspended load of meltwater streams also contributes to lacustrine sediments. In general, there is a spectrum of sedimentation from fluvial outwash, through deltaic to deep-water deposition, the last being characterized by the finest sediments (Smith & Ashley, 1985).

3.1.3 Depositional landforms and landsystems

The deposition of glacial sediments produces a wide variety of landforms. These are of importance to the ecology of recently-deglaciated terrain because of the major and diverse ways in which topographic factors influence plant habitats in the field. As Kellman (1975) has explicitly pointed out, there is a hierarchy of environmental influences on plants within which topography (operationalized by, for example, slope position and aspect) occupies an important place. Although such factors have no direct affect upon plants, they are often closely correlated with factors having specific physiological effects (such as air and soil temperatures, moisture, insolation and wind). Furthermore, as topographic factors are relatively easy to measure, they provide a useful means of approximating environmental conditions in the field and may explain, statistically, a high proportion of plant distributional patterns. A similar pragmatic solution to spatial variation in the soil landscape has long been recognized under the guise of the catena concept (Milne, 1935; Bushnell, 1942; Hall, 1983).

In this section, emphasis is given not only to individual landforms, which have been described in detail by geomorphologists (e.g. Price, 1973; Embleton & King, 1975a; Sugden & John, 1976; Goldthwait, 1988), but also to glacial landsystems. These are defined as distinctive areas of the glacial landscape characterized by particular assemblages of landforms together with their associated sediments (Boulton & Paul, 1976; Eyles, 1983a; Shaw, 1985). They are particularly useful for classifying and mapping groups of landforms which consistently recur and are genetically related. Glacial landsystems are therefore considered as an appropriate framework for summarizing much of the initial heterogeneity in the physical environment of glacier forelands which, for ecological purposes, have too often been regarded as homogeneous.

Clayton & Moran (1974) recognize four 'elements' in the deglaciated landscape – 'preadvance', 'subglacial', 'superglacial' and 'postglacial' – each characterized by different process–form relationships. Boulton & Paul (1976) and others have since widely utilized the idea of subglacial and supraglacial landsystems (which correspond to Clayton & Moran's subglacial and superglacial elements) and have added the glaciated valley landsystem (see also Boulton & Eyles, 1979). This three-fold subdivision of glacial landsystems is used here as the basis for discussion, whilst acknowledging that all such models are selective (Price, 1973), that different combinations of landforms and sediments may occur in reality, and that more comprehensive classifications are possible (e.g. Kurimo, 1980; Brodzikowski & Van Loon, 1987).

The subglacial landsystem represents the bed over which the glacier moved and its characteristic landforms reflect the processes operating at the glacier bed (Eyles, 1983a). The subglacial topography, together with glacial erosion and the activity of subglacial streams, ensures that till does not usually mantle the bedrock surface completely. Where bedrock knobs are important, bedrock-drift complexes may be found with lodgement till or basal melt-out till on gently sloping bedrock surfaces, on stoss (up-glacier) sides, or interspersed with subglacial glacio-fluvial sediments in sites of former lee-side cavities (Eyles & Menzies, 1983). In flatter areas, a till plain of lodgement till is characteristic with, according to Boulton & Paul (1976), three principal landforms – drumlins, fluted moraines and 'push' moraines – all of which generally occur in groups or swarms (Fig. 3.2).

Fluted moraines and drumlins are streamlined forms aligned parallel to ice-flow direction. Individual flutes are elongated ridges, usually up to 2 m in width and rarely exceeding 1 m in height, which may be hundreds of metres long. These may be superimposed on larger 'megaflutes' and probably form where water-soaked till is intruded into cavities in the lee of boulders or other obstacles at the base of moving ice (Hoppe & Schytt, 1953; Andersen & Sollid, 1971; Boulton, 1976b). Drumlins are rounded hills of variable size, usually 5–50 m in height and typically hundreds of metres in length, with a characteristic elliptical plan and asymmetric (gentler lee-side) long profile. They are clearly ice-moulded but there remains much controversy over the extent to which their formation can be attributed to specific depositional or erosional processes (Menzies, 1979).

Annual 'push' moraine ridges are commonly <2 m high and <10 m wide and may be hundreds of metres long, although they are often discontinuous and poorly preserved. They are generally produced from till by push, shear or deformation mechanisms in response to a winter advance

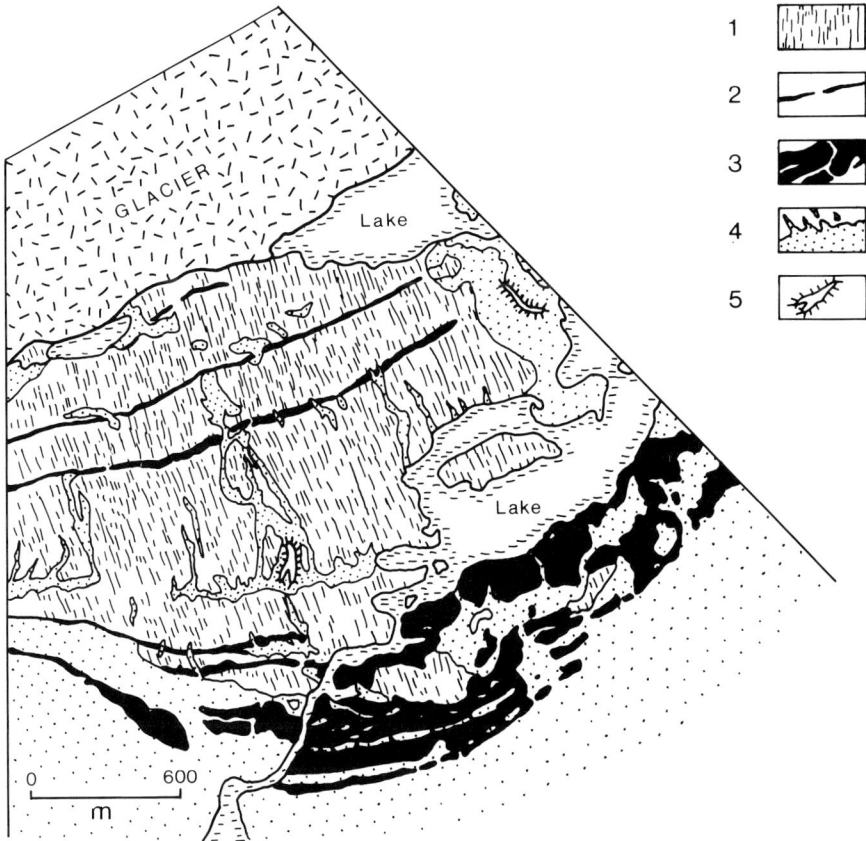

Fig. 3.2. Geomorphological map of a sector of the foreland of Breidamerkur-jökull, Iceland, showing some of the features of a subglacial landsystem: (1) drumlinized till plain with fluted moraines; (2) annual (push) moraines; (3) terminal (push-moraine) complex; (4) sandar and braided streams; (5) eskers (after R. J. Price, from Eyles & Menzies, 1983).

during the retreat of an active glacier snout (Worsley, 1974b; Birnie, 1977; Sharp, 1984). They are therefore oriented transverse to the ice-flow direction and their plan form mirrors details of the glacier margin. Larger-scale terminal and recessional moraine ridges may also be built up by similar mechanisms during more persistent glacier advances or stationary phases (Bothamley, 1987).

   Additional features of the subglacial landsystem include certain types of transverse moraine ridge formed subglacially (Lundqvist, 1981), meltwater channel fills and eskers (the latter being sinuous ridges of glacio-fluvial sands and gravels deposited along the length of subglacial streams). Subglacial landsystems are commonly found in front of temperate ice caps and glaciers not confined by valley walls. These glaciers slide over their beds and transport a relatively large quantity of debris close to the bed (Boulton & Paul, 1976). Subglacial landsystems also tend to characterize those glacier forelands where deglaciation has proceeded by ice-marginal retreat rather than downwasting (Prince, 1973; Krüger & Humlum, 1981).

Supraglacial landsystems, on the other hand, predominate in areas of inactive or slowly-moving ice masses, which release a considerable amount of englacial and supraglacial debris near their margins, and are more typical of sub-polar and polar glaciers. They also tend to develop in association with stagnating ice after a glacier surge (Wright, 1980; Sharp, 1985a). Although most of the sediment in the supraglacial landsystem may actually be subglacial in origin, it is deposited from the ice surface. Such material may be incorporated by freezing on to the base of the glacier and redistributed within the ice by thrust faulting. As the ice melts from above the debris is released at the glacier surface. The final chaotic pattern of landforms results from deposition above and between downwasting ice bodies (Paul, 1983).

The landforms range from relatively featureless supraglacial till plains to belts of irregular hummocky terrain. The latter in turn vary from hummocky moraines, in which supraglacial melt-out till may overlie basal melt-out till, to mounds of trough fillings now forming 'inverted' relief due to the melting of ice cores in the former surrounding slopes (Paul, 1983). These mounds (sometimes termed kame moraines) comprise interdigitating depositional complexes, including 'flow till' and glacio-fluvial sediments, and form a continuum with true kames (mounds of glacio-fluvial sands and gravels). Buried ice continues to melt out, prolonging the instability of the landscape (see sections 3.2.2 & 3.2.3) and, as in the case of the subglacial landsystem, considerable complexity is added by the deposition of ice-marginal and proglacial outwash (see section 3.2.1).

A change in the nature of deglaciation, due to climatic change or other influences on glacier dynamics, can lead to a shift in the dominant landsystem. Krüger & Humlum (1981) cite the zonation on the glacier foreland of Höfdabrekkujökull, Iceland. Within the outermost marginal moraine complex, probably formed in a glacial advance at the end of the nineteenth century, fluted moraines (with only a sparse covering of angular boulders of supraglacial origin) indicate deglaciation by frontal retreat. After the deposition of an inner moraine complex, Höfdabrekkujökull appears to have retreated by downwasting with a considerable addition of supraglacial deposits, including hummocky moraine and sheets of 'flow till'.

The third landsystem is the glaciated valley landsystem, in which is recognized the characteristic landform assemblages produced by valley glaciers. An enhanced supply of coarse-grained debris to the glacier surface from headwalls and steep valley sides is crucial to the understanding of this landsystem (Boulton & Eyles, 1979; Eyles, 1979, 1983b). During transport towards the glacier snout, this debris is concentrated by ice flow and ablation to form supraglacial lateral and medial moraines (Small, 1983; Gomez & Small, 1985). Together with a variable quantity of debris raised from the basal transport zone of the glacier to englacial and supraglacial positions (Matthews & Petch, 1982; Matthews, 1987; Small, 1987a,b; Vere & Benn, 1989), this material is deposited at the margin of the glacier to form a wide variety of landforms, depending on the quantity of debris, position relative to the valley sides, and ice dynamics.

On the valley floor, many features of the subglacial and supraglacial landsystem may be found. In the trough between the glacier and the valley sides, characteristic ice-marginal lateral moraine ridges, which can grow by accretion or superimposition to many tens of metres in height (Röthlis-berger & Schneebeli, 1979), are produced by dumping and related pro-cesses, particularly during glacier advances. With the destruction and burial of landforms by meltwater activity along the valley floor, these moraine ridges often dominate the landscape. In addition, as deglaciation occurs, meltwater streams may occupy the trench between the glacier and the lateral moraine ridge, depositing sands and gravels, which form terraces (kame terraces) as the glacier retreats (Boulton & Paul, 1976; Eyles, 1983b).

3.2   Proglacial landscape modification

Ecological studies on glacier forelands have tended to proceed on the assumption of a stable physical landscape. However, some of the dynamic physical processes that produce the main features of a newly-deglaciated landscape continue after deglaciation, and numerous new subaerial processes are initiated once the land surfaces and the various substrates are exposed. To return to the analogy employed earlier, these create a shifting stage on which the biological players are beginning to appear. Later in the performance, there may also be interruptions between scenes and the props may change between acts.

3.2.1   Glacio-fluvial activity

The largest and fastest changes in most newly-deglaciated land-scapes follow from the establishment and maintenance of a drainage network (Price, 1980). The initial routes followed by meltwater across the glacier foreland are determined by the general form of the bedrock surface and the overlying depositional landforms. Continuing a process that began under the glacier, glacio-fluvial erosion and deposition often produce major modifications to existing features, and create new ones. Meltwater channels are cut and abandoned, landforms buried or dissected, sediments stripped from bedrock, and lakes infilled. New suites of landforms and sediments develop from washed, rounded and sorted sediments, and continue to be reworked.

Most impressive amongst the glacio-fluvial landforms is the sandur (Icelandic: 'sand plain'); an expanse of coarse glacio-fluvial sediments (particularly cobbles, gravels and sands) characterized by shallow, anasto-mosing (braided) channels (Smith, 1985). Sandar are well-developed in front of glaciers in Iceland, Arctic Canada and Alaska, where they have been described by, amongst others, Price (1971), Church (1972), Boothroyd & Ashley (1975) and Haraldsson (1981). Two types were recognized by Krigström (1962), both of which may form in the immediate proglacial area or farther downstream, depending on the local topography. Plain sandar (or outwash plains) are usually associated with ice caps and ice sheets. They are much broader than valley sandar (or valley trains), which are restricted in their lateral development by valley sides and normally originate from one main channel issuing from an individual cirque or valley glacier (e.g. Fahnestock, 1963; Bluck, 1974; Fenn & Gurnell, 1987).

The characteristic braided channel pattern of a sandur is a response to three main factors: the highly variable discharge of meltwater from the glacier; the large quantity of sediment carried as bedload; and the non-cohesive channel banks (Miall, 1985). These same factors also enable large areas of the surface of the sandur to remain active, and hence unstable, almost indefinitely.

General levels of discharge associated with sandar depend closely on the seasonal pattern of ice (and snow) melt. Superimposed on this are diurnal effects and aperiodic runoff variations (Röthlisberger & Lang, 1987). Church & Gilbert (1975) defined five main periods of runoff, two of which have high discharges and account for most sediment transport and channel instability: (1) the nival flood period, with irregular, high discharges; and (2) the late summer period, with a closer correspondence between the melting rate of glacier ice and discharge. At other periods, discharges are low or the water is frozen.

Diurnal fluctuations are particularly well developed in streams fed by small temperate glaciers in the late summer period, and in the absence of rain-generated runoff. Aperiodic runoff variations fall into two main categories, those caused by synoptic weather changes (such as storms) and those related to the temporary blockage and sudden release of water from subglacial or ice-marginal locations. Jökulhlaups (Icelandic: 'glacier bursts') are particularly large-scale cases of the latter, caused by the catastrophic drainage of ice-dammed lakes (see, for example, Blachut & Ballantyne, 1976; Shakesby, 1985). These can cause catastrophic modification to sandar surfaces and to glacier foreland landscapes generally. Such modifications have been described by Heim (1983), who observed and mapped the effects of jökulhlaups in front of Kötlujökull, Iceland, between 1945 and 1980.

Large bedloads are derived from the sediment transported from the glacier terminus, supplemented by the erosion of loosely-consolidated sediments exposed on deglaciation (Church & Ryder, 1972; see also, Richards, 1984). Except in flood, braided rivers typically are able to transport only a small fraction of the available bedload (Østrem, 1975). At other times, gravel and sand is deposited in the channels producing a variety of bedforms and bars (Miall, 1985). These further constrict and divert flow, a process facilitated by banks that are easily eroded unless vegetated (Smith, 1976; Sidle & Milner, 1989). As channel gradients decline and channels become increasingly distributed downvalley, water depth decreases together with stream competence and particle size (Krigström, 1962; Church, 1972; Ballantyne, 1978). Such trends, particularly the irregular decrease in particle size, have led to the recognition of broad zones or gradients in the sedimentary characteristics of a sandur from proximal to distal reaches (Klemek, 1972; Boothroyd & Nummedal, 1978; Miall, 1985; Smith, 1985).

In proximal reaches, relatively few, deep and narrow channels are dominated by diagonal and longitudinal bars (elongated parallel to flow direction) composed primarily of massive to crudely horizontally-stratified gravels. Blocks of ice may be buried during rapid aggradation near the glacier, later to form 'pitted outwash' or 'kame and kettle' topography on melting (Price, 1973). Maximum clast size may reach 50 cm near the ice margin. Finer-grained deposits, characteristic of waning flows, are relatively rare, as are small-scale bedforms.

Several topographic levels may be recognizable on an active sandur,

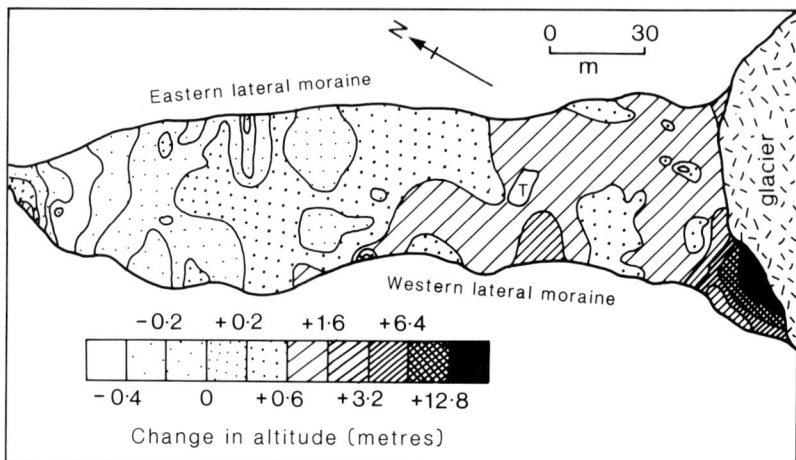

Fig. 3.3. Aggradation and degradation on the valley sandur of the Glacier des Bossons, French Alps, from April 1973 to April 1974, during a period of glacier advance (T, abandoned terminal moraine) (from Maizels, 1979).

each higher level being occupied by flows of increasing magnitude and decreasing frequency, and sometimes associated with increasing vegetation cover (Williams & Rust, 1969). Three or four such levels may be present near the glacier. Downstream, over a distance varying from < 1 km to > 10 km, the bedload changes to pebbly sand and, eventually, to sand with little or no gravel. These distal reaches are characterized by linguoid or lobate transverse bars with planar crossbedding as the principal internal structure. Large areas of the sandur may here be at a single topographic level, reflecting the aggradation and lateral migration of numerous, relatively ill-defined channels.

An additional complication is introduced by changes in the position of the glacier snout. This may be accompanied by shifting stream origins. It also appears that a glacier advance is likely to produce proximal aggradation, whereas a glacier retreat produces channel incision. In Iceland, Thompson & Jones (1986) reported a maximum downcutting of 7.9 m over 114 years in reponse to a recession of 320 m by the glacier Svinafellsjökull. This reflected the change from proximal to distal conditions and the associated local reduction of channel gradient. Maizels (1979) monitored the pattern and rate of aggradation over a six-year period following a rapid advance of the Glacier des Bossons in the French Alps. Although annual variation was high, net proximal aggradation occurred (Fig. 3.3). At the same time, there was an appreciable increase in the density of channels and a corresponding decrease in average channel width.

Inactive terraces with palaeochannel systems may therefore be isolated from the active sandur by the incision of active channels. Thompson & Jones (1986) identified two distinct causes of terrace formation in Iceland. At Svinafellsjökull, the timing and rate of downcutting was directly related to the rate of glacier recession, whereas at Kotarjökull the terraces represented stages in the recovery of the stream from the aggradational effects of a jökulhlaup dating from A.D. 1727.

From an ecological viewpoint, it is vital to recognize differences in instability between the different levels of the active sandur, and to distinguish these from the relatively stable surfaces of the terraces. Similarly, terrain age of a terrace (determined as the timing of the cessation of activity) may be considerably less than that of a neighbouring moraine (the age of which is likely to be a closer reflection of the date of deglaciation). A useful example is provided by the valley sandur on the Austerdalsbreen glacier foreland, southern Norway, where Maizels & Petch (1985) dated the stabilization of moraines and sandur surfaces independently. On the sandur surfaces they were able to demonstrate a progressive increase in terrain age downvalley and laterally (towards the valley side) on progressively higher terraces. This reflected the location of active surfaces downvalley of terminal moraines at each phase of deglaciation, and the migration over time of active channels towards the centre of the valley.

Once channels are abandoned as courses for glacial meltwater, they may still remain active as the courses of streams fed by snow melt and summer rain. A developmental sequence exhibited by such streams has been described by Sidle & Milner (1989) from Glacier Bay, Alaska. Sediment supply from stream banks decreases, largely as a result of stabilization by roots from developing riparian vegetation. Later in the sequence, debris dams, produced by the collapse of vegetated banks, in turn promote the establishment of woody vegetation on gravel bars and floodplains.

3.2.2   Consolidation and slope stabilization

The rapidity of post-depositional change in till has been stressed by Boulton, Dent & Morris (1974) and by Boulton & Dent (1974). Indeed, major changes in the structure and texture of till may take place after deposition but prior to deglaciation. Further changes then occur once the till surface is exposed to subaerial processes.

Till exposed in front of the retreating margin of Breidamerkurjökull, south-east Iceland, has been shown by Boulton, Dent & Morris (1974) to possess a two-layer structure (Table 3.1). A similar structure has been reported by Boulton & Paul (1976) from Nordenskiöldbreen in Svalbard

Table 3.1. *Till characteristics on deglaciation at Breidamerkurjökull, south-east Iceland (from Boulton, Dent & Morris, 1974).*

Depth (cm)	Mean void ratio	Mean coefficient of consolidation (cm^2 min^{-1})	Mean % silt + clay ($<4\phi$)	Field shear strength (kg cm^{-2})
Topmost 50 cm	0.68	0.046	39	0.075
Below 50 cm	0.38	0.0064	24	0.325

(Fig. 3.4), and by Sharp (1984) from Skálafellsjökull in Iceland. The upper layer, about 0.5 m thick, has a higher concentration of silt and a lower bulk density (high void ratio) attributed to the effects of subglacial crushing and shearing. The lower layer is thought to lie below the level at which shear stresses imposed by the overriding glacier exceed the strength of the till. This layer is a denser, less fine-grained horizon with a platy structure, possibly the product of consolidation (Boulton, Dent & Morris, 1974).

The loose and poorly-consolidated upper layer is probably caused by dilation during deformation in the water-saturated subglacial environment. Similar destruction and remoulding of a previous dense structure to one of higher void ratios has also been found in fluted moraines, small push moraines and annual moraines (Boulton & Paul, 1976). However, there is no evidence for the existence of such a layer in front of Eyjabakkajökull, a surging glacier in south-east Iceland (Sharp, 1985b), where relatively high shear strengths are found near the surface of the recently-exposed tills. This is presumed due to the absence of shearing and crushing beneath a downwasting glacier snout.

Once deglaciated, consolidation occurs, particularly due to changing drainage conditions. This effect can be seen in Fig. 3.4. Near to the glacier, the upper horizon of the lodgement till has a high void ratio (0.69) in contrast to the lower horizon (0.35). Farther away from the glacier, there is a reduction in the void ratios in the upper horizon (to 0.39), which Boulton & Paul (1976) attributed to drying out in the Arctic climate (see also section 3.3.2). This, and data from Breidamerkurjökull (Boulton & Dent, 1974) suggest that much of the consolidation takes place very rapidly within the first year of deglaciation.

Initially steep slope angles are lowered by a variety of other processes, including slope wash, creep and gelifluction, slumping and sediment flow.

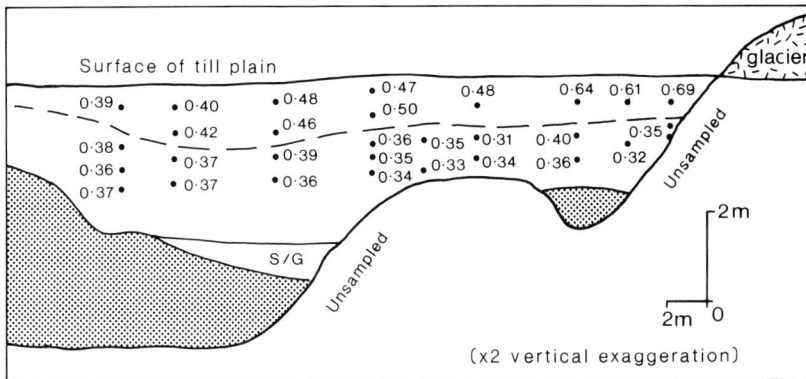

Fig. 3.4. Cross section of a till plain showing void ratios with increasing depth and increasing distance from Nordenskiöldbreen, Svalbard. The broken line indicates the two-layered structure of the till sheet. Bedrock is shaded; S/G is sand and gravel (from Boulton & Paul, 1976).

Many of these processes are particularly active on newly-deglaciated terrain with high water content and negligible vegetation cover. Furthermore, the possibility of the melting of buried glacier ice and the thawing of frozen sediments add considerably to the instability of the terrain (see section 3.2.3).

The slopes of large moraines and other large landforms, slopes continually supplied by rockfall, debris flow, slush flow or avalanche, and slopes undercut by meltwater streams may remain unstable for long periods. However, smaller-scale slopes and slopes not influenced by continual disturbance or undercutting tend to stabilize rapidly. This has been clearly demonstrated by Welch (1970) who investigated the slope characteristics of recessional moraines in front of the Athabasca Glacier, Alberta, Canada.

Based on the measurement of slope profiles across moraines deposited between A.D. 1880 and A.D. 1966, Welch (1970) found that most morphological change occurs within the first six to ten years after deposition. Initially, slopes are highly variable with slope angles varying from 1° to 79°. Maximum slope angles rapidly stabilize at approximately 30° (Fig. 3.5). Similarly, surface roughness values show an approximately linear decrease with respect to time for the first ten years. Declining values through time reflect the progressive elimination of surface irregularities (such as large stones, slump scars and features of initial deposition).

Similar results have been obtained by Sharp (1984) working on annual moraine ridges at Skálafellsjökull, Iceland. Immediately after ridge formation he found a rapid decline in mean maximum slope angle from 34.2

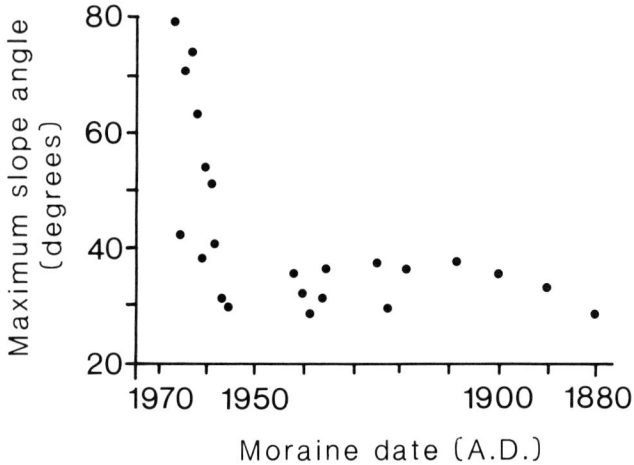

Fig. 3.5. Maximum slope angles on recessional moraines of increasing age, Athabasca Glacier, Alberta, Canada (from Welch, 1970).

$\pm 1.89°$ (± 1 standard deviation) on a one-year-old ridge, to 21.95 $\pm 3.2°$ on a ten-year-old ridge. Maximum slope angles on ridges between 17 and 43 years old average between 18.1 $\pm 2.7°$ and 19.7 $\pm 1.15°$. An anomalously steep-sided 51-year-old ridge is composed almost entirely of boulders. Although the different 'equilibrium' values recorded by Welch (1970) and by Sharp (1984) may reflect differences in the sediment composition or environment at the two sites, they may also be accounted for by different measurement techniques.

3.2.3   Pervection

Pervection, the mechanical movement or downwash of solid particles, especially silt, by subsurface flow (Paton, 1978), is a further characteristic process of post-depositional modification on newly deglaciated terrain. On exposure, the downward percolation of water may transport sufficient fine-grained particles to change fundamentally both the grain-size distribution of the till and its structure. For instance, Boulton & Dent (1974) show that lodgement till in front of Breidamerkurjökull loses up to 10% of its silt + clay fraction from its surface layers during the first year of exposure. A third-order trend surface relating to the percentage silt + clay with depth and distance along a transect from the glacier snout (Fig. 3.6) shows not only the reduction in the amount of fines near the surface but also a progressive reduction in the rate at which the zone of

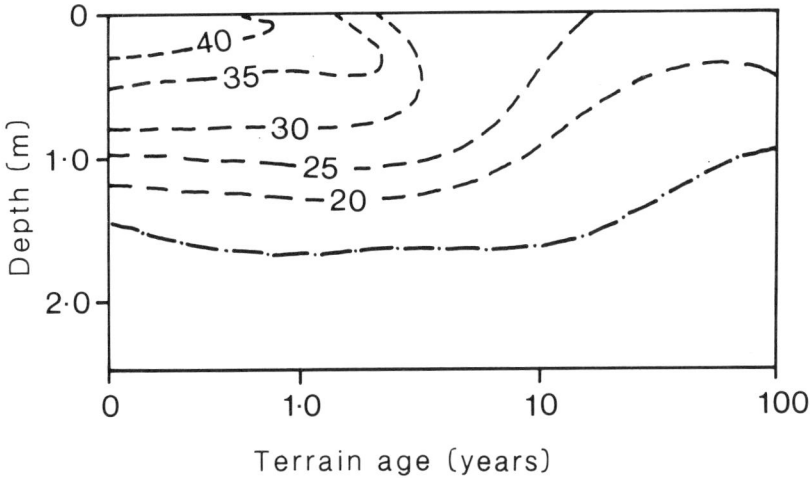

Fig. 3.6. Third-order trend surface of the percentage fines (silt + clay) in a till sheet with depth and increasing distance from the margin of Breidamerkurjökull, Iceland (from Boulton & Dent, 1974).

highest concentration of fines moves down the profile: 45 mm yr^{-1} between 0 and 10 years, 3 mm yr^{-1} after about 20 years, and 1 mm yr^{-1} after about 55 years (Boulton & Dent, 1974).

The more compact lower layers of the till impede the downward movement of fines, which leads to the development of a dense silty layer in the upper part of the compact till. This silt-rich layer reaches its maximum development after 10 to 30 years and is subsequently degraded by the continued downwashing of fines (cf. Romans, Robertson & Dent, 1980). At the surface, a combination of pervection with other processes, such as deflation by wind (see section 3.2.8) and frost sorting (section 3.2.5) may produce a coarse stony lag. Thus, Boulton & Dent (1974) report 30 to 40% of the surface to be covered by subangular stones only one year after exposure; after four years, 90% of the surface is so covered.

Similar patterns have been detected in the surface layer of tills in front of glaciers in Jotunheimen, southern Norway (Matthews & Beckett, unpublished). In front of all seven glaciers investigated, statistically significant reductions were detected in the finer than 0.125 mm fraction (fine sand, silt and clay) with distance from the glacier snout (Fig. 3.7). Except for Søre Illåbreen and Hurrbreen, where glacier retreat records are not available, distances of 200 m from the snouts of these glaciers correspond to terrain ages of 10 to about 20 years.

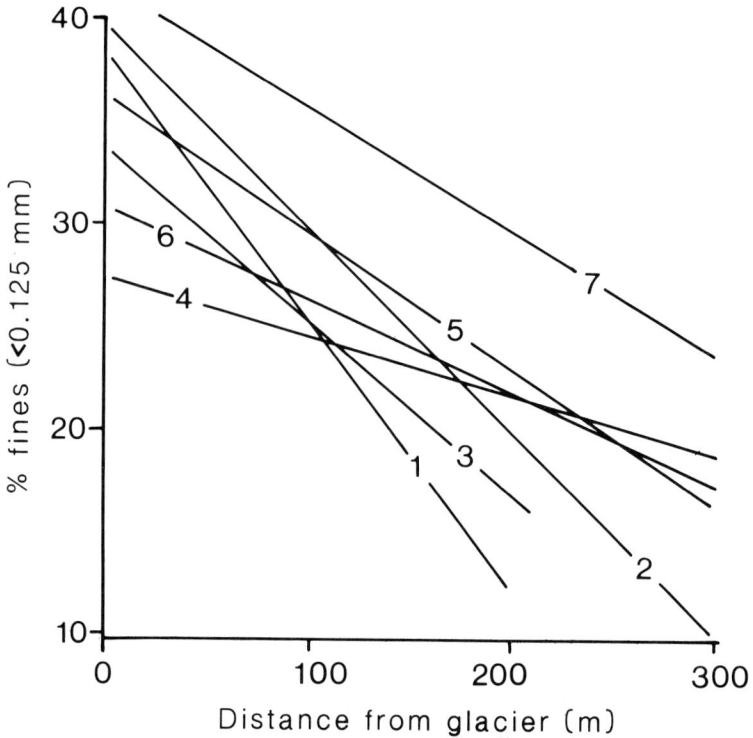

Fig. 3.7. Trends of decreasing percentage fines (fine sand + silt + clay) in the uppermost 5 cm of till sheets from seven glacier forelands, Jotunheimen, southern Norway: (1) Böverbreen, (2) Hurrbreen, (3) Leirbreen, (4) Søre Illåbreen, (5) Storbreen, (6) Storgjuvbreen, (7) Tverråbreen (P. J. Beckett & J. A. Matthews, unpublished data).

3.2.4   Cryogenic processes: frost weathering

A wide variety of disturbances to recently-deglaciated terrain arise from the freezing and thawing of the ground and, in some cases, the melting of buried glacier ice. Here, emphasis is given to several important processes that have been specifically investigated on glacier forelands or are likely to influence significantly soil development and plant growth. These processes include: frost weathering (gelifraction); frost-heave and frost-sorting; frost creep and gelifluction (solifluction); rock glacier movement; and some more rapid forms of periglacial mass movement, including those associated with the melting of buried glacier ice. For an account of the full range of cryogenic processes and their resulting effects on landforms and sediments, readers are referred to several books devoted specifically to periglacial geomorphology (Embleton & King, 1975b; French, 1976; Washburn, 1979; Harris, 1981; Clark, 1988). Although it has for long been recognized that such processes are of major importance to arctic-alpine vegetation and soils (e.g. Benninghoff, 1952; Raup, 1951, 1971; Sigafoos, 1951) studies in the context of glacier forelands are rare.

The breakdown of rock by freeze–thaw action (frost weathering) involves a range of poorly-understood processes ranging from granular disintegration (microgelifraction), through superficial frost scaling due to the formation of thin ice plates parallel to the rock surface, to frost shattering (macrogelifraction), which includes the growth of ice in cracks (frost wedging) and the breaking of previously unweakened rock by a true bursting process (frost splitting) (McGreevy, 1981; Lautridou & Ozouf, 1982; Lautridou, 1988). Whilst hydration shattering (White, 1976) may be a contributory process, particularly in fine-grained sedimentary rocks, the mechanical action of ice expansion and ice crystal growth are believed to be the most important physical weathering processes in arctic-alpine environments (Washburn, 1979; Price, 1981). However, as Thorn (1988) has pointed out, frost weathering is subject to some very specific constraints, and is probably much more restricted in periglacial environments than is usually believed.

Many factors influence the nature of frost weathering. The frequency of freeze–thaw cycles is an important control on the effectiveness of various kinds of frost action, but this factor is not all-important (Washburn, 1979). The rate, degree and depth of cooling, the composition of the rock, the frequency and character of discontinuities and other weaknesses, the distribution of snow and the availability of water are also of significance. The last factor has been found particularly important on and near glacier forelands in Jotunheimen, southern Norway, where intense frost shattering tends to be most effective near water bodies. For example, the shoreline of a former ice-dammed lake that existed near Böverbreen during the 'Little Ice Age' is littered with large angular slabs of rock prised from bedrock outcrops (Matthews, Dawson & Shakesby, 1986). The process has continued around the present lake shoreline since the lake surface fell to its present level *ca* A.D. 1876 (Shakesby & Matthews, 1987).

Coarse and angular debris is the typical product of frost shattering, the exact size and shape depending on the rock type and the intensity of frost action (Price, 1981). Massive rocks, such as many limestones and granite, shatter more randomly than stratified rocks, such as slate or schist. However, the rate and extent of freezing can produce different results in the same rock type. Thus, high tropical mountains, which are dominated by diurnal freeze–thaw cycles that penetrate to only shallow depths in the rock, reputedly produce relatively small rock fragments (Troll, 1958; Hastenrath, 1973). In middle and high latitudes, however, where pronounced seasonal freezing occurs, the smaller number of relatively intense freeze–thaw cycles allows the production of a greater quantity of large-calibre debris.

Providing other factors are suitable, frost comminution may increase once the bedrock has been broken, due to an increase in the exposed surface area of the rock and the water holding capacity of the debris (Price, 1981). Lautridou (1988) and others have studied the fine-grained products of gelifraction in detailed laboratory studies. The threshold of fragmentation (size of the smallest particles that can be split by frost) varies from 0.1 mm (very fine sand) for weathered granites to 1.0–2.0 mm (coarse sand), although occasional much smaller particles down to clay size may be scaled off during the process. Together with the products of glacier erosion, the products of frost weathering therefore help account for the scarcity of clay-sized particles in glacier foreland environments.

3.2.5   Frost-heave and frost-sorting

Frost-heave processes are widespread on glacier forelands in middle and high latitudes, although these processes too are subject to constraints, particularly those imposed by moisture supply, particle-size characteristics and the distribution of snow and vegetation. Frost heave is the upward displacement of the ground in response to the development of distinct lenses of clear ice parallel to the ground surface (Harris, 1981). Provided a water supply is present, ice lenses develop as water is drawn towards the freezing front in fine-grained sediments and soils. Susceptibility to frost-heave increases from near zero in coarse sand (0.6–2.0 mm) to a maximum in fine silt (0.006–0.02 mm), from which it slowly declines to near zero in heavy clay. Growth of ice lenses during freezing disrupts the soil structure and results in an increase in the moisture content and void ratio. On thawing, excess water released by melting ice lenses, in conjunction with substrate consolidation (and a reduction in void ratio) leads to over-saturation, the development of high pore-water pressures, and loss of strength. In combination, repeated freezing and thawing may therefore lead to extensive churning or cryoturbation causing damage to plant roots and the formation of non-sorted patterned ground (French, 1988). On a smaller scale, the common occurrence of gaps around stones is probably due to the frost-heaving of the immediate surface of the surrounding material (Washburn, 1979).

A related process is the development of needle ice (pipkrake), which consists of columnar ice crystals that elongate perpendicular to the soil surface and are capable of lifting a cap of fine sediment and small stones several centimetres above the ground. The process may well be responsible for the formation of nubbins described by Washburn (1979) as small round-to-elongate earth lumps, one to several centimetres in diameter. These and related microrelief features caused by frost action may be important in providing suitable sites for seed germination and establishment on glacier forelands. The formation of lenticular to spherical moss balls, which may be unattached to the substratum, have also been attributed to frost disturbance by needle ice (Beck, Mägdefrau & Senser, 1986). According to Price (1981) needle ice development is the major frost process operating in tropical mountains and in middle-latitude maritime climates (see also Pérez, 1987; Lawler, 1988).

A further set of processes produces particle-size sorting in hetero-geneous, unconsolidated sediments and soils. Frost-heave and needle ice development, together with other processes, including the upfreezing of stones and boulders, gravity sorting, the mass displacement of unfrozen fines into coarser material by cryostatic pressures, the migration of fines ahead of a freezing front, and the preferential removal of fines by thaw water, result in vertical and lateral particle-size sorting (Washburn, 1979). Such frost-sorting processes lead to the creation of sorted patterned ground which, like non-sorted patterned ground can be effectively classified according to geometric form into circles, polygons, nets, steps and stripes (Washburn, 1956; Nicholson, 1976). It is also useful to differentiate large-scale and small-scale forms. Other aspects may also be of significance, such as the presence or absence of domed centres and marginal troughs, and whether or not cracking of the ground surface is involved in their formation.

The initiating process in polygonal patterns is generally cracking due to desiccation or frost cracking (Price, 1981: 187), whereas circles appear to form in the absence of cracking (Washburn, 1979: 158). Elongated patterns – including nets, steps and stripes – are induced on slopes of only a few degrees. Large-scale polygons and circles (> 2 m diameter) are probably characteristic of a permafrost environment (Goldthwait, 1976), whereas smaller forms can develop where the ground freezes seasonally. Thus, the small-scale patterning occuring in middle-latitude mountains with mari-time climates, and in tropical mountains, must be due to shallow freezing and probably involves needle ice, which affects only the uppermost 2–3 cm and, according to Price (1981), cannot move clasts larger than 5–10 cm in diameter.

Several types of patterned ground have been studied on glacier forelands in Jotunheimen, southern Norway (Ballantyne & Matthews, 1982, 1983; Harris & Matthews, 1984). Boulder-cored frost boils (Fig. 3.8a) are caused by the seasonal ratchet-like uplift of individual boulders through the surrounding sediment and soil. This probably results from the frost-pull mechanism whereby freezing sediment adheres to the upper part of the boulder, which is raised as ice lenses grow and heaving of the frozen sediment occurs (Harris & Matthews, 1984). On thawing, consolidation of the surrounding sediment prevents the resettling of the boulder to its former position and a silt and vegetation cap may remain on the upper surface of the boulder. Matchsticks inserted into the ground on the Storbreen glacier foreland have been similarly uplifted, suggesting that the process probably affects the larger roots of both seedlings and older colonizing plants. Indeed, experimental studies on alfalfa (*Medicago sativa*) have demonstrated its effectiveness in the upheaval of this tap-rooted species (Perfect, Miller & Burton, 1987, 1988).

Sorted circles up to 3.3 m in diameter and comprising centres of fine sediment surrounded by marginal troughs filled with cobbles and boulders (Fig. 3.8b) have been described by Ballantyne & Matthews (1982) on the glacier foreland of Slettmarkbreen. Here, this type of sorted patterned ground appears to originate immediately after deglaciation and to stabilize sufficiently to allow the colonization by crustose lichens on the clasts and vascular plants on the fine centres within 50 years of deglaciation. These sorted circles are best developed on or near the crests of broad moraine ridges where the snow is likely to be less deep in winter and hence frost is likely to penetrate deeper into the ground.

Sorted polygons up to 0.4 m in diameter have been examined by Ballantyne & Matthews (1983) on a till sheet exposed over a 35-year period by the retreat of Storbreen. The polygons are initiated from desiccation cracks (Fig. 3.8c) which, as they enlarge and form a network on terrain of increasing age, become progressively filled with clasts (Figs. 3.8d, 3.9a,b). Major changes in the fine (<2 mm) fraction of the sediments are also involved: the migration of coarse sand (>500 μm) into the cracks (Fig. 3.9c,d) being accompanied by the displacement of silt (<63 μm) into the centres (Fig. 3.9e,f) and the depletion of fine sand (63–250 μm) from both cracks and centres (Fig. 3.9g,h). The first two trends can be explained in terms of frost-sorting processes, coarse sand moving into the cracks while silt moves into the centres in front of a horizontally-advancing freezing plane (lateral sorting). The depletion of fine sand from centres and cracks suggests the removal of this size grade from the surface layers by pervection.

Fig. 3.8. Cryogenic and related disturbance on glacier forelands in Jotunheimen, southern Norway: (a) Boulder-cored frost boil, Storbreen; (b) Sorted circle, Slettmarkbreen; (c) Polygonal desiccation cracking, Storbreen; (d) Sorted polygon, Storbreen; (e) Non-sorted polygons, Böverbreen.

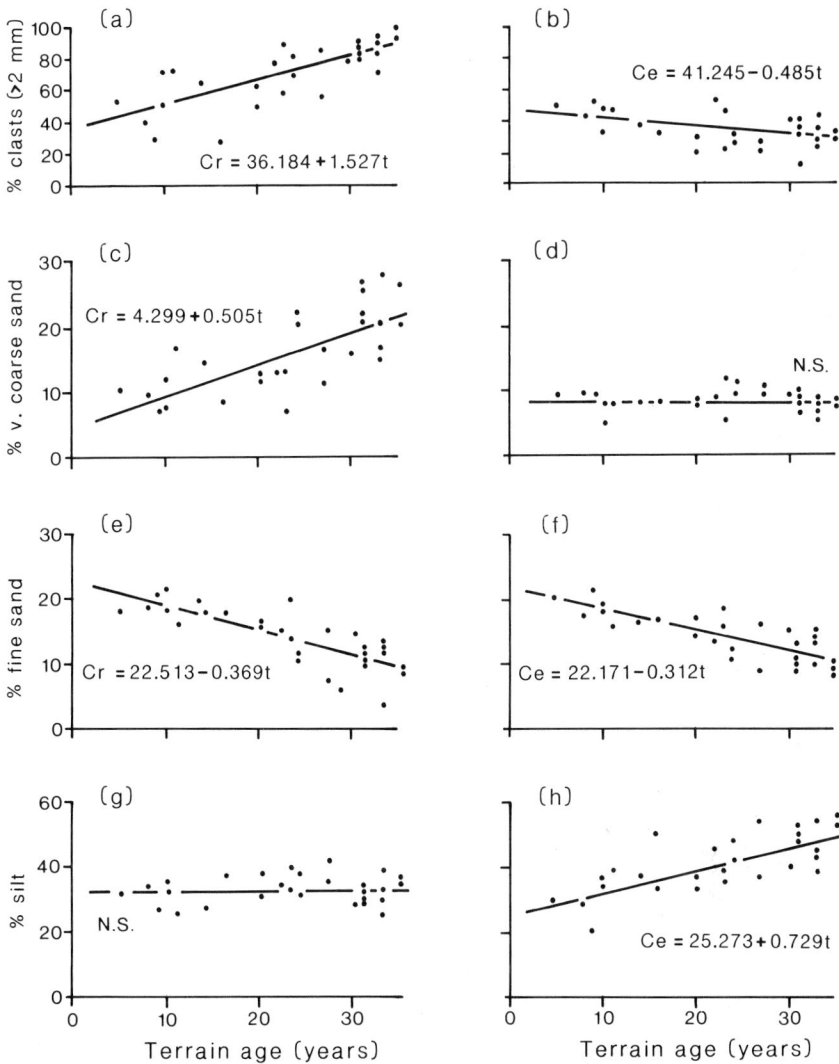

Fig. 3.9. Sedimentary evidence for frost-sorting and related processes associated with the development of sorted polygons on terrain of increasing age at Storbreen, Jotunheimen. Percentage by weight of various fractions for 'cracks' (Cr) and cell centres (Ce); (a) % clasts > 2 mm in cracks; (b) % clasts > 2 mm in cell centres; (c, d) % very coarse sand (1000–2000 μm) in the fine (< 2 mm) fraction; (e, f) % fine sand (125–250 μm) in the fine fraction; (g, h)) % silt (< 63 μm) in the fine fraction. All regression equations are statistically significant (p < 0.001) (from Ballantyne & Matthews, 1983).

Again, the appearance of plants, particularly in the form of mosses covering the crack walls on the oldest terrain, suggests that after 35 years these polygons may be approaching the end of their development.

3.2.6   Solifluction and other periglacial slope processes

Solifluction, the slow downslope movement of the uppermost metre or so of soils and sediment on gradients as low as 1–2°, is the most widespread type of mass movement above and beyond the tree line in both arctic and alpine environments (Harris, 1981). It is likely, therefore, to be common on glacier forelands even though it has rarely been reported. Although other processes are sometimes involved (Matthews, Harris & Ballantyne, 1986), frost creep and gelifluction are the most important constituent processes. Frost creep results from the uplift of particles due to frost-heave or needle ice development (perpendicular to the slope), followed by vertical settling, whereas gelifluction is the slow flow that occurs due to the loss of strength as ice-rich sediments thaw (Washburn, 1979). Frost creep is primarily controlled by the number of freeze–thaw cycles and achieves its greatest importance in tropical and middle latitudes (Benedict, 1976). Its effects, such as the production of small steps and terracettes, and the downslope curvature of tree-trunks, are best seen on drier slopes, where flowage is not a factor (Price, 1981). Gelifluction is the main process in the production of extensive non-sorted or sorted sheets, terraces and lobes, which generally require soil saturation during thaw and slopes less than 20–25° (above which they tend to become unstable and more rapid mass movements are characteristic).

Although solifluction lobes have been recorded on glacier forelands in Norway (Harris, 1982), in the Nepal Himalaya (Watanabe *et al.*, 1989), and in the Alps where they have been used in the dating of Neoglacial moraines (e.g. Röthlisberger, 1976), the only specific study of them on glacier forelands known to the author is that of Archer, Simpson & Macmillan (1973) in New Zealand. These authors recognize rectilinear terraces as well as lobate forms on a lateral moraine of Tasman Glacier and note likely important effects on soil and vegetation development (see section 5.4.2).

Another type of slow periglacial mass movement is that exhibited by rock glaciers. These are lobate or tongue-shaped bodies of ice-cored or ice-cemented sediment, which move downslope under their own weight (Corte, 1976; Giardino, Shroder & Vitek, 1987). They are of relevance to glacier forelands because they may develop downslope of debris-covered glaciers (Whalley, 1974; Johnson, 1987) or from ice-cored moraines (Lindner & Marks, 1985; Dzierzek & Nitychoruk, 1987). They provide a distinctive and inhospitable substrate for plants characterized by blocky surface debris as well as instability. Although the upper surface of active rock glaciers is often *relatively* stable, with considerable lichen cover on boulders (Haeberli, King & Flotron, 1979; Vere & Matthews, 1985), pockets of fine debris suitable for the growth of vascular plants are sparse (Hartman & Rottman, 1986; Johnson & Lacasse, 1988). In a survey of five rock glaciers in the San Juan Mountains, Colorado, U.S.A., Hartman & Rottman (1986) located 89 species growing as isolated populations of single species or as highly localized communities on more stable sites.

Resedimentation (section 3.1.2) and the stabilization of slopes (section 3.2.2) on glacier forelands involve rapid as well as slow mass movements, both of which are likely to be most important immediately following deglaciation. Lawson (1982) states that sediment flow is the predominant process that reworks deposits at the terminus of the Matanuska Glacier, Alaska. Here the material flows after its strength is reduced by excess pore pressures and seepage pressures generated by meltwater from thawing ice. Theakstone (1982) has described the periodic formation over a 10–15 year period of extensive sediment fans generated by snow melt on the glacier foreland of Austerdalsisen (Svartisen), northern Norway. Similar debris flows or skinflows are common types of mass movement in periglacial environments, where they are often confined to the active layer (Rapp, 1960; Harris, 1981; French, 1988; Lewkowicz, 1988).

A second type of rapid mass movement, the thaw slump (Washburn, 1979), is of importance on glacier forelands where buried glacier ice or extensive ground ice melts. Thaw slumps produce collapsed topography, including topographic depressions, sometimes termed 'thermokarst' (French, 1976). Many examples of collapsed and collapsing glacial deposits occur in Adams Inlet, Glacier Bay, Alaska, where the back-wasting of a kame terrace has been recorded over a 17-year period (McKenzie & Goodwin, 1987). The terrace front has retreated at an average rate of 4.3 m yr^{-1}, destroying the vegetated terrace top and forming a field of low hummocks. Surface lowering has proceeded at an average rate of about 0.25 cm yr^{-1}. However, the resulting increase in the depth of depressions on the terrace top has been insufficient to destroy the vegetation cover. In front of the surging Klutlan Glacier, Yukon Territory, Canada, where interdisciplinary studies have been carried out on massive ice-cored moraines (Wright, 1980), it has been calculated (Driscoll, 1980) that nearly 1200 years may be required to melt completely all the ice in the moraines. The general development of this ice-cored landscape is one of downwasting in which a major topographic reversal takes place: an original ice mass with a gentle convex surface melts to leave a basin floored by a concave mantle of morainic debris (Watson, 1980).

As a result of their episodic and spatially discontinuous nature, rapid mass movement processes have not been investigated as fully as slow mass movement in periglacial environments. However, Lewkowicz (1988) has speculated that sporadic, rapid mass movements are of secondary import-ance compared to continuous, slow mass wasting. Beyond the most recently deglaciated zone and outside areas with ice-cored deposits, this is also likely to be the case on most glacier forelands.

3.2.7   Nivation

The term 'nivation' is used here to describe the suite of geomorpho-logical processes associated with late-lying snow (Embleton & King, 1975b; Embleton, 1979; Thorn, 1988). Although snow is widely regarded as one of the most important environmental influences affecting arctic-alpine plants (see section 3.3.3), it tends to be viewed by ecologists as a passive environmental factor. In contrast, the geomorphological literature sug-gests that certain processes may be enhanced beneath or at the margins of late-lying snowbanks. It seems clear that as a result of the increased moisture availability associated with melting snow, slope wash and soifluc-tion are enhanced (Thorn, 1976; Thorn & Hall, 1980; Hall, 1985). Snow-banks are well-known as traps for mineral and organic particles blown by the wind (see section 3.2.8). Physical and chemical weathering may also be intensified. In addition, certain rapid mass movement processes, such as snow avalanches and slushflows, are commonly associated with melting snow (Washburn, 1979) and, beneath thick snowbeds, considerable forces can be exerted by snow creep (Costin *et al.*, 1973; Mathews & Mackay, 1975).

On glacier forelands, the erosion, transport and deposition of surficial material associated with nivation is likely to be most intense following deglaciation, when sediments are unconsolidated and a vegetation mat is absent (Thorn, 1976, 1988). Possible changes include the enlargement of hollows, the removal of fine sediment from beneath snowbeds producing stone pavements (White, 1972), the development of solifluction lobes downslope, and the continuation of niveo-aeolian deposition. Indeed, Oliver, Adams & Zasoski (1985) indicate that various disturbances asso-ciated with snow affect about half of the foreland of Nooksack Glacier, Washington. Such disturbances are particularly important beneath steep valley sides, where snow accumulates and avalanches occur (see section 6.3.1).

3.2.8 Aeolian processes

Boulton & Dent (1974) have analysed the effects of wind action on a till plain in front of Breidamerkurjökull, Iceland. The stony lag that develops rapidly at the surface of the newly-exposed till (see section 3.2.3) appears to result primarily from the deflation of silt and fine sand by frequent strong winds blowing off the glacier. The stony layer thickens with time in relation to the exposure of the site to wind and is best developed adjacent to the central axis of the foreland where the winds are strongest. Some of the wind-blown material is trapped by vegetation on older sites, but its accumulation is patchy because the first plant community to achieve a complete ground cover is a moss–lichen heath (*Rhacomitrium canescens–Stereocaulon* spp.), which is easily stripped off by the wind following disturbance. Thus, 'armoured' surfaces may be formed in exposed locations close to glaciers, and aeolian horizons may be deposited in sheltered sites downwind.

Aeolian deposition of organic and mineral material has been examined by Baranowski & Pękala (1982) on the glacier foreland of Werenskioldbreen, south-western Spitzbergen. Measurements were made of these two components of deposition on snow patches and comparisons were made with the quantities deposited on snow patches in nunatak and in coastal tundra zones. At a total of 75 sites, the mean annual deposition is 29 g m^{-2}, including 24% organic material, but on the glacier foreland it is considerably higher (85 g m^{-2}) than on the tundra (12 g m^{-2}) or on the nunataks (17 g m^{-2}). The organic component is much less important than the mineral component on the glacier foreland (6% as opposed to 55% and 7% on the tundra and on the nunataks, respectively). The largest quantities, up to 174 g m^{-2} including 4% organic material, are deposited in the end moraine zone of the glacier foreland, where exposed moraine crests provide local source areas of fine sediment. Friedel (1936) has described the erosion of moraine crests and the deposition in vegetated areas of loess-like material up to 3 m thick on the foreland of the Pasterzegletscher in the Austrian Alps.

Deflation also occurs on other exposed or desiccated surfaces, such as glacio-fluvial outwash deposits and wherever cryogenic or other processes disrupt or prevent the development of vegetation to expose a wind-susceptible substrate (Åkerman, 1980). Bogacki (1973) has described the frequent occurrence in front of Skeidarárjökull, Iceland, of dust storms. These are composed of sand and finer grain sizes derived from volcanic dust on the glacier surface as well as from sandur surfaces. The major part of the dust is dropped within 1.0–1.5 km of its source in moist sites and depressions, particularly in old channels, and finer material tends to be transported farthest. In temperate and maritime regions deflation may be strongly seasonal in character, being most likely during dry periods in summer or autumn, when snow cover has largely disappeared, rivers and streams are low, and the surface layers of sediments and soils dry out (Koster, 1988). However, in polar or continental regions with low snowfall, or in particularly exposed topographic situations, it may be more important in winter when winds are strongest (McKenna-Neuman & Gilbert, 1986; Riezebos *et al.*, 1986).

Other relevant geomorphic processes associated with wind include the enhancement of desiccation cracking, enrichment of soils with salts, wind abrasion and niveo-aeolian deposition (Åkerman, 1980; Baranowski & Pękala, 1982; Bryant, 1982; Seppelt *et al.*, 1988). In vegetated areas, desiccation cracks may be confined to the vegetation mat or may penetrate into the soil beneath. Particularly in areas of low rainfall and high groundwater tables, drying of surface soils may be accompanied by the formation of salt crusts. Atmospheric inputs of salt can also be important in windy and maritime regions. As blowing snow and mineral particles produce faceted and polished stones (ventifacts) in windy periglacial environments, they are clearly capable of damaging plant tissue. Niveo-aeolian deposits involve the deposition of a combination of wind-transported mineral particles, snow and organic detritus; these too may directly affect vegetation and soil development.

3.3   The climatic environment

Climate is the third major component of the physical landscape to be examined in this chapter. In view of the general importance of climate in the explanation of plant distribution and ecology (e.g. Woodward, 1987), it is surprising that so little attention has been given to this factor in ecological studies on glacier forelands. As in the case of the substrate, distinct characteristics have often been recognized but homogeneity and stability have been assumed. In reality, the climatic environment on recently-deglaciated terrain exhibits considerable variations in space and in time.

3.3.1   Regional climate

Various combinations of snowfall and temperature regimes are capable of producing an excess of snowfall over snowmelt sufficient to allow the consolidation of snow into ice. Hence glaciers and glacier forelands exist in a large range of locations and climates from polar to equatorial and from maritime to continental. As pointed out in the context of arctic and alpine plants generally, such a great geographical range precludes generalizations on the environment (Bliss, 1971).

At present 10% of the world's land area is glacierized (14.9 million km^2), of which 96% is in Antarctica and Greenland (Flint, 1971). About half of Antarctica has a mean annual accumulation of < 100 mm (Embleton & King, 1975a) and the mean annual accumulation for the whole of Greenland has been estimated at only 340 mm (Hattersley-Smith, 1974). Such low snowfalls produce glaciers only in areas with extremely low temperatures. The mean annual air temperature is − 23 °C at Camp Century, Greenland, and may reach − 50 °C in parts of Antarctica (Sugden & John, 1976). On a world scale, therefore, the distribution of glacier ice is primarily dependent on low temperatures. Extremely high snowfalls are required to maintain glaciers in regions with a mean annual temperature close to 0 °C (Sugden, 1982).

Temperature is also the major factor in the restriction of glaciers to progressively higher altitudes with decreasing latitude, as suggested by snowline altitudes (Barry & Ives, 1974; Price, 1981). The climatic snowline, defined as the average position of the line between seasonal and perennial snow, lies some 100–300 m below the glaciation level, defined as the critical summit elevation necessary to produce a glacier (Charlesworth, 1957; Østrem, 1974; Østrem, Haakensen & Eriksson, 1981). Thus, glaciers form only above about 5000 m in the tropics but may form close to sea level at high latitudes. The main factor accounting for the increase in the altitude of glaciers from maritime coasts towards continental interiors, is decreasing snowfall. This is clearly shown in the Arctic and sub-Arctic. The effect is accentuated by the presence of coastal mountain ranges, as in Alaska and Scandinavia.

Summer temperature conditions at the climatic snowline are strongly related to winter snowfall in polar and temperate latitudes. In southern Norway, for example, mean summer temperature at the climatic snowline varies from about + 4 °C to − 1 °C (Liestøl, 1967; Sutherland, 1984) from west to east. On most of the glacier forelands in this region, temperatures are considerably higher than this. Some of the outlet glaciers from the Jostedalsbreen ice cap descend into the surrounding valleys to altitudes below 300 m. In eastern Jotunheimen, on the other hand, several glacier snouts lie above 1800 m, much closer to the climatic snowline and glaciation limit. Superimposed on the west-east trend is considerable variation in glacier snout and hence glacier-foreland altitudes, accounted for mainly by glacier size (the larger glaciers descending to lower altitudes). In the west, many glacier snouts descend below the tree line, here composed largely of mountain birch (*Betula pubescens*). The glacier forelands therefore lie wholly within sub-alpine or boreal zones (Dahl, 1986; Moen, 1987). In Jotunheimen, however, almost all the glacier forelands lie within the alpine zone. Indeed, in eastern Jotunheimen, most lie within the mid- or high-alpine belts and many fall within the zone of discontinuous alpine permafrost (King, 1983, 1984; Harris & Cook, 1986). The mean annual air temperature on the forelands varies from about $+4\,°C$ in the west to about $-5\,°C$ in the east (Matthews, 1987). Mean annual precipitation also varies considerably between these forelands, within an approximate range of 700–2000 mm (Erikstad & Sollid, 1986).

Such large differences in the regional climatic environment over short distances are associated only with transects across mountain ranges. However, even greater climatic differences can occur between glacier forelands widely separated along latitudinal gradients or gradients in continentality. In particular, many ecologists have stressed the diverse ways in which the climatic environment differs between arctic and alpine ecosystems (e.g. Wright & Osburn, 1967; Billings & Mooney, 1968). Such differences involve not only mean values but also seasonality and diurnal climatic patterns. Seasonal changes in solar radiation, day length, and temperature become relatively unimportant at low latitudes where diurnal climatic rhythms predominate. For example, Barry (1981) contrasts the diurnal and seasonal temperature ranges on Mt. Wilhelm, Papua New Guinea, with those of Niwot Ridge, Colorado, U.S.A., which lie at comparable altitudes (3480 m and 3750 m, respectively). On Mt. Wilhelm, the diurnal range is about 7–8 °C throughout the year, compared with a seasonal range of 0.8 °C for mean monthly temperatures. The corresponding values for Niwot Ridge are 6–8 °C and 21 °C, respectively. Thus glacier forelands near the equator experience 'summer every day and winter every night' (Hedberg, 1964) and snow may fall on any day of the year, a remarkable contrast to climatic conditions in front of glaciers in middle and high latitudes where much of the precipitation occurs as summer rain.

Although not specifically concerned with glacier forelands, the classification and ordination of the climates of selected tundra and tundra-like sites carried out by French & Smith (1985) is useful in emphasizing that there are differences as well as similarities between hemispheres. Using a principal components analysis of a variety of climatic data, they demonstrate that five sub-Antarctic islands form a loosely coherent group which are distinguished from Northern Hemisphere tundras by their extreme oceanicity. General wetness and strong winds combine to produce a significant chill factor and a vegetation structure that resembles that of much higher latitudes in the Northern Hemisphere. The latitudinal difference between northern and southern sites with equivalent vegetation types can be as much as 20° or even 30°. However, the primary production of the southern sites is higher than the northern sites because of soil enrichment (from salt spray, seals and seabirds) and the longer growing season.

The regional climatic environment of any glacier foreland may therefore be considered as a point within a multidimensional continuum of variation. Extremes within this continuum include the polar deserts of the Arctic and Antarctic (Svoboda & Henry, 1987; Aleksandrova, 1988), and temperate climates supporting broad-leaved evergreen forests in New Zealand and South America (Wardle, 1980; Rabassa, Rubulis & Suarez, 1981). Other dimensions of the continuum involve alpine, tropical alpine, continental and oceanic climates. A varied rather than a uniform biological response to the climatic environment should perhaps be expected, although it is possible that similar biological effects may result from interactions between different environmental components (Webber, 1974).

3.3.2   Meso-scale climatic gradients

On any glacier foreland, there exist climatic gradients due to (1) major topographic features (topoclimate) and (2) the proximity of a glacier (glacier climate). These meso-scale climatic patterns are superimposed on the regional climate of the locality.

Topoclimatic gradients result from the influence of slope, relief and aspect on the heat and water balance and also on secondary wind circulations (Geiger, 1969). The slope of the ground affects the angle of incidence of solar radiation, and the flow of water on and near the surface. Relief determines the extent of shadow effects, and can greatly affect the amount and distribution of rain and snow. Aspect results in differences between north- and south-facing slopes and an asymmetry of easterly and westerly slopes in terms of warmth and moisture. In addition, secondary wind circulations, such as up- and down-slope breezes and valley and mountain winds, are commonly produced (Flohn, 1969). These in turn influence temperature and moisture distributions, as in the case of valley-bottom temperature inversions and thermal belts on valley sides (Geiger, 1971; Oke, 1987). Glacier forelands are therefore likely to exhibit important topoclimatic gradients, particularly in the case of valley glaciers, where significant transverse (cross-valley) gradients as well as longitudinal (down-valley) gradients are almost certain to exist.

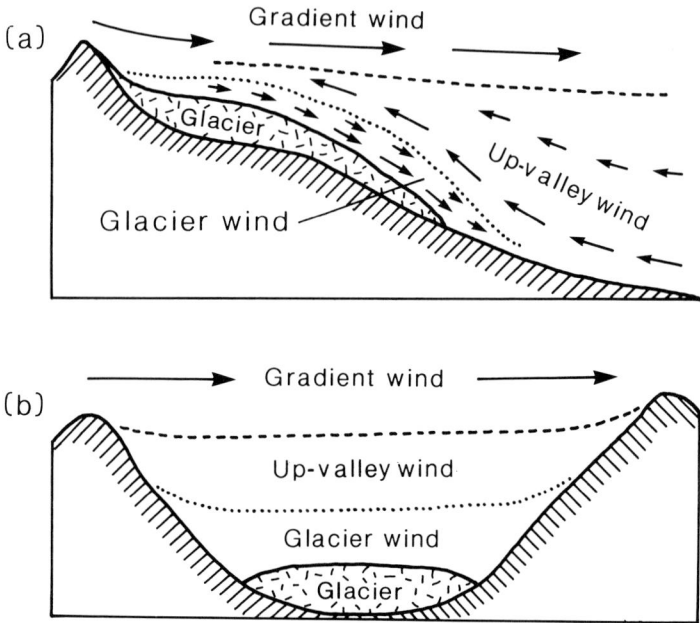

Fig. 3.10. Model of a glacier wind and its relation to gradient and up-valley winds in fine-weather: (a) long profile; (b) cross profile. Dotted line represents the upper limit of the glacier wind (after E. Eckhart, from Geiger, 1971).

Glacier climates have rarely been examined in any detail. However, their existence is immediately brought home to field workers by the cold winds which, almost invariably, are encountered at glacier snouts. Such glacier winds or firn winds were first studied in detail in the European Alps by Tollner (1931, 1935). Geiger (1971) provides a simple model relating the glacier wind to local and regional wind systems (Fig. 3.10). The cold ice surface gives rise to a steep vertical temperature gradient and a cool skin of air slides down the glacier surface with a maximum velocity of about 3.0 m s^{-1} at a height of about 2.5 m (Hoinkes, 1954). During the day, in the fine weather situation shown in Fig. 3.10, the down-valley influence of the glacier wind is restricted by the anabatic or valley wind (blowing up-valley) which develops in response to heating of the valley-side slopes. At night, however, the glacier wind may be reinforced by a katabatic or mountain wind (blowing down-valley) which develops as the valley sides cool. A double velocity maximum characterizes the three Austrian glaciers investigated by Hoinkes (1954): the first before sunrise, corresponding to maximum cold air flow; and the second before sunset, when the temperature difference is greatest between the ice and the surrounding area. Similar double maxima have been observed on the Worthington and Castner Glaciers in the Brooks Range of central Alaska (Streten & Wendler, 1968). In contrast, the McCall Glacier (north-eastern Alaska) exhibits a single nocturnal maximum and resembles more closely the summer flow at the margins of the Greenland and Antarctic ice sheets (Streten, Ishikawa & Wendler, 1974).

Glacier winds may extend to a vertical height of only a few tens of metres on small glaciers. In front of Nigardsbreen, a large valley glacier in southern Norway, well-developed glacier winds involve a thicker air layer, extending to vertical heights of 100 m or more (Evers, 1951). At the margins of the Greenland and Antarctic ice sheets vertical thicknesses of several hundreds of metres can develop with average annual velocities of 19 m s^{-1} and averages as high as 45 m s^{-1} on individual days (Flohn, 1969). Although the strengths of glacier winds are obviously influenced by the regional gradient-wind and by topography, they are normally remarkably persistent. For example, at the snout of the Pasterze Gletscher in the Grossglockner massif (Austria), the relative frequency of glacier winds at 14.00 hrs during July and August reaches 90% (Flohn, 1969). However, they also tend to die out within a relatively short distance of the glacier snout because the air is slowed by friction and because it is thermally modified in its lower layers (Oke, 1987).

Useful though limited data on glacier climates are available from temporary meteorological stations set up during the summer months. For example, Wójcik (1973) obtained data on sunshine duration, solar radiation, wind velocity, precipitation, relative humidity and temperature from a site on the crest of the outermost end moraine in front of Skeidarárjökull, Iceland. The station was located at a distance of 3 km from the glacier snout and 20 km from the sea, and at an altitude of 114.5 m above sea level. His data show that during the day in fine weather, atmospheric pressure falls over the foreland and the sandur beyond as the ground surface becomes hotter; air temperatures rise from as low as 2 °C to as high as 19 °C, and relative humidity falls to below 80%. At the same time a sea breeze gradually increases its range inland and reduces the influence of the glacier wind until about noon when the latter reaches no farther than the glacier snout. At the convergence of the sea breeze and the glacier wind a wall of convective cloud develops. During the night, the influence of the glacier wind extends beyond the foreland where it is reinforced by a developing land breeze. The main limitations of such temporary meteorological stations are that they may be unrepresentative of longer time periods and may also produce data that are highly site specific (cf. Derbyshire & Blackmore, 1975).

Temperature conditions have been investigated in more detail in front of Skaftafellsjökull, Iceland, at eight sites located between 8–1200 m from the glacier margin (Lindröth, 1965). All sites were located on horizontal ground in unshaded positions, a continuous vegetation cover present at the two most distal sites having been removed. Thermometers were placed horizontally, at 150 cm and 5 cm above the till plain and at 3 cm below the surface of the ground, and were shaded when measurements were made. Readings at hourly intervals through 24-hour periods reveal a major effect of the glacier within about 50 m of the ice margin, diminishing to negligible effect at a distance of some 600 m. Under clear skies, maximum air temperatures at the glacier snout are depressed by about 5 °C at 150 cm and by about 9 °C at 5 cm above the ground (Fig. 3.11a). The gradients are much weaker under cloudy conditions, and/or at night, when a difference of only 2–3 °C develops between the most proximal site (site 1) and the most distal site (site 8) at 5 cm above the ground (Fig. 3.11b). It should be noted, however, that the night-time gradient in minimum air temperatures under clear-sky conditions is in the opposite direction, with temperatures about 2 °C warmer at the glacier snout than at site 8. This is caused by the interaction of the glacier wind with rapid radiative cooling on the till plain. Near the ground, the latter creates a stable layer of cool air, which the glacier wind prevents from developing close to the glacier.

One consequence of all the gradients shown in Fig. 3.11 is the relatively narrow range of air temperatures characteristic of sites close to the glacier, both maximum and minimum temperatures becoming more extreme with increasing distance from the glacier snout. Furthermore, at 3 cm below the ground surface, maximum temperatures may be 5 °C or more higher than the corresponding air temperature at 5 cm above the ground, even near the glacier snout. Minimum ground temperatures are also higher and the differences between the daily extremes are greater. Although the strong ground temperature gradients developed on sunny days may be maintained during the night, they do not extend as far from the glacier (site 4 at 190 m from the glacier being closely similar to sites on older terrain).

Similar effects have been measured in the Alps (Schreckenthal-Schimits-chek, 1933; Maizels, 1973), Norway (Ballantyne & Matthews, 1982; Vetaas, 1986), British Columbia (Fraser, 1970), Baffin Island (Rannie, 1977) and Yukon Territory (Nickling & Brazel, 1985). Whilst these limited data suggest that Lindröth's (1965) results are fairly representative of conditions elsewhere, it is clear that climatic gradients associated with glacier climate

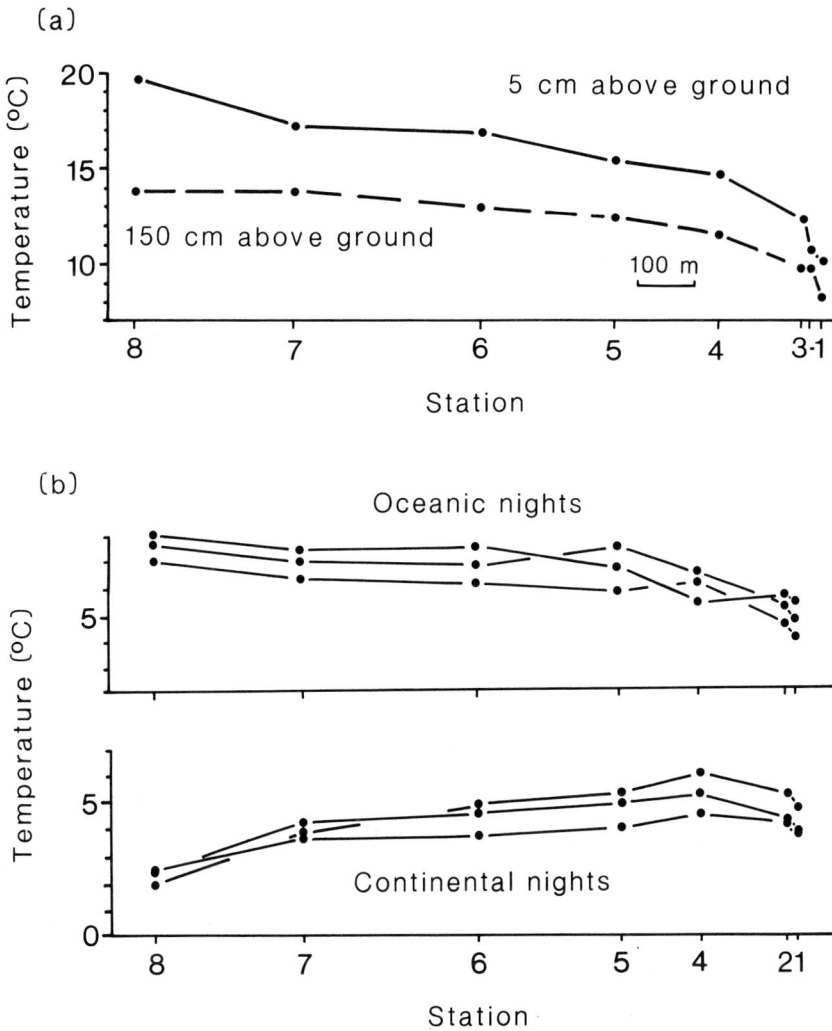

Fig. 3.11. Air temperature variations with distance from the glacier margin of Skaftafellsjökull, Iceland: (a) air temperature at 12.00 noon, August 8th 1962, at 5 cm and at 150 cm above the ground surface; (b) minimum air temperature in July during three oceanic and three continental nights at 5 cm above the ground. Glacier is to the right (from Lindröth, 1965).

vary considerably from foreland to foreland and are strongly related to glacier size. In all cases the steepest gradients are found close to the glacier, particularly within the first 50 m or so. In relation to cirque and small valley glaciers, such as the Glacier des Bossons in the French Alps (Maizels, 1973), the glacier climate appears to be insignificant at distances greater than about 200 m from the glacier snout. However, in front of large valley glaciers, ice-caps and ice-sheets, or where topography and gradient-wind conditions are favourable, the influence of the glacier climate can extend for many kilometres.

One further glacier-related environmental gradient should be mentioned, although it is not strictly climatic as it is closely related to glacier hydrology. This is the ground moisture gradient with distance from the glacier snout. Ballantyne & Matthews (1982) give the moisture content within the uppermost 5 cm of the till plain in front of Storbreen under clear-sky conditions at 16.00 hrs in August. Whereas the till on most parts of the glacier foreland is well drained, meltwater at the ice margin continually recharges groundwater in the adjacent substrate so that a steep gradient is found within about 40 m of the ice margin. The percentage moisture content by weight falls from about 70% at a distance of 10 m from the glacier snout to < 10% at a distance of 50 m, a somewhat steeper gradient than the corresponding ground temperature gradient (Fig. 3.12). With the low water-holding capacity of sediments and soils, well-drained sites on glacier forelands are likely to experience high moisture deficits during periods of low rainfall in summer. Thus, except where the ground water table is close to the surface, or at sites fed by streams, snowmelt or other water sources, glacier foreland habitats are likely to be relatively xeric (cf. Fraser, 1970). On older terrain, the moisture content may increase slightly in response to changes in the soil and/or vegetation cover (Crocker & Major, 1957).

3.3.3   Microclimate

Microclimate involves the characteristics of climate which vary over short distances, both horizontally and vertically, in response to small-scale topographic variation and differences in the nature of the ground surface and substrate. There is clearly no sharp dividing line between the mesoclimatic gradients discussed in the previous section and microclimatic gradients, which are as diverse as the possible combinations of factors that control them (Geiger, 1971; Rosenberg, 1974; Oke, 1987). Here, attention will be confined to several recurring effects which are found on glacier

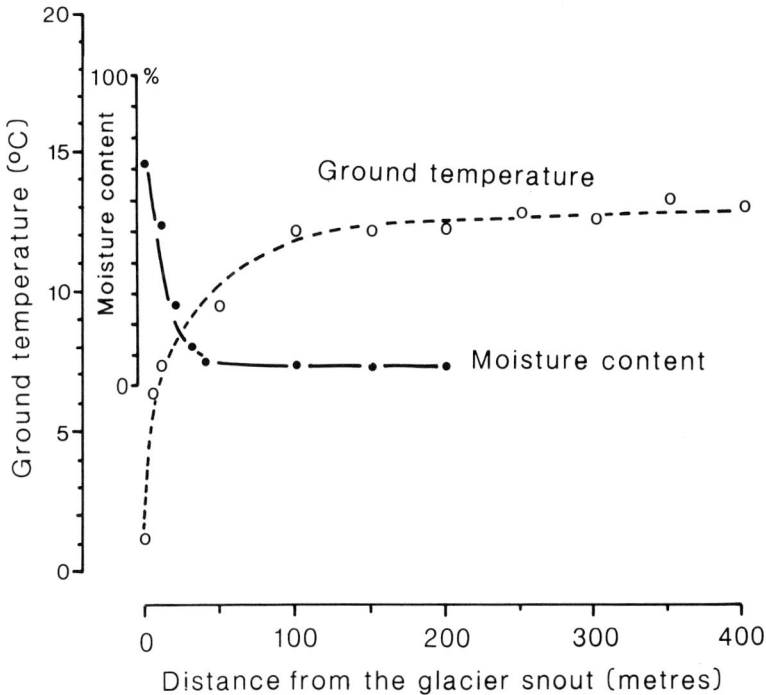

Fig. 3.12. Ground temperature at 5 cm depth and soil moisture content (% by weight) within the uppermost 5 cm, with increasing distance from the glacier margin of Storbreen, Jotunheimen (at 16.00 hrs under clear-sky conditions in August) (after J. A. Matthews & P. J. Beckett, from Ballantyne & Matthews, 1982).

forelands and are likely to be important in soil development and plant succession.

Differences in the air temperature at 5 cm above the ground surface have been compared for proximal (south-facing), crest and distal (north-facing) sites on five moraine ridges in front of Bödalsbreen, southern Norway (Vetaas, 1986). Measurements were made during two contrasting time periods, late morning (11.00–11.30 hrs) and early evening (18.00–18.30 hrs), on several occasions, mostly under cloudy conditions. Strong temperature gradients with distance from the glacier develop by the late morning, when relatively high temperatures characterize the proximal slopes on four of the five moraines. Only on the moraine nearest the glacier (distance about 1 km) is the proximal slope cooler. On one occasion, with a clear sky and a strong glacier wind, the temperature on the proximal slope of the

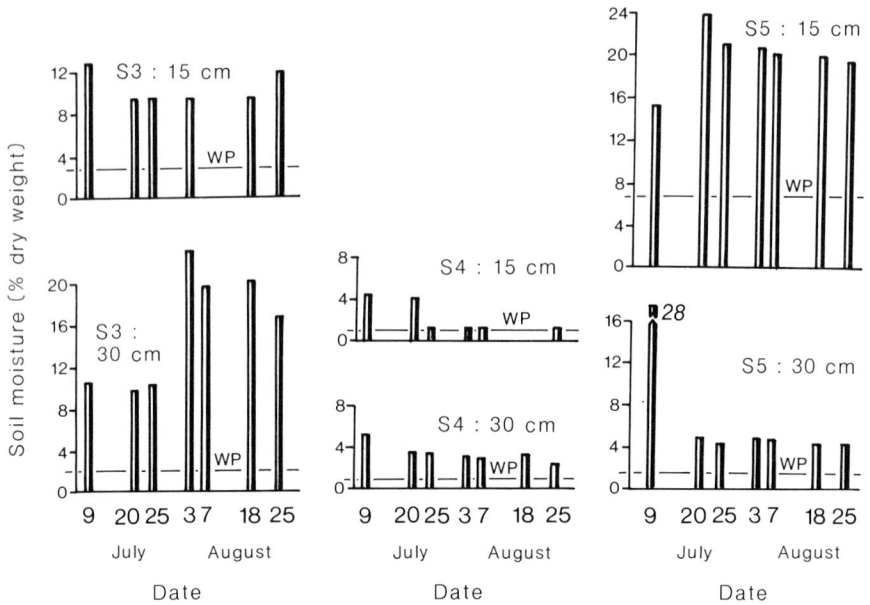

Fig. 3.13. Summer soil moisture regime (% by weight) in relation to wilting point (WP) at depths of 15 cm and 30 cm for three sites (S3, S4 and S5) on the foreland of Sentinel Glacier, British Columbia, Canada (from Fraser, 1970).

oldest moraine (about 1.75 km from the glacier) was > 6.0 °C higher than on the distal slope. At the same time, moraines closer to the glacier exhibited progressively smaller temperature differences between the proximal and distal slopes; and on the youngest moraine, the proximal slope was 1.8 °C cooler than the distal slope. Thus, on the older moraines of this glacier foreland, a south-facing aspect can outweigh the glacier wind effect.

Data from a number of microenvironmental stations established on the forelands of Sentinel and Warren Glaciers in the subalpine zone of Garibaldi Park, British Columbia, have been analysed by Fraser (1970). Five stations were located on each foreland, within or adjacent to sample plots used to define vegetation associations. Soil moisture contents were measured at various times during one field season using soil moisture resistors at depths of 15 and 30 cm. Evaporation (using a modified Piche evaporimeter) and minimum ground temperatures (based on unshaded maximum-minimum thermometers) were also recorded.

Soil moisture records for three sites in front of Sentinel Glacier are shown in Fig. 3.13 (Fraser, 1970). Permanent wilting points of the soils were determined for the 2 mm fraction using a pressure-membrane apparatus.

Site S3 (Junco drummondii-Lupino-Salicetum barclayi association) is influenced by the Garibaldi Lake water table, which rises in early August due to high seasonal snowmelt. Soil moisture content therefore remains at or above 10% throughout the summer. In contrast, after initial snowmelt water drains from the profile, site S4 (Junco parryi-Lupino-Salicetum barclayi association) receives no auxilary water supply other than precipitation. Soil moisture levels at a depth of 15 cm therefore fall rapidly to below 2% by the end of July and remain close to the wilting point. Site S5 (Lupino-Abieto-Tsugetum mertensianae association) consists of stratified glacio-fluvial deposits with a considerable fine silt fraction at 15 cm but a very coarse sand layer beneath. At the beginning of the field season, lateral seepage occurs from a meltwater stream along the layer of coarse sand. As soon as this water source disappears in early July, soil moisture falls to a relatively constant level of about 4%. Higher moisture contents at 15 cm depth can be attributed at least in part to the greater water-holding capacity of the finer-textured sediments. Although all the stations experience summer frosts, the thermometer records show variations in frost frequency, which depend in particular on topographic position and air drainage. Variations in rate of evaporation appear to be closely related to exposure to down-valley winds (Fraser. 1970).

Lindröth (1962) examined the temperature gradient with distance from a perpendicular rock wall composed of basalt on the Skaftafellsjökull glacier foreland, Iceland. Air temperatures were measured at 5 cm and at 150 cm above the ground surface at distances of 0.5 to 25.0 m from the rock wall. During persistent sunshine, the effect of heating and cooling of the rock wall on the air temperature is detectable at distances of up to about 2.0 m. With varying sunshine conditions, air temperatures at a distance of 0.5 m from the rock wall and at 5 cm above the ground fluctuate rapidly within a 4 °C range.

The effects of soil characteristics and a variable vegetation cover on air and soil temperatures were also examined by Lindröth (1962) who considers that variations in particle size and moisture content account for most of the variability about the main trends in Fig. 3.11. His data point to the interactions between these variables as well as their independent effects (see also section 6.1.3). In coarse gravel, air circulation from the surface goes deeper, whereas finer-grained substrates register relatively high temperatures in the surface layers and retain their heat longer. However, the water-holding capacity of soils rapidly increases where particle sizes are < 0.2 mm and evaporation from such surfaces (e.g. after rainfall) produces a cooling effect.

The measurements referred to above relate only to summer conditions, and take no account of snow depth and distribution. The small-scale pattern of snow accumulation depends on the interaction of wind and topography. Snow is one of the most important microclimatic factors in arctic-alpine environments, in relation to soils (Retzer, 1974; Ellis, 1980; Rieger, 1983), animals (Remmert, 1980; Marchand, 1987), and plants (Gjaerevoll, 1956; Billings, 1974; Chernov, 1985). Its significance for plant growth lies mainly in its influence on air temperature, wind and moisture conditions near the ground surface, and on the length of the growing season (Friedel, 1952; Dahl, 1956; Billings & Bliss, 1959; Detwyler, 1974; Scott, 1974a; Miller, 1982; Wijk, 1986a,b).

The remarkable similarity in the pattern of melting of snow from year to year has been the subject of detailed work by Friedel (1952) for an area of 32 km² in the eastern Alps near the Pasterze Gletscher. Based on a photogrammetric survey of the melting snow made in 1935, supplemented by direct measurements on the ground, the area was subdivided into 620 sections enabling the preparation of charts showing the percentage of the ground becoming snow-free by certain dates as a function of altitude, aspect and slope. Smaller-scale investigations of the depth and duration of snow cover have been made by Elven (1974) on the glacier foreland of Blåisen, Hardangerjökull, southern Norway. In this location in the alpine zone, some ridge crests are snow-free for virtually the whole year, whereas snowbed sites are commonly snow-free for three months or less. On three glacier forelands in the sub-alpine zone of Garibaldi Park, British Columbia, various plant communities are associated with snow-free seasons of 2–5 months (Fraser, 1970).

3.3.4   Climatic change

The existence of recently-deglaciated terrain in front of glaciers from all parts of the globe is indisputable evidence of recent world-wide climatic change. Many lines of evidence indicate that the 'Little Ice Age' glacier expansion episode was primarily a response to a global cooling (Williams & Wigley, 1983; Lamb, 1984). Similarly, the retreat from the 'Little Ice Age' glacier maximum appears to have been caused primarily by a global warming of some 1–2 °C.

Superimposed on this trend have been shorter-term climatic fluctuations. These include the climatic deteriorations that produced glacier advances (and resulted in the formation of end moraine ridges). On many glacier forelands, such as Storbreen (Figs. 1.1 & 2.4), the general pattern has been one of accelerating retreat with fewer glacier advances towards the present day. During the twentieth century, both the newly-exposed pioneer zone and the older parts of the glacier foreland have experienced relatively high average temperatures. The twentieth century warming appears to have been more marked at higher latitudes. For example, on Svalbard, mean annual temperatures rose by about 6 °C between A.D. 1900 and A.D. 1930 (Kelly *et al.*, 1982; Brázdil, 1988). Thus, during their relatively short history, most glacier forelands have experienced a changing regional climate with a dominant warming trend and minor fluctuations.

Particular sites are also affected by climatic change resulting from their changing positions relative to the local glacier climate (see section 3.3.2). The existence of a glacier climate varying systematically with distance from the ice margin means that any area of terrain experiences a different climate on exposure from beneath the retreating glacier than it experiences later in time. The rate of glacier retreat determines the rate of climatic amelioration at newly-exposed sites, while the size of the glacier is an important determinant of the scale of this amelioration. In addition, as has been clearly recorded in the reduced growth of certain trees (see section 2.2.2), sites may experience a deteriorating climate during the approach of an advancing glacier.

Although either element can exhibit short-term reversals, the long-term warming trends in the regional and the glacier-climatic elements of change tend to be mutually reinforcing, leading to a strong amelioration of local climatic conditions following deglaciation. In terms of a chronosequence approach, this type of change is a problem because it is correlated with terrain age and other environmental gradients.

3.4 Spatial variation and change in the physical landscape

It is clear from the above survey that the physical environment of recently-deglaciated landscapes is characterized by spatial variation and change rather than uniformity and stability. Topography, parent materials and climate all exhibit spatial patterns at various scales. They also experience temporal changes of different types and of varying duration. This section summarizes this variability and dynamism with particular reference to generalizations relevant to a discussion of soil development and vegetation succession (chapters 4 to 6).

3.4.1 Spatial patterns at various scales

First, at the largest scale, there are regional variations in the nature and extent of glacier impact on the landscape and proglacial landscape

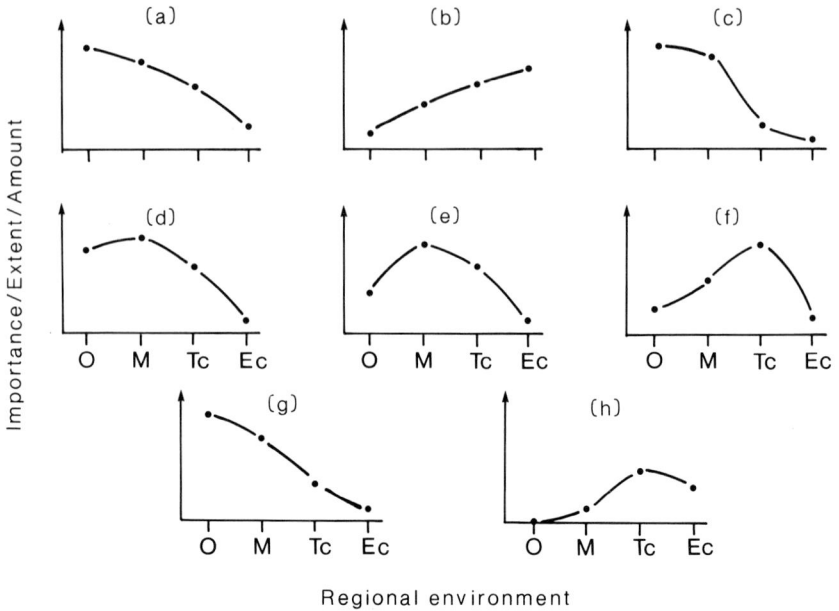

Fig. 3.14. Variations in some glacier characteristics and in the nature and extent of glacier impact on proglacial landscapes in relation to four regional environments (types of glacier landscape): oceanic (O), maritime (M), temperate-continental (Tc) and extreme continental (Ec). (a) Accumulation at the equilibrium line and precipitation on the glacier foreland; (b) altitude of the equilibrium line and glacier snout; (c) rate of glacier erosion and glacier-bed transformation; (d) extent of glacier variations; (e) degree of development of supraglacial and lateral moraines; (f) degree of development of terminal moraines; (g) degree of development of glacio-fluvial landforms, sediments and activity; (h) quantity of buried ice beneath the foreland (from Golubev & Kotlyakov, 1978, with additions).

modification. These produce systematic variations between glacier forelands in terms of their size, landforms and parent materials, and the nature and extent of subsequent reworking and disturbance. Some of these regional variations are summarized in Fig. 3.14, which compares glaciers and glacier foreland landscapes from *oceanic* (O), *maritime* (M), *temperate-continental* (Tc) and *extreme-continental* (Ec) regions from temperate and sub-polar latitudes (Golubev & Kotlyakov, 1978). At least three of these types are present within most large glacierized regions of the temperate zone. Examples are provided by: the Olympic Mountains, U.S.A., and the eastern coast of Kamchatka (O); the central Alps and western Caucasus (M); the central Altai and Tien Shan (Tc); and the Suntar Khayata and eastern Pamirs (Ec).

As continentality increases, so precipitation decreases (Fig. 3.14a), glacier foreland altitudes rise along with equilibrium line altitudes (b), ice velocity and the rates of subglacial erosion and glacier-bed modification decrease (c), and the rate of glacier variations and hence glacier foreland size are reduced (d). The *extreme-continental* type, associated with relatively inert glaciers (which tend to freeze to their beds) and least subglacial meltwater activity, is characterized by minimal development of supraglacial and lateral moraines (e), terminal moraines (f) and glacio-fluvial landforms and sediments (g). Although conditions for the formation of all three types of landform are inversely related to continentality, moraines are poorly developed in *oceanic* regions because their survival is unlikely where there is a combination of large glacier variations, high glacier velocities and large meltwater runoff. Thus lateral moraines are best developed in *maritime* regions. Terminal moraines, which are more liable to erosion by meltwater streams, tend to be best developed in *temperate-continental* regions. Further regional differences are indicated in terms of an increase in the quantity of buried ice within moraines, and in glacier foreland landscapes generally, from *oceanic* to *temperate-continental* regions (h). This suggests a parallel increase in major disturbance due to thawing. However, other kinds of post-depositional modification, including other kinds of cryogenic activity, are likely to respond to different aspects of the regional climate already outlined in section 3.3.1.

Another useful framework is provided by Gorbunov (1978), who classifies mountain regions into eight major geocryological types named after the mountain ranges where they are best developed. This framework emphasizes the predominant importance of latitude and altitude on periglacial morphoclimatic zonation. However, it takes into account the fact that latitudinal and altitudinal morphoclimatic zones can only be crudely equated (Tricart & Cailleux, 1973) and allows for the distinctiveness of areas such as the equatorial mountains (Büdel, 1982; Young, 1989). In Gorbunov's scheme, each type is characterized by a distinctive series of altitudinal belts, defined on the basis of the presence or absence of permafrost, seasonally-frozen ground or shorter-term periodic freezing (Fig. 3.15). Thus, depending on their location and size, glacier forelands may exist in and exhibit the cryogenic characteristics of up to four altitudinal belts.

At the glacier foreland scale, numerous spatial patterns are again recognizable in terms of the glacial legacy, proglacial modification and climate. These include zonations and gradients that are in the same direction as the terrain-age gradient, and hence may be confounded with it

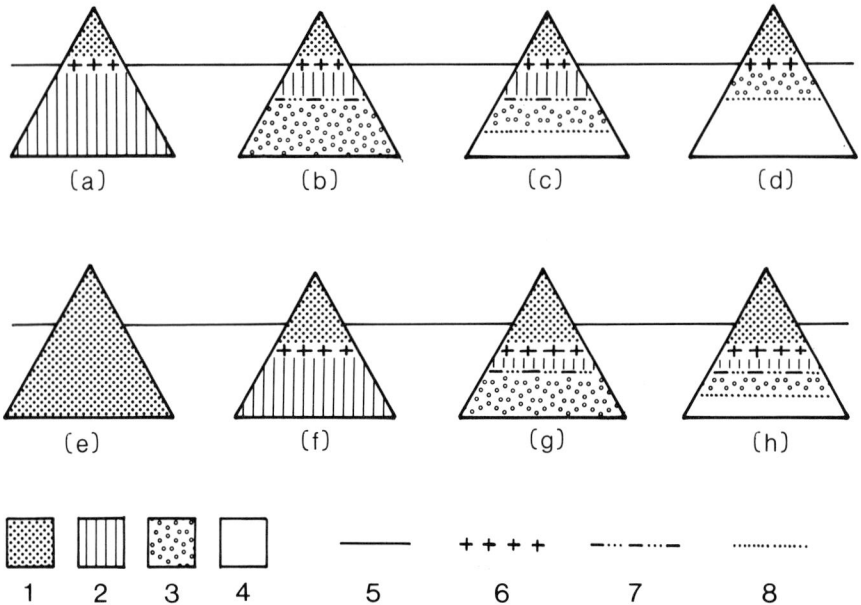

Fig. 3.15. A classification of mountain periglacial environments (types of mountain geocryological region) based on the nature and altitudinal pattern of periglacial belts: (1) permafrost belt; (2) belt of seasonally-frozen ground; (3) belt of short-term periodically-frozen ground; (4) frozen ground absent; (5) snowline; (6) lower limit of permafrost; (7) lower limit of seasonal freezing; (8) lower limit of short-term freezing. (a)-(d) Oceanic types: (a) Chugach, (b) New Zealand, (c) Himalayan, (d) Ecuadorian. (e)-(h) Continental types: (e) Verkhoyansk, (f) Tien Shan, (g) Tibetan, (h) Central Andean. (from Gorbunov, 1978).

in ecological studies. They also include other systematic patterns that are unrelated to the terrain-age gradient. Both are illustrated in relation to landforms and parent materials on the geomorphological map of the glacier foreland of Breidamerkurjökull, Iceland (Fig. 3.2). Broad morphogenetic zones can be recognized with increasing distance from the glacier margin (see also Galon, 1973).

Although the precise patterns may be very different, most glacier forelands exhibit systematic spatial variations in various aspects of the physical environment. Numerous examples of spatial variation in the same direction as the gradient of terrain age are provided in the previous sections of this chapter. These include: the particle-size characteristics of glaciofluvial deposits (see section 3.2.1); the slope angles of moraine slopes and the degree of consolidation of till (section 3.2.2); the degree of development of frost-sorted patterned ground (section 3.2.3); the extent of aeolian deposits (section 3.2.5); and the existence of a glacier climate (section 3.3.2). Systematic variation parallel to the ice margin is most likely in association with valley glaciers where, for example, there may be a greater concentration of glacio-fluvial deposits near the valley axis (section 3.2.1), glacially-eroded bedrock may be more commonly encountered on valley sides (section 3.1.1), and aspect can produce a considerable topoclimatic contrast between opposite valley side slopes (section 3.3.2).

Superimposed on the systematic patterns of the type just described are smaller-scale physical environmental patterns that include the patchy effects of snow banks on nivation processes and microclimate (sections 3.2.7 & 3.3.3), the linear effects of moraine ridges or stream channels on drainage and microclimate, and the repeated erosional and depositional patterns associated with bedrock outcrops. For example, Hallet & Anderson (1980) have described micro-scale patterns on limestone surfaces exposed by the retreat of the Castleguard Glacier, at the southern edge of the Columbia Icefield, Canadian Rocky Mountains. Areas dominated by former intimate ice–rock contact are either abraded or covered with subglacial carbonate precipitate and solutional features. The remainder of the bedrock surface comprises an interconnected network of irregular cavities, which represent areas of former glacier-bedrock separation. Similar features have been described in front of the Glacier de Tsanfleuron, Switzerland (Sharp, Gemmell & Tison, 1989).

Whittaker (1987, 1989) has made a unique attempt to quantify the interrelationships among a variety of environmental variables (including soil variables) on the Storbreen glacier foreland and to examine their relationship to the terrain-age gradient. At 108 sites where vegetation composition was also recorded, he quantified 11 environmental variables in addition to terrain age (see section 5.4.3 for precise definitions). Kendal's tau (τ), a non-parametric correlation coefficient, was used to measure the strength of relationships between variables. Non-metric multidimensional scaling was then used to provide a visual representation of the correlation matrix in the form of an environmental plexus diagram (Fig. 3.16).

The principal feature of the environmental plexus is the group of variables comprising terrain age, frost churning, disturbance, altitude, soil depth, litter depth and depth of rooting. The remaining variables are clearly separate from this group, termed the *terrain-age factor complex* by Whittaker (1987). Three (moisture regime, time of snowmelt and exposure) suggest a second group with somewhat weaker interconnections. The latter group can be interpreted as a *microtopographic factor complex* (Matthews & Whittaker, 1987). Although it represents a simplification, this analysis is

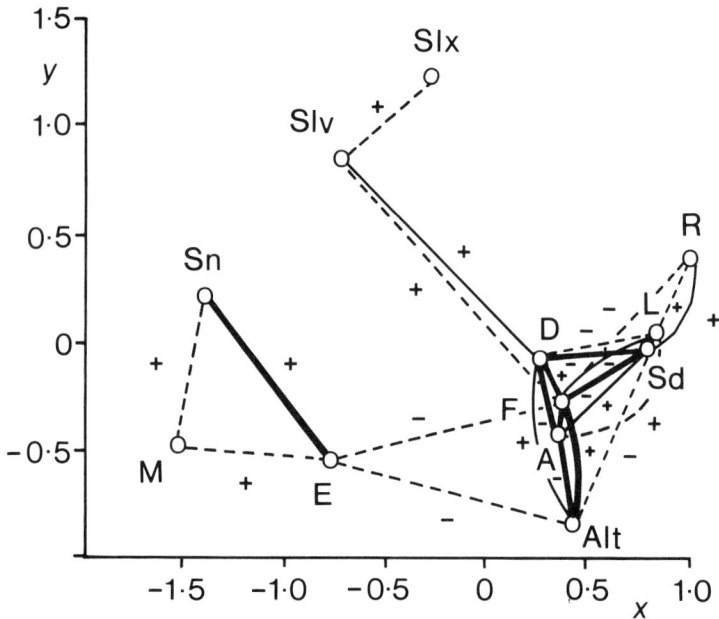

Fig. 3.16. Environmental plexus defined by non-metric multidimensional scaling of intercorrelations between 12 variables on the Storbreen glacier foreland, Jotunheimen: (A) terrain age, (Alt) altitude, (D) disturbance, (E) exposure, (F) frost churning, (L) litter depth, (M) moisture, (R) root depth, (Sd) soil depth, (Sn) snow melt, (Slv) slope movement, (Slx) slope maximum. Correlation coefficients (τ): $> \pm 0.3$ (- - -), $> \pm 0.4$ (——), $> \pm 0.5$ (▬▬) (from Whittaker, 1987).

extremely valuable as it demonstrates two things. First, there are many environmental patterns strongly related to the pattern of terrain age. It is possible that these variables are causes or effects of the time-dependent processes of soil development and vegetation succession. Second, the other environmental variables, which are not correlated with terrain age, may be truly independent variables.

3.4.2   Physical processes and landscape change

On exposure from beneath the retreating glacier, major physical environmental changes occur in the proglacial landscape. Boulton & Dent (1974) provide a model of the processes affecting parent materials on the till plain in front of Breidamerkurjökull, Iceland (Fig. 3.17). Major changes include the deflation and downwashing (pervection) of silt from the near-surface layers creating a lag, the production of a horizon of maximum silt accumulation that in turn moves down the profile, and the accumulation of

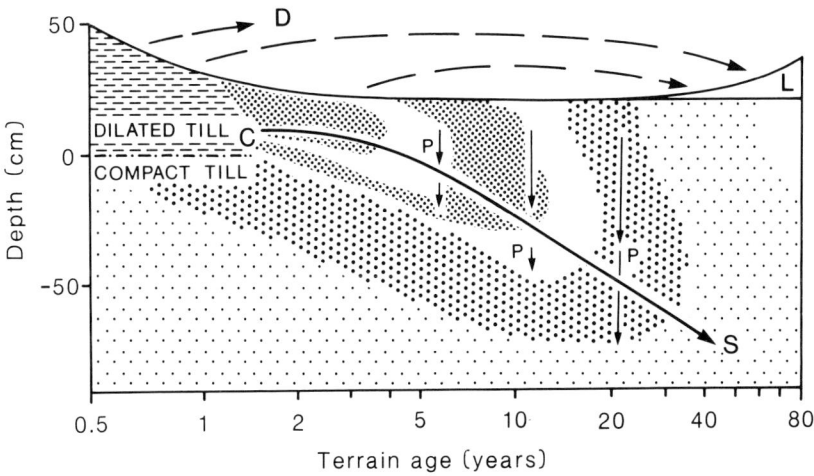

Fig. 3.17. Model of processes affecting till texture with time since deglaciation. Shaded areas give a schematic representation of changes in % fines (fine sand, silt and clay) with time and depth; C, collapse of dilatant structure; P, pervection; S, horizon of maximum silt accumulation; D, deflation; L, loess-like deposition (from Boulton & Dent, 1974).

loess-like wind-blown material at greater distances from the glacier margin (see sections 3.2.2, 3.2.3 & 3.2.7). The rapidity and magnitude of the physical environmental changes that parallel the early stages of ecosystem development indicate the inadequacy of any concept of pedogenesis or vegetation succession that assumes a stable physical environment (cf. Matthews & Whittaker, 1987).

The sum of the effects of the physical environmental processes acting through time might be expected to produce an overall topographic simplification of the landscape, as positive landscape features (such as ridges and mounds) are degraded and negative features (e.g. depressions) become filled in. However, preliminary work by Elliot (1989) suggests that within the first 40 years of deglaciation at Austre Okstindbreen, northern Norway, the fractal dimension, and hence the roughness of the landscape, may increase as a result of other processes (e.g. dissection or the upheaval of large blocks by frost heave).

These and related changes may be viewed as representing the adjustment of the newly-deglaciated landscape to subaerial environmental conditions. In the context of longer timescales and glacio-fluvial sedimentation, Church & Ryder (1972: 3049) term such processes *paraglacial* and define them as: 'nonglacial processes that are directly conditioned by glaciation'.

They refer 'both to proglacial processes, and to those occurring around and within the margins of a former glacier that are the direct result of the earlier presence of the ice'. Church & Ryder also refer to the paraglacial period as the time interval during which paraglacial processes occur. Glacial deposits are unstable in the proglacial fluvial environment; they provide easily-accessible material for fluvial erosion, transport and deposition, and the quantity of material available bears no relationship to concurrent production of debris by weathering. A similar paraglacial concept has been applied to mass movement processes (Johnson, 1984) and cryogenic (periglacial) processes (Thorn & Loewenherz, 1987).

Nitychoruk & Dzierzek (1988) suggest that present-day slope processes (including debris falls, debris flows, snow-debris avalanches, deflation, solifluction and frost weathering) are most active in the zone of contact between slopes and glacier snouts in Wedel Jarlsberg Land, Svalbard. They attribute the enhancement of these processes partly to the glacier climate. However, as debris flows and snow-debris avalanches are the most important processes, and they are caused by heavy rains and snowmelt, the increase in activity in the ice-contact zone is better regarded as primarily a paraglacial effect, caused by the availability of debris for transport by rapid mass movements.

The paraglacial period is one in which rapid and large-scale initiating processes are important. Subsequently, the *normal* periglacial regime is characterized by sustaining processes, such as sheet wash, rock fall and frost creep (Thorn & Loewenherz, 1987). Thus, the paraglacial period has been described as exhibiting relaxation as the landscape attempts to equilibrate. The notion of an asymptotic decline in the influence of paraglacial processes with distance from a glacier margin and time since deglaciation is represented in Fig. 3.18. Although different processes operate at different rates, a zone and period of *dominance* by paraglacial processes may therefore be recognized. However, it should be pointed out that some processes are catastrophic and/or are initiated by complex trigger mechanisms, particularly during the early stages of the paraglacial period. In the context of mass movement phenomena, this has led Johnson (1987) to question the applicability of steady state or dynamic equilibrium models in the paraglacial landscape.

Whether or not paraglacial or other physical environmental processes are tending towards equilibrium, their presence on recently-deglaciated terrain has largely been ignored by ecologists in their enthusiastic pursuit of biological processes to explain vegetation succession and soil development.

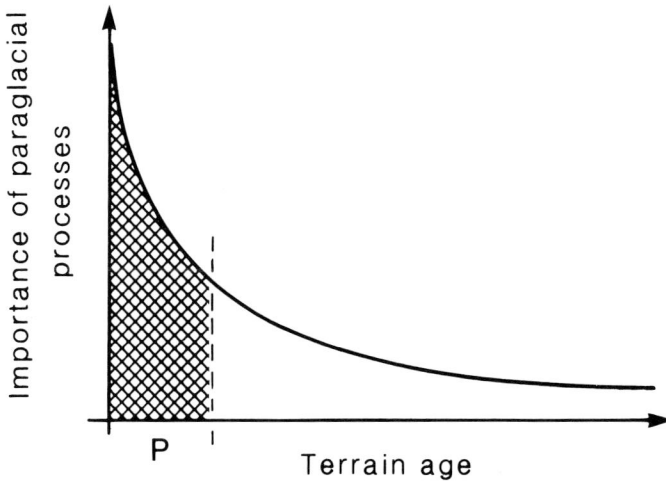

Fig. 3.18. Schematic representation of the asymptotic decline in the importance of paraglacial processes with increasing distance and time from the margin of a retreating glacier. Shaded area suggests the paraglacial zone/stage.

This chapter has presented a detailed account of the physical landscape, and particularly the physical processes, in order to counteract this trend. Several of these processes are of a magnitude and type to be of at least potential ecological relevance. The following chapters make the case for including physical processes as an *essential* part of glacier foreland ecology.

4

Soil development

Soil, by definition, is both a physical and a biological phenomenon. It can therefore be viewed as a vital link between the physical and biological components of the landscape. In this chapter, soil development on recently-deglaciated terrain is considered in its own right. Later chapters will consider its role in relation to the development of the vegetation cover and the landscape as a whole.

4.1   Soil chronosequences and chronofunctions

4.1.1   Conceptual framework

The recognition of a chronosequence, and its quantitative expression as a chronofunction, presupposes that time can be isolated as a factor in soil formation. This notion, which is intimately linked with the concept of *independent* soil-forming factors, appears to have been originated in the late nineteenth century by the Russian soil scientist Dokuchaev. In a series of publications between 1877 and 1900, Dokuchaev recognized climate, subsoil (parent material), vegetation, fauna, man, relief and age of the land surface as significant pedogenic factors (cited in Crocker, 1952). Although Dokuchaev stressed that the soil is the expression of all these factors, he seems also to have believed that they could vary in their relative importance and hence that they could be regarded as statistically independent (Crocker, 1952).

Clarification of the nature of soil-forming factors, the explicit recognition of soil chronosequences and chronofunctions, and the formalization of the functional factorial methodology has been based largely on the work of Jenny (1941, 1946, 1958, 1961, 1980), although other representations have been proposed (see, for example, Stephens, 1947; Crocker, 1952; Olsson, 1958; Stevens & Walker, 1970; Runge, 1973; Johnson, Keller & Rockwell, 1990) and the whole approach is not without limitations that have sometimes given rise to heated debate (Chesworth, 1973, 1976a,b; Yaalon, 1975, 1976; Huggett, 1976). Jenny's formulae will be adopted here, beginning with his 'fundamental equation of soil-forming factors':

$$s = f(cl, o, r, p, t \ldots) \tag{1}$$

In the fundamental equation, a particular soil property (s), or indeed the soil as a whole, is a function of, or is dependent on, the various soil-forming factors, which include: climate (cl) in the sense of regional climate; organisms (o) in the sense of the available biota; relief (r) in the sense of the original topography and incorporating the water table; the nature of the parent material (p); and time in the sense of the duration of soil formation (t). Other factors, such as fire, air-borne salts or pollutants, may be important locally. All the factors are regarded by Jenny as statistically independent. However, vegetation is regarded as distinct from the available flora and, together with the soil, is viewed as part of a larger interdependent ecosystem (Jenny, 1958, 1980). Thus:

$$l, v, a, s = f(cl, o, r, p, t \ldots) \tag{2}$$

wherein properties of the whole ecosystem (l), vegetation (v), animal communities (a) and soil (s) are represented as being a function of the five main soil-forming factors.

A complementary formulation (Jenny, 1961, 1980) substitutes three state factors for the five soil-forming factors:

$$s = f(S_o, I, t) \tag{3}$$

This recognizes that soil-forming factors may be of different kinds and may play different roles in soil formation. First, they may be 'passive', a part of the initial state of the system (S_o), e.g. the parent material and original topography. Second, they may be 'active', influx variables or external flux potentials (I), which affect the soil as it develops, e.g. some climatic and biotic variables. Third, there is the time factor (t); that is the age of the system. The idea of state factors is useful because it provides a point of contact between the functional factorial approach and the systems approach to soil genesis (Huggett, 1976). It also emphasizes the unique nature of the time factor, which has no influence on soils in itself. This had been recognized earlier by Stephens (1947), who used a version of the fundamental equation, based on Wilde (1946), which may be written as:

$$s = f(cl, o, r, p) \, \mathrm{d}t \tag{4}$$

Soil is regarded as an integral of soil-forming factors (each of which has independent and dependent aspects) and all of which change through time. This equation recognizes time as the only truly independent factor and is therefore more realistic than equation (1).

The most recent formula is that of Johnson, Keller & Rockwell (1990):

$$s = f\left(D, P, \frac{dD}{dt}, \frac{dP}{dt}\right) \tag{5}$$

This too recognizes the unique quality of the time factor and groups the soil-forming factors into two classes: a set of 'passive vectors' (P) and a set of 'dynamic vectors' (D). The dynamic vectors of pedogenesis fluctuate through time (dD/dt) to a greater extent than the passive vectors (dP/dt). Lists of factors comprising the two sets have been provided by the authors. The lists are useful for parameterization but the terms involved possess no real advantage over Jenny's respective use of the initial state of the system (S_o) and influx variables (I).

The existence of dependent as well as independent relationships between the soil-forming factors, together with problems of definition and measurement, mean that it has not proved possible to solve these various representations of the fundamental equation. However, the effects of any one of the factors on a particular soil can be investigated in situations where either (1) the range of the other factors is small, or (2) the other factors have negligible effect on the soil property; that is, where the other factors are more-or-less invariant or ineffective. Hence, a soil chronosequence has been defined as '... a sequence of soils developed on similar parent materials and relief under the influence of constant or ineffectively varying climate and biotic factors, whose differences can thus be ascribed to the lapse of differing increments of time since the initiation of soil formation' (Stevens & Walker, 1970: 333).

The majority of chronosequences that have been reported in the literature, and the type most relevant to soil development on glacier forelands, involve soils that started to develop at different times in the past and are still developing in the present-day landscape. This has been termed a post-incisive chronosequence (Vreeken, 1975), which is characterized by a time-transgressive starting point and an isochronous end point to soil development.

In terms of an equation, an *ideal* soil chronosequence (Jenny, 1980) can be written as:

$$s = f(t)_{cl, o, r, p \ldots} \tag{6}$$

or $$s = f(t)_{S_o, I} \tag{7}$$

where the soil, or a soil property, is a function of terrain age, with all the other factors held constant. If the differences between soils in such a sequence depend only on the passage of varying periods of time since the initiation of soil formation, rate equations of soil formation (chronofunctions) can be derived where the terrain ages in the chronosequence are precisely known.

Ideal chronosequences probably do not exist in nature. In particular, post-incisive chronosequences of the type characteristic of glacier forelands have several limitations. These relate mainly to the extent to which it is possible, in practice, to hold constant the factors other than time. There are two main limitations, which have been widely discussed (Crocker, 1952; Stevens & Walker, 1970; Birkeland, 1974; Vreeken, 1975; Yaalon, 1975; Jenny, 1980; Pickett, 1988). First, the initial state of the system is unlikely to have been identical for all sites in the chronosequence. For example, both the parent materials and the original topography are normally character-ized by considerable spatial variation (see section 3.1). Even where the initial conditions were closely similar this may be impossible to verify because of changes at the sites since the initiation of soil development. Second, as there has been only partial overlap in the history of soil development at the various sites in a chronosequence, it is likely that the influx variables have differed, at least to some extent, between sites. In other words, climatic or other changes in the environment are likely to have affected differently the various sites on terrain of differing age (see sections 3.2, 3.3.4 & 3.4 for many examples).

To these limitations must be added those imposed by the nature of the timescale itself, particularly the problems of incomplete chronosequences and imprecise dating (chapter 2). Few chronosequences are free from major error terms introduced from this source. Thus, although a chronosequence tends to be conceptualized as a perfect representation of soil development at a particular site, *actual* chronosequences fall short of this ideal.

In order to take account of the limitations outlined above, Jenny modified his chronosequence equation to denote an *actual* chronosequence as:

$$s = f(t, cl, o, r, p \ldots) \tag{8}$$
$$\text{or} \qquad s = f(t, S_o, I) \tag{9}$$

where time is the *dominant* factor (t) and the other factors are *subordinate*. Relief and parent material, defined in terms of the initial state of the system, are controlled only to the extent that they are invariate between sites at time zero. The influx variables are controlled only to the extent that they are expressed in similar site histories.

Actual chronosequences must therefore be regarded as approximations to sequences of soil development at particular sites. With careful site selection in the field, and accurate dating, close approximations are

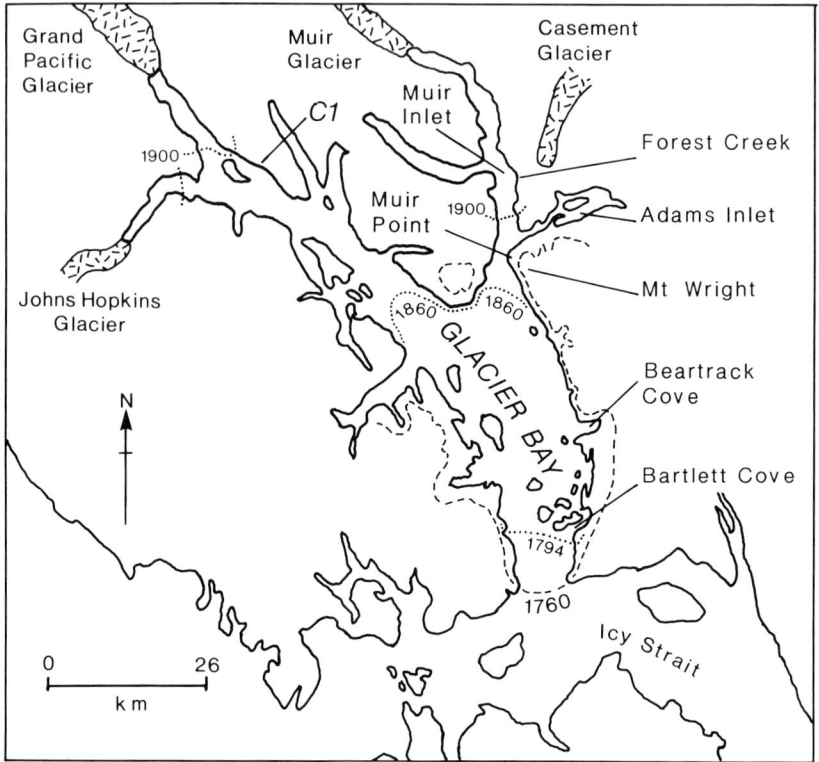

Fig. 4.1. Glacier Bay, Alaska, U.S.A.: location of important studies of soil and vegetation chronosequences. The 'Little Ice Age' glacier limit of *ca* A.D. 1760, selected dated glacier retreat positions, and selected glacier snouts are shown. C1 indicates the location of W. S. Cooper's permanent quadrat No. 1 (from Lawrence, 1979, and others).

possible, at least on the relatively short timescales and the restricted spatial scales appropriate to glacier forelands. According to Yaalon (1975), difficulties in collecting appropriate data for solving the equations, rather than the theoretical approach, remain the main constraint.

4.1.2   An example: Glacier Bay, Alaska

This section describes one chronosequence in sufficient detail to exemplify the methodology outlined above. The example draws on the classic studies in the Glacier Bay area of Alaska, which was mentioned in the introduction (see section 1.3). The potential of this area for the investigation of soil development was first exploited by Crocker & Major (1955). The present brief account is based on this and on later work in the

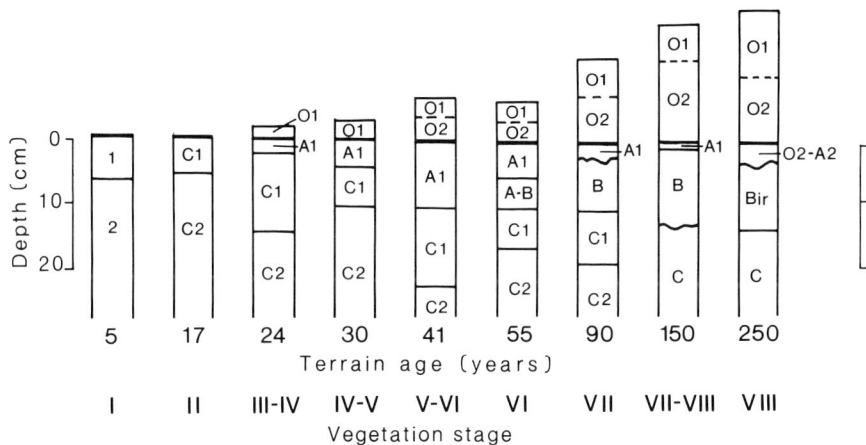

Fig. 4.2. Soil profile development in relation to terrain age and vegetation stage, Muir Inlet, Glacier Bay, Alaska. Soil horizons are lettered (from Ugolini, 1968).

same area, largely by Ugolini (1966, 1968) but also by Bormann & Sidle (1990). Complementary studies have also been made at the neighbouring Herbert and Mendenhall glaciers (Chandler, 1942; Crocker & Dickson, 1957).

The chronosequence is based largely on sites located on the east side of Glacier Bay, between Muir and Casement Glaciers, and Bartlett Cove (Fig. 4.1). Terrain ages vary from 0 to 200–300 years (Goldthwait, 1966). The mild, maritime climate is characterized by a small annual and diurnal temperature range, high relative humidity and cloudiness, heavy precipitation and rather strong winds (Loewe, 1966). Consequently, both vegetation and soil development proceed rapidly. Parent materials, relief, regional climate, and probably the available biota, are relatively constant between sites. Ugolini (1966) stresses that only sites deglaciated during the last 55 years (from the same till sheet in front of Casement Glacier) are likely to provide a close approximation to an *ideal* chronosequence. At the older sites (from moraine ridges beyond Muir Inlet), the soil-forming factors other than time increasingly diverge. Indeed, F. R. Stephens, in an unpublished manuscript, suggests that sites on the relatively old terminal moraines utilized by Chandler (1942), Crocker & Major (1955), Crocker & Dickson (1957) and Ugolini (1966, 1968) at Glacier Bay and neighbouring forelands, may represent different chronosequences characteristic of drier conditions.

The *actual* chronosequence, in terms of soil profile development, is

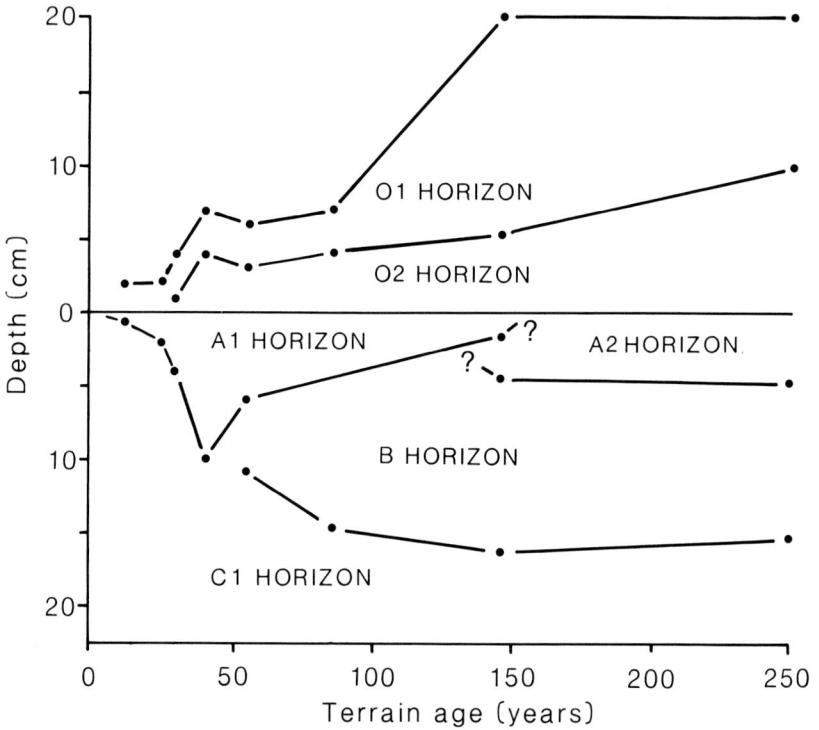

Fig. 4.3. Soil horizon development in relation to terrain age, Muir Inlet, Glacier Bay, Alaska (from Ugolini, 1968).

shown in Figs 4.2 & 4.3. The first visible sign of soil formation is the appearance of a discontinuous, grey-brown A1 horizon. This is found in areas deglaciated for approximately 10 years and is covered by a crust of mosses and a discontinuous patchy carpet of *Dryas drummondii* (vegetation stage I according to Decker, 1966; see section 5.1.2). An organic O1 horizon appears next, in areas deglaciated for about 13 years and still sparsely vegetated. On terrain ages of about 30 years, a thin, humified O2 layer develops below partly decomposed leaves, mainly of *Alnus crispa* (vegetation stages V and VI). Within the first 40 years, the A1 horizon reaches its maximum thickness of about 10 cm, thereafter decreasing in thickness until it becomes indistinct after about 150 years. The soils developed during the first 40 years are classified as regosols.

An incipient, illuvial B horizon (distinguishable by colour) occurs in the area deglaciated for 55 years, and first appears where dense thickets of alder, overtopped by black cottonwood (*Populus balsamifera*), are beginning to be colonized by seedlings of Sitka spruce (*Picea sitchensis*). The B horizon appears at the bottom of the A1 horizon and seems to grow upwards. In areas deglaciated 85 years prior to the study, beneath dense alder thickets with interspersed spruce trees, the entire A1 has been replaced by a B horizon; other profiles of the same age still show an A1 or a transitional A/B horizon. Ugolini tentatively classifies these soils as podzolic. An eluvial A2 horizon first appears in areas deglaciated for about 150 years, or possibly more, and its thickness of about 5 cm remains fairly constant for at least the next 100 years. The terrain deglaciated for about 150 years is characterized by mature spruce forest (vegetation stage VII). On older terrain, deglaciated for up to 200–300 years, where a mature forest of spruce, sometimes with western hemlock (*Tsuga heterophylla*) (vegetation stage VIII) is found, profiles with and without A2 horizons persist. Both brown podzolic and podzol soils are therefore present at this stage. The surface organic layer (O1 and O2 horizons) appears to attain its maximum thickness of some 20 cm within about 150 years of deglaciation at sites where alder trees and other understorey species are still contributing significantly to the forest floor litter.

Physical and chemical soil properties also change within the chronosequence as shown by the graphical chronofunctions in Fig. 4.4. Although some textural differences in the profiles have been noted, and attributed to variations in the parent material, or to surface wash and cryogenic activity shortly after deglaciation, Ugolini (1966) considers that textural differences resulting from soil-forming processes are only minor. However, samples of both whole soil and the fine fraction (< 2 mm) from these predominantly sandy soils exhibit a consistent decline in bulk density values with increasing terrain age (Fig. 4.4a). Discrepancies between the values reported by Crocker & Major (1955), Ugolini (1966, 1968) and Bormann & Sidle (1990) can be attributed to the different techniques used, within-site variability and the lapse of time between samplings. The trend of decreasing values with time is attributed to the progressive addition of litter to the soil (Fig. 4.4a), development of a weak crumb structure, root channelling and insect tunnelling (Ugolini, 1968).

The pH–time function (Fig. 4.4b), based on samples from the surficial layer of the mineral soil, shows a rapid drop in pH values from about 8.2 for fresh till to about 6.8 after 17 years under alder, with a further decline to between 4.5 and 5.0 under the spruce-hemlock forest. Litter residues are consistently more acid than the soil, and there is a concomitant fall in the soil carbonate content. As organic carbon steadily accumulates in the surface organic layer (forest floor litter), the organic content of the

(a)

(b)

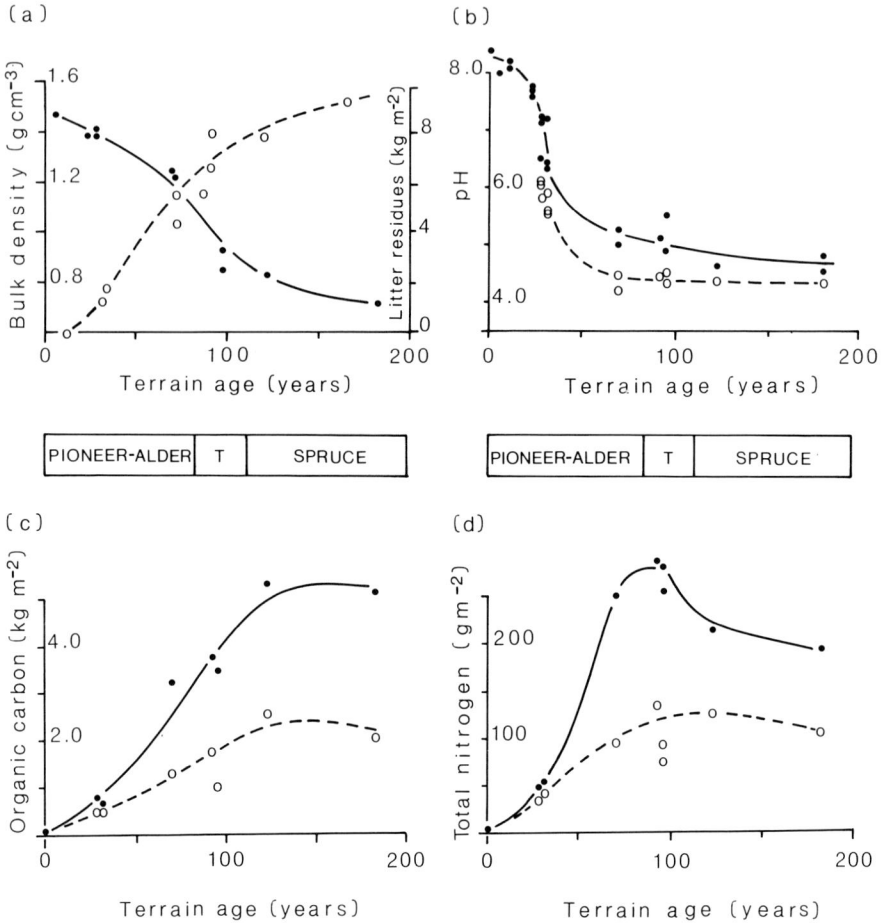

| PIONEER-ALDER | T | SPRUCE |

| PIONEER-ALDER | T | SPRUCE |

(c)

(d)

Fig. 4.4. Physical and chemical soil chronofunctions from Glacier Bay, Alaska:
(a) bulk density of the fine fraction (<2 mm) in the uppermost 5 cm of the
mineral soil (solid line), and the accumulated litter residues (broken line) on the
soil surface; (b) pH of the uppermost 5 cm of the mineral soil (solid line), and pH
of the litter residues (broken line); (c) organic carbon in the soil profile (solid
line), and in the mineral soil (broken line); (d) total nitrogen in the soil profile
(solid line), and in the mineral soil (broken line). Vegetation stages are also
shown (T, transitional stage) (from Crocker & Major, 1955).

underlying mineral soil builds up (Fig. 4.4c). The nitrogen content also
increases first in the organic horizons and later in the mineral soil, and then
is depleted as alder is replaced by spruce (Fig. 4.4d).

The chronosequence represents stages in the development of a podzol
that has yet to approach full maturity. This contention is supported by the

accumulation of free iron oxides in the B horizon (Ugolini, 1968; Bormann & Sidle, 1990) as well as the presence of the thin bleached A2 horizon (see section 4.2.5). Although Crocker & Dickson (1957) hesitated to use the term podzol for similar soils in front of the Herbert and Mendenhall glaciers, other workers, including Chandler (1942) and Stevens (1963) were in no doubt. Based on their work farther to the north in the vicinity of the Nelchina and Tazlina glaciers, Kubota & Whittig (1960) suggested that similar soils might be nanopodzols. However, their suggestion was rejected by Ugolini (1968) in favour of regarding these soils as shallow iron-podzols. Whereas the brown podzolic soils seem to be a necessary stage in the ontogeny of these podzols in this area, it is uncertain whether they can coexist indefinitely under the same environmental factors. The prevalence of podzols over brown podzolic soils at sites near Bartlett Cove (eight of the ten profiles excavated by Ugolini were podzols) suggests that the brown podzolic soils may eventually disappear.

Although there is not one single explanation to account for all the observed trends in the Glacier Bay chronosequence, it appears that the establishment of plants and intensive leaching play key roles. Soil-development processes proceed downwards more rapidly in the early stages. Then, as time progresses, the morphological, physical, chemical and biological properties of the various horizons are intensified. Further discussion of the processes involved in this chronosequence, the importance of plants in the acceleration of pedogenic processes, and the concept of stability in the context of glacier foreland soils, will be presented elsewhere.

4.2　Soil properties and pedogenic processes

The example from Glacier Bay clearly illustrates the chronosequence approach to the study of soil development on recently-deglaciated terrain. In this section, precise values of particular soil properties from a variety of glacier foreland chronosequences are described systematically. These patterns are used to infer change and as the basis for discussion of the pedogenic processes involved in the transformation of parent material into soil.

4.2.1　Texture

Changes in physical properties involve not only the relatively slow decrease in bulk density detected at Glacier Bay and attributed to organic matter additions (section 4.1.2, and Fig. 4.4a), but also other, often relatively rapid, changes in soil texture (Lüdi, 1945; Viereck, 1966; Fitter &

Parsons, 1987). Several processes responsible for these textural changes, including pervection, cryogenic and aeolian processes have already been considered under the heading of proglacial landscape modification (see section 3.2). The apparent lack of textural trends reported in some chronosequences (e.g. Persson, 1964; Ugolini, 1966, 1968; Jacobson & Birks, 1980; Mellor, 1984, 1985) may well be the result of a combination of a small number of samples, a heterogenous parent material and the rapidity of change.

Textural changes similar to those found with distance from Breidamer-kurjökull, Iceland, by Boulton & Dent (1974) (see section 3.2.3, and Figs. 3.6 & 3.17), with distance from seven Jotunheimen glaciers by Beckett & Matthews (Fig. 3.7), and in association with the development of patterned ground with distance from Storbreen, Jotunheimen, by Ballantyne & Matthews (1982) (see section 3.2.5 and Fig. 3.9) are likely to be a common feature of chronosequences elsewhere. Similar underlying processes are likely to account for the stony pavement characteristic of the earliest stages of the Glacier Bay chronosequence (Crocker & Major, 1955; Crocker & Dickson, 1957).

Two general tendencies are found on the glacio-fluvial outwash deposits of the Muldrow Glacier, Alaska (Viereck, 1966). First, the texture of the surface layers tends to be finer than the unaltered parent material at depths greater than about 20 cm. Second, the clay (<0.002 mm) and silt (0.002–0.02 mm) fractions at any depth increase slightly through time at the expense of the sand fraction (0.02–2.0 mm) (Fig. 4.5a). A similar trend is described by Lüdi (1945) from soils on the lateral moraines of the Aletsch Gletscher, western Alps. Although Viereck attributes both changes to a combination of near-surface weathering and the addition of wind-blown fines to the surface, it is unlikely that weathering would produce such a substantial effect on texture after only 200–300 years.

Depth plots for the clay and sand fractions from terraces of different age exhibit two departures from the above trends (Fig. 4.5b). The first is the significantly higher concentration of clay (and silt) near the ground surface at the late meadow stage (about 100 years), which is not present on younger terrain. This is apparently caused by the deposition of fines blown on to the newly-formed sandur surface from freshly-deposited moraines in the vicinity. A related explanation in terms of aeolian deposition was advanced for an unusually high content of fine sand and silt at one soil site within a chronosequence involving moraine ridges at Engabreen, northern Norway (Worsley & Alexander, 1976b; Alexander, 1982). In this case, however, the source of the fines is thought to have been the local sandar, deposited prior

(a)

(b)

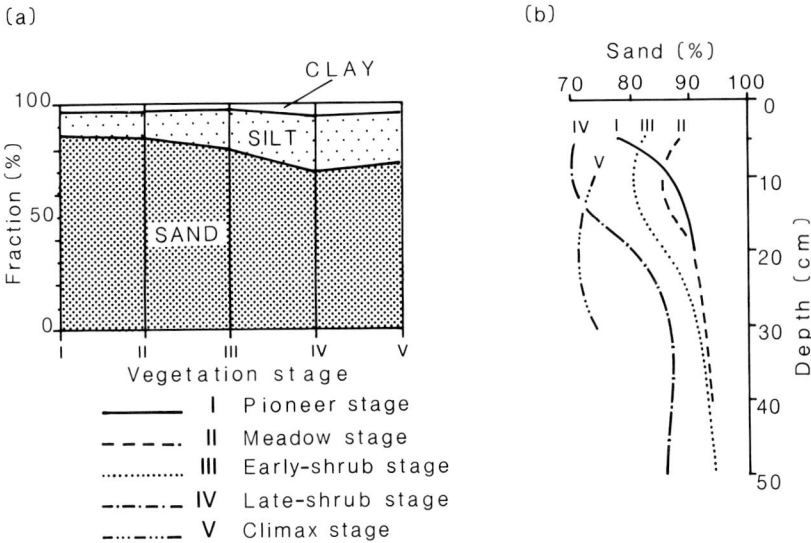

Fig. 4.5. Soil texture variations in glacio-fluvial deposits of the Muldrow Glacier, Alaska, in relation to terrain age and vegetation stage (I–V): (a) proportion of various fractions at 10 cm depth; (b) proportion of sand (0.2–2.0 mm) and clay (< 0.002 mm) with depth in the soil profile (from Viereck, 1966).

to and immediately after an advance of Engabreen which failed to reach the site.

The second anomaly in Fig. 4.5b is the relatively high concentration of fines at a depth of around 10 cm in the meadow, early shrub and late shrub stages, which is not apparent in the pioneer stage (25–30 years). Although cryogenic activity may possibly be involved (Viereck, 1966), this pattern is consistent with the pervection of fines described in section 3.2.3. Pervection may also account for small, but in some cases statistically significant, increases in clay content with depth at each site, despite there being no clear trend between sites in the chronosequence. Such a pattern has been reported from moraines at both Engabreen (Alexander, 1982) and the Robson Glacier, British Columbia, Canada (Sondheim & Standish, 1983).

One of the most interesting textural patterns found from a soil chronosequence is the polynomial relationship shown in Fig. 4.6. This chronofunction is based on sampling the upper 5 cm of soil in front of the Athabasca Glacier, Alberta, Canada (Fitter & Parsons, 1987). Under *Dryas drummondii*, the percentage of fines (< 2.0 mm) falls from about 60% to about 40% in the first 40 years and then steadily rises to about 80% on moraines deglaciated for some 120 years. In unvegetated areas there is no significant

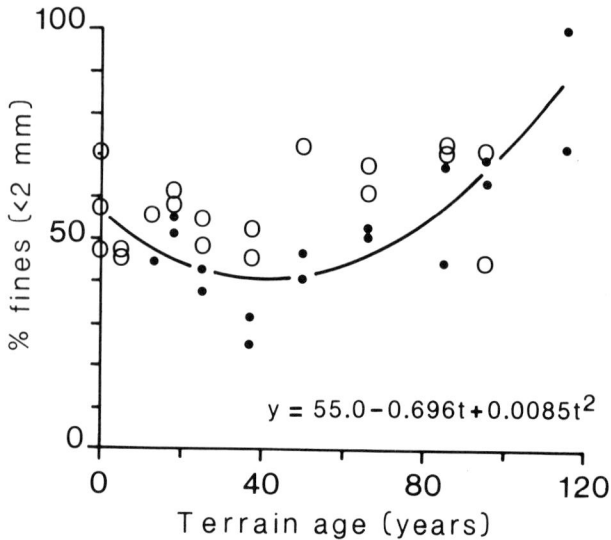

Fig. 4.6. Soil texture (< 2 mm fraction, uppermost 5 cm) from moraines of increasing age, Athabasca Glacier, Alberta, Canada. The fitted regression line applies to samples from under *Dryas* (filled circles; $p < 0.05$); there is no significant relationship for unvegetated terrain (open circles) (from Fitter & Parsons, 1987).

relationship. Even though Fitter & Parsons offer no explanation for these trends, the initial decline is likely to be the result of pervection, possibly aided by cryogenic processes and deflation, while the later increase can be assigned to aeolian deposition. An explanation for the absence of any trend in the unvegetated areas probably lies in the instability of such areas, particularly their likely susceptibility to cryoturbation, which would prevent the development of a surficial layer depleted of fines. The vegetated areas, on the other hand, are likely to be in the most stable areas. Furthermore, the presence of a vegetation cover may not only suppress cryoturbation but may also enhance aeolian deposition.

4.2.2   Micromorphology

The progressive development of the soil microfabric on the glacier foreland of Breidamerkurjökull has been investigated by Romans, Robertson & Dent (1980), who examined thin sections prepared from undisturbed samples collected by Boulton & Dent (1974). The results confirm the textural changes observed by Boulton & Dent, and aid their interpretation. The initial fabric of freshly-exposed dilated till is porous with occasional bubble pores near the surface, which are absent in subglacial till. The formation of such bubble pores or vesicles probably occurs during thawing of the sediments (Harris, 1983; Van Vliet-Lanoë, 1985), although other origins, such as the wetting of dry soils (Ugolini, 1966; Harris, 1985) are also possible. After one year of exposure at freely-draining sites, some of the primary pore space at depths of 20–50 cm is filled with fines, particularly silt, producing a spongy fabric. After two years, further translocation of silt produces an open fabric with the spaces between stones bridged by patches and lenses of finer matrix material. By this time there is a sharp boundary at about 50 cm, below which silt fills most of the pores. After 20 years this dense silt layer has been partly destroyed, while after 80 years it has been washed out of the profile.

Silt droplets (flattened or lenticular patches of fines which lie roughly horizontal) develop within 30 years, although they are not found on terrain deglaciated for 12 years (Romans, Robertson & Dent, 1980). This type of microfabric generally develops under the influence of both pervection and freeze–thaw processes (cf. Harris & Ellis, 1980; Ellis, 1983; Van Vliet-Lanoë, Coutard & Pissart, 1984; Harris, 1985; Van Vliet-Lanoë, 1985). During freezing and ice segregation, pore water is drawn from surrounding sediments to growing ice lenses. Thus, as the ice lenses expand, the surrounding matrix loses water and is compacted. During thaw, drainage of abundant water washes fines down the profile. The growth of ice lenses parallel to the ground surface, in addition to the infilling of voids vacated by the melting ice lenses, produces the patchy silt-droplet pattern. Such patterns therefore require a suitable combination of moisture availability and thermal regime, combined with a permeable and frost-susceptible substrate. Their absence on the youngest terrain at Breidamerkurjökull is attributed by Romans, Robertson & Dent (1980) to the collapse of structure under permanently wet conditions close to the glacier margin. Silt droplets form after the first 20–30 years, when pervection has increased the permeability of the surface layers and internal profile drainage improves.

The ubiquitous platy structure found at various depths in the soils at Glacier Bay may be of related origin. While the platy structure persists throughout the soil chronosequence, it appears closer to the surface in the more recently deglaciated areas and it becomes less distinct in the older soils, where it tends to be found at greater depths (Ugolini, 1966). This pattern may result from cryogenic activity becoming progressively subdued by an increasing vegetation cover; alternatively, it may represent a characteristic of the newly-deposited till and therefore of the parent material (see section 3.2.2).

Very rapid changes involving the interaction of pervection and freeze–thaw processes have been inferred by Bothamley (1987) from micromorphological analyses at Styggedalsbreen, Jotunheimen, of undisturbed samples from depths of up to 40 cm from moraine ridges of three ages (about 5 years, 50 years and 230 years). Amongst the trends detected is an increase with depth in the frequency of cutans (distinct cappings or coatings of fine material on the surface of individual larger grains). The proportion of top-surface cutans (capping the upper surface of a grain) to other-surface cutans (covering the whole grain or a different surface) also increases with depth, which suggests that the rotation of grains, associated with cryoturbation, decreases with depth. These trends are clear even in tills exposed for only a few years, yet there is no apparent trend with terrain age.

In soil chronosequences derived from moraines near two other southern Norwegian glaciers, the presence of lenticular and alveolar microstructures, together with matrix granules, suggest that cryogenic processes are operative in subsurface (Ae, B, C) horizons (Mellor, 1986a). Of the chronofunctions constructed for the frequency of illuvial coatings (predominantly composed of silt and fine-sand particles), the relationships are statistically significant only for the B and C horizons from the Storbreen glacier foreland, Jotunheimen (Fig. 4.7). In both cases, there is a decrease through time in the area of coatings in thin sections. Rapid formation during the earliest post-depositional stages is the most likely explanation for the abundance of coatings in the youngest profiles examined (terrain age *ca* 50 years). Subsequent cryoturbation, bioturbation, and wetting and drying may have caused progressive disruption and a corresponding decline in the frequency of coatings in the older soils (Mellor, 1986a).

4.2.3   Organic content
Organic content has been measured in many glacier foreland soil chronosequences, in terms of either organic carbon (e.g. Crocker & Major, 1955; Crocker & Dickson, 1958; Burger & Franz, 1967; Fitze, 1980; Alexander, 1982; Sondheim & Standish, 1983; Mellor, 1985; Mahaney, 1990), loss-on-ignition (e.g. Stork, 1963; Persson, 1964; Viereck, 1966; Jacobson & Birks, 1980; Messer, 1988), or both (Ugolini, 1966, 1968). This reflects the importance of organic processes in soil development. However, the number of samples have rarely been large enough for detecting anything more than a general trend of increasing values.

The accumulation of organic matter at Glacier Bay to form a litter layer on the ground surface and its subsequent incorporation in mineral horizons

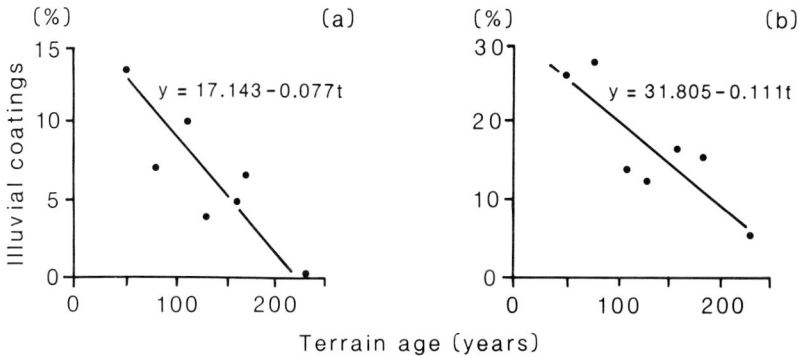

Fig. 4.7. Micromorphology point-count data chronofunctions for illuvial coatings within (a) the B horizon and (b) the C horizon, Storbreen Jotunheimen. Fitted regression lines are statistically significant ($p < 0.05$) (from Mellor, 1986a).

was briefly described in section 4.1.2. This has been examined in detail by Crocker & Major (1955) and by Ugolini (1966, 1968), and comparable data on organic carbon in relation to depth and terrain age are also given by Crocker & Dickson (1958) for the chronosequences at the neighbouring Herbert and Mendenhall Glaciers (Fig. 4.8). The dry weight of forest floor material does not differ significantly between the three chronosequences during the earlier shrub-dominant stages. However, much greater accumulations of plant residues build up at the Mendenhall Glacier, levelling out at about 16 kg m^{-2} after 120–150 years, during the period of spruce dominance (Figs. 4.4a & 4.8a). Depth functions of organic carbon (Fig. 4.8b) indicate that the bulk of the organic matter in the mineral soil resides close to the surface. The absence of a second maximum in these profiles emphasises the incipient nature of the B horizon and the lack of a diagnostic illuvial humus horizon even after 250 years (Ugolini, 1966). Although there appears to be a reduction in the rate of accumulation of total organic carbon (forest floor + mineral horizons) with age, organic carbon content continues to increase throughout each chronosequence, at least for the first 150 years. After about 150 years, organic carbon within the Mendenhall soils reaches about 9.0 kg m^{-2}, whereas at Glacier Bay the corresponding value is about 5.0 kg m^{-2} (Crocker & Major, 1955; Ugolini, 1966).

In front of Engabreen, the organic carbon content of the surface organic (F) horizon reaches about 30% after about 100 years (Alexander, 1982). After this, under birch (*Betula pubescens*) woodland, values remain fairly constant. In the upper mineral (A or E) horizon a similar trend is apparent, with values reaching 3–4%. At Austerdalsbreen in southern Norway,

(a)

(b)

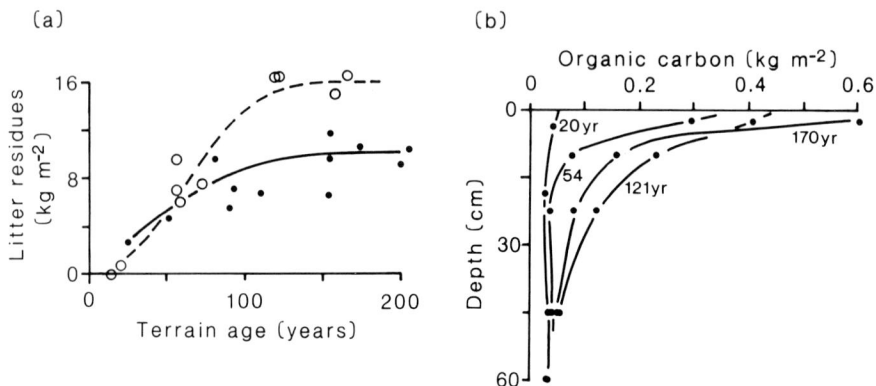

Fig. 4.8. Organic matter in soil chronosequences, Herbert and Mendenhall Glaciers, south-eastern Alaska: (a) chronofunctions for litter accumulation (solid line represents Herbert Glacier); (b) depth functions for organic carbon (per 2.5 cm column) with increasing terrain age for the Mendenhall sequence (from Crocker & Dickson, 1957).

where podzolic soils similar to those at Engabreen are developing on terminal moraines under birch, the organic carbon content of the surface organic (LF) horizon reaches only 15% after 230 years (Mellor, 1985). A chronofunction constructed by Mellor gives no indication of any reduction in the rate of increase of the organic carbon content through time. In this area, as is the case near Engabreen, organic-rich horizons beyond the Neoglacial glacier limits are characterized by organic carbon contents in excess of 30%.

Studies by Messer (1984, 1988) of 18 chronosequences from southern Norway, most of which are from Jotunheimen and involve alpine brown soils above the birch tree line, indicate that organic content (loss-on-ignition) within the uppermost 10 mm of the mineral soil on moraines dating from *ca* A.D. 1750 varies between about 3 and 32%. She found, moreover, that the most common pattern exhibited by the corresponding chronofunctions is an accelerating rate of increase in organic content through time (Messer, 1984: 109). Weaker trends are characteristic of the 10–30 mm depth interval. In most of these chronosequences the organic content is not approaching a steady state after 230 years, even in the uppermost 10 mm of the soil. In contrast, on moraines in front of the Robson Glacier, British Columbia, maximum values of *mean* organic carbon of about 0.87, 0.32 and 0.21% are reached after about 100 years at depths of 0–15, 15–30 and 30–35 cm, respectively (Sondheim & Standish, 1983). Here, a non-podzolized brown soil (orthic dystric brunisol) is developed on a calcareous substrate and organic carbon values appear to reach a temporary steady state.

Micromorphological examination of the organic surface horizon at Austerdalsbreen after about 230 years shows a predominance of fresh to slightly decomposed organic material forming a spongy microstructure (L layer), below which occurs a more humified (F) layer, with greater fragmentation, a darker colour, a granular microstructure and faecal pellets (Mellor, 1986a). It is not surprising, therefore, that the penetration of organic carbon into the mineral horizons at Austerdalsbreen is slight, reaching about 2% in the E horizon and about 0.5% in the B horizon after about 230 years (Mellor, 1984).

After the same time interval above the tree line at Storbreen, organic matter in the surface (Ah) horizon is moderately to strongly decomposed and well mixed with mineral matter and faecal pellets, forming a granular/ crumb microstructure (Mellor, 1986a). Mellor suggests that the difference between the nature of the surface horizons at Austerdalsbreen and at Storbreen might be explicable in terms of varying rates of organic matter production and decomposition. The likelihood of organic matter accumulation being controlled by the *balance* between production and decomposition is also used by Messer (1988) to account for the lack of a correlation between rate of organic accumulation and selected climatic variables in the soil chronosequences from the Jotunheimen and Jostedalsbreen region (see section 4.3.4).

4.2.4   pH and base status

A decline in pH with time appears to be an almost universal feature of glacier foreland chronosequences. In the 18 southern Norwegian chronosequences investigated by Messer (1988), initial pH values in the uppermost 10 mm of the mineral soil range from 5.3 to 8.5, falling to values between 3.8 and 5.9 after 230 years (see also section 4.3.4 and Fig. 4.22). Generally, the reaction of newly-deposited till is close to neutral, or well above this in areas of calcareous parent material. Values fall rapidly in the surface layers of the mineral soil, only failing to become clearly acidic in areas characterized by highly calcareous bedrock, such as in the vicinity of the Robson Glacier (Sondheim & Standish, 1983).

Depth functions from the Mendenhall Glacier chronosequence, developed in a parent material predominantly composed of quartz diorite, are quite typical of non-calcareous substrates (Fig. 4.9). The pH falls from > 7 to about 5 within 50–60 years, with relatively small changes in the later stages. At Glacier Bay, where the parent material contains both calcite and

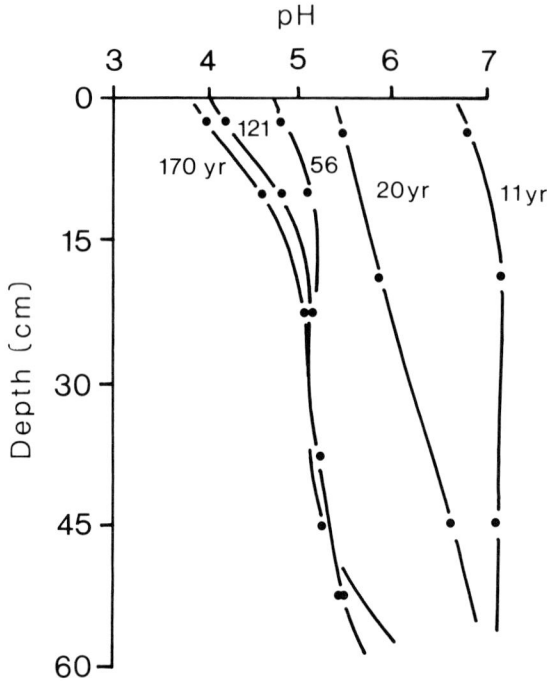

Fig. 4.9. Depth functions for soil pH with increasing terrain age, Mendenhall Glacier, south-eastern Alaska (from Crocker & Dickson, 1957).

dolomite, Ugolini (1966, 1968) has demonstrated the selective removal of carbonates; the calcite leaching more readily than the dolomite and the calcite:dolomite ratio in the uppermost mineral horizon decreasing from 1.7 after 5 years to about 0.2 after about 40 years. The rapid decline in calcium carbonate content, which is accompanied by an equally rapid decline in pH, is shown in Fig. 4.5b. Although there is considerable evidence that leaching becomes more effective under a vegetation cover (see section 4.3.3), Ugolini states that there is also a distinct loss of carbonates at the surface of the soil in unvegetated areas.

Messer (1988) found that initial values of cation exchange capacity (CEC) range from 1.99 to 5.42 meq 100 g^{-1} in front of southern Norwegian glaciers. Such low values are typical and are accounted for by the lack of a clay-humus complex. Organic matter is absent at first and the clay content of glacier foreland deposits is usually very low (see section 3.1.2). Thus, in the example shown from Glacier Bay, the high pH reflects the very low values for exchangeable hydrogen (Fig. 4.10) and a correspondingly high

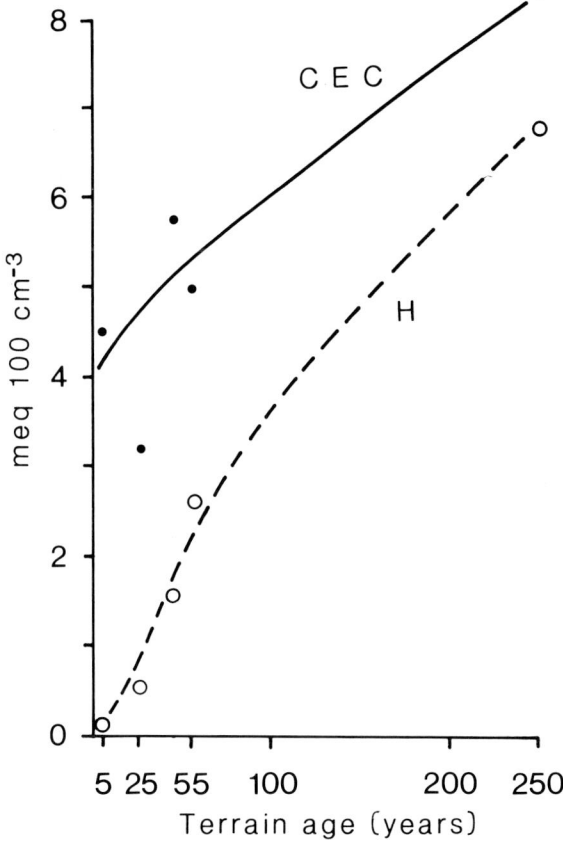

Fig. 4.10. Chronofunctions for cation exchange capacity and exchangeable hydrogen near the soil surface, Muir Inlet, Glacier Bay (from Ugolini, 1966).

degree of base saturation in freshly-deposited till (see also, Persson, 1964; Gennadiyev, 1978; Alexander, 1982; Vetaas, 1986). In the early stages of soil development, therefore, rapid rates of pH decline may occur partly because of the relative ease with which the small quantity of exchangeable bases can be leached and replaced by hydrogen ions (see section 4.3.1).

The increase in CEC through time (Fig. 4.10) is dependent on the accumulation of organic matter in the surficial mineral horizon, which is generally quite slow. In southern Norway, CEC values vary from < 1.0 to 9.0 meq 100 g^{-1} after 100 years (Messer, 1988). After 230 years in Jotunheimen, CEC values still do not exceed 10.0 meq 100 g^{-1} in the uppermost 10 mm of the mineral soil; at the lower altitudes of the

Jöstedalsbreen, however, where higher rates of organic matter incorporation are characteristic, values rise to > 70 meq 100 g^{-1} after the same time interval (see section 4.3.4).

A similar pattern is reported in the early stages of the development of mountain meadow soils from moraines in the Elbruz region of the Central Caucasus (Gennadiyev, 1978). Here, CEC increases from 9–10 meq 100 g^{-1} to 24 meq 100 g^{-1} in the upper horizon after 300 years. During further development of the profile, Gennadiyev reports only a slight increase in CEC, although exchangeable hydrogen continues to increase. At Glacier Bay, the quantity of exchangeable bases remains relatively constant across the chronosequence; only the podzol beneath the spruce-hemlock forest (stage VIII, *ca* 250 years) shows a marked depletion of bases (Ugolini, 1968). Although there are relatively few data points, the increase in CEC appears to be largely taken up by an increase in exchangeable hydrogen (Fig. 4.10). Because CEC is controlled by the content of organic matter and because, in turn, the exchangeable hydrogen is a function of CEC, Ugolini argues that, in this case, the organic matter appears to be the source of the acidity in the soil. However, in the soils at Engabreen in northern Norway, Alexander (1982) recognises an irregular decline in the total exchangeable bases which, although most marked on terrain ages younger than about 45 years, continues on older moraines. He attributes the *reduced* rate of loss to the establishment of a vegetation cover and the development of nutrient cycling. In addition, it is recognized that some input of bases is likely from precipitation and sea spray, which would tend to replace those lost.

Nutrient cycling also appears to be important in the upper soil horizons of sandar surfaces in front of Franz Josef Glacier, New Zealand (Stevens, 1963, 1968; Burrows, 1990). In this long chronosequence, initial pH values of 8.0 fall by 3.5 units in the first 55 years. Subsequently, they fall a further 1.5 units in about 10 000 years. Over the same time interval, CEC and organic carbon levels rise, the latter reaching values of 20 kg m^{-2} beneath mature kamachi-mixed podocarp forest (see section 6.1.5). Calcium content at first declines rapidly due to leaching; later it rises slowly, together with potassium levels. As there is relatively little unweathered parent material accessible to roots after this length of time, the plants must eventually rely almost totally on recycling of nutrients from the organic component of the ecosystem.

Significant increases with age of CEC and exchangeable potassium and magnesium are found in the bleached E horizon at Austerdalsbreen, southern Norway, but corresponding increases in calcium and sodium are not observed. This may be attributed to the release of bases by pedogenic

weathering processes from the gneissic parent material, which is comparatively rich in potassium and magnesium but poor in calcium and sodium
(Mellor, 1985). In the Kebnekajse region of northern Sweden, both calcium
and potassium levels fall with increasing surface age, the calcium appearing
to be leached at a faster rate than the potassium (Stork, 1961). Increasing
sodium and potassium values are found together with declining calcium
values across the chronosequence at Skaftafellsjökull, Iceland, which is also
characterized by remarkably high initial CEC values attributed to the large
surface area provided by particles of pumice and rhyolitic ash (Persson,
1964). Jacobson & Birks (1980) suggest that relatively high values of
calcium, magnesium and potassium in soils at least 100 years old at the
Klutlan Glacier, Yukon Territory, may result in part from nutrient
addition from loess.

Soils in front of several glaciers in Antarctica exhibit high pH values and
high salt concentrations to the point of exhibiting salt-enriched horizons,
surface salt efflorescences and salt crusts (Everett, 1971; Claridge &
Campbell, 1977; Pastor & Bockheim, 1979; Campbell & Claridge, 1987;
Bockheim, 1990). The salts, which originate from the weathering of
bedrock, from the atmosphere, and from saline lakes, are largely sulphates,
chlorides and nitrates of sodium, potassium, magnesium and calcium.
Most of the moraines in the chronosequences from Antarctica are much
older than those under discussion here and, in the most arid areas, leaching
is negligible. However, it is salutory to reflect on the possibility of dominant
upward movements of soil water in dry and/or windy environments
elsewhere (see section 3.2.8).

Thus, although detailed and comparable information on the base status
of soils in glacier foreland chronosequences is largely lacking, the data
available suggest a complex set of physical, chemical and biological
processes is responsible for apparently similar simple patterns of pH decline
through time.

4.2.5   Iron and aluminium

Morphological evidence for the movement of iron in the soils at
Glacier Bay is provided by the development of a B horizon (Fig. 4.3) in
which, eventually, there is a detectable increase in free iron oxides (the iron
fraction not bound within the crystal lattice of soil minerals). According to
Ugolini (1968), the incipient B horizon recorded in the 55-year-old soil does
not show a distinct accumulation of free iron oxides. More accumulation is
evident after 150 years, and definitely in the 250-year-old soil.

More detailed results are available from the Engabreen podzolic chrono-

(a)

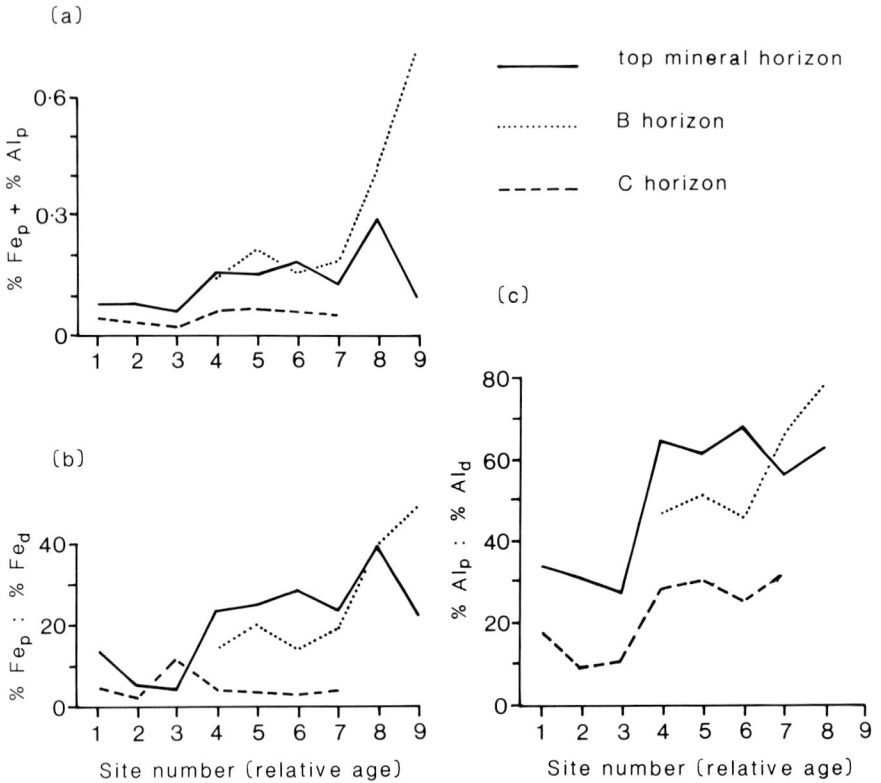

(b)

(c)

Fig. 4.11. Indices of iron and aluminium from three soil horizons in a chronose-quence, Engabreen, northern Norway: (a) $\%Fe_p + \%Al_p$, (b) $\%Fe_p : \%Fe_d$, (c) $\%Al_p : \%Al_d$ (from Alexander, 1983).

sequence, where Alexander (1982) has investigated the free oxides and the organically-complexed sesquioxides of iron and aluminium. The amount of organically-complexed iron and aluminium is given by the pyrophosphate-extractable fractions (Fe_p and Al_p). The free iron and aluminium, which includes finely-divided inorganic oxides, amorphous inorganic oxides and the organically-complexed sesquioxides, is given by the dithionite-extractable fractions (Fe_d and Al_d). Thus, levels of organically-complexed sesquioxides ($\%Fe_p + \%Al_p$) are shown in Fig. 4.11a, and the organically-complexed proportions of the total free oxides are shown in Fig. 4.11b ($\%Fe_p : \%Fe_d$) and Fig. 4.11c ($\%Al_p : \%Al_d$)), respectively.

The organically-complexed iron and aluminium (Fig. 4.11a) increase progressively from sites 1 to 8 (*ca* 255 years) in both the A/E and B horizons while values in the C horizon remain low. There is a marked increase between sites 3 (*ca* 70 years) and 4 (*ca* 115 years), which corresponds with

(a)
$$\log y = 2.837 + 0.73 \log t$$

(b)
$$y = 0.798 - 0.306 \log t$$

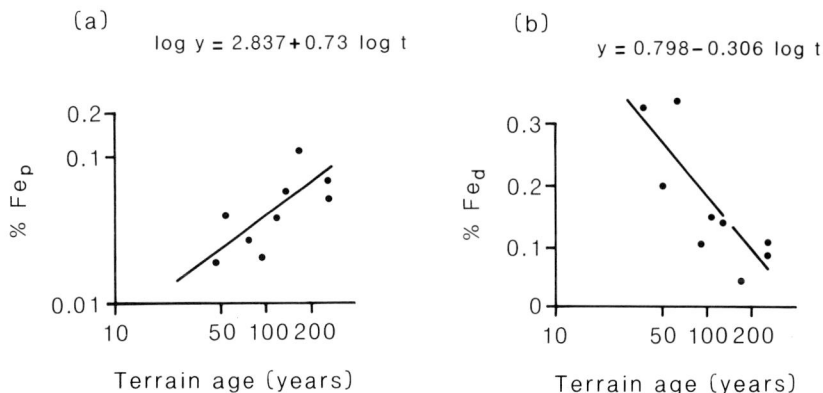

Fig. 4.12. Selected chronofunctions for iron in the E horizon at Austerdalsbreen, southern Norway: (a) % Fe_p, (b) % Fe_d (from Mellor, 1985).

a major increase in the soil organic content. According to Alexander (1982), this suggests a corresponding increase in the quantity of organic-complexing agents. Eventually, the amount of organically-complexed iron and aluminium in the B horizon exceeds that in the E horizon, reflecting the translocation of iron and aluminium sesquioxides from the latter to the former. As accumulation in the B horizon continues, depletion of complexed iron and aluminium appears to occur in the E horizon between sites 8 and 9 (> 2500 years). Similar trends are shown in the proportion of the free oxides that are complexed (Figs. 4.11b,c). However, a higher proportion of the aluminium is complexed (maximum 80%, as opposed to a maximum of < 50% of the iron).

At Austerdalsbreen, southern Norway, similar podzolic soils have yielded compatible results, although the changes appear to be more gradual (Mellor, 1984, 1985). Selected chronofunctions for iron in the E horizon are shown in Fig. 4.12, and similar trends are exhibited by aluminium. The increase through time exhibited by organically-complexed iron (Fig. 4.12a) accompanies a corresponding increase in organic carbon. At the same time, there is a significant decline in the free iron oxides (Fig. 4.12b), which accumulate in the B horizon. The inverse relationship between the trend of the latter chronofunction and that of organic matter suggests a source other than organic matter for the free iron oxides. It also supports the concept of iron translocation in an inorganic form (Anderson *et al.*, 1982; Childs, Parfitt & Lee, 1983) after initial release during till formation (Ellis, 1980b; Mellor, 1985), although *in situ* chemical weathering is also a possible mechanism of release (see below).

Significant increases in organically-complexed iron and aluminium in the A and B horizons of alpine brown soils at Storbreen, may be explained in similar terms to the increase in the E and B horizons at Austerdalsbreen (Mellor, 1984, 1985); that is, through increasing inputs from organic matter. However, no significant age-related trends are found in the free iron oxides ($\%Fe_d$) in either horizon. These results differ from those presented by Sondheim & Standish (1983) from the brunisolic soils at the Robson Glacier. They demonstrate significant changes in $\%Fe_d$ and in ($\%Fe_d + \%Al_d$) with age and depth, but only changes with depth in the case of Al_d, and no significant changes with either age or depth in the pyrophosphate-extractable fractions. Much lower quantities of organically-complexed iron in these soils are nevertheless consistent with their much lower organic content. Sondheim & Standish (1983) suggest that increases in free iron and aluminium oxides through time, and decreases with depth on the older moraines, can be explained in terms of *in situ* weathering of, in this case, an exceptionally calcareous parent material.

4.2.6   Chemical weathering processes

The whole question of the nature and effectiveness of chemical weathering in arctic-alpine environments in general and in arctic-alpine soils in particular is highly controversial. It is the opinion of Darmody, Thorn & Rissing (1987) that research on chemical weathering has been neglected in such regions, in the mistaken belief that weathering in periglacial environments is overwhelmingly dominated by mechanical (frost) weathering. However, Ellis (1980c, 1983) has stressed the dearth of evidence in northern Norwegian mountain regosols, brown soils, podzols and bog soils of not only chemical but also physical weathering. In this section, most attention will be given to evidence produced by recent work in southern Norway, where the nature and scale of pedogenic weathering has been investigated by Mellor (1984, 1986a, b, c) and by Darmody, Thorn & Rissing (1987). Their studies have involved the examination of textural variations, micromorphology under optical and scanning electron microscopes (SEM), and mineralogy by X-ray diffraction.

In principle, chemical weathering of silt-and sand-rich parent materials should produce an increase in the clay content and an increase towards the surface in the clay:silt ratio. Neither trend is found at any of the four alpine and sub-alpine glacier foreland chronosequences examined by Mellor. At Storbreen, samples from five depths within the uppermost 20 cm on four till surfaces ranging in age from 9 to about 230 years, yielded an average of about 2.5% clay at the youngest site (clay:silt ratios (0.04–0.10) and < 1.5% at the other sites (clay:silt ratios < 0.09) (Darmody, Thorn &

Rissing, 1987). The slightly lower values at the older sites are counter to any predicted weathering trend and probably reflect initial site differences. In analysing the texture of soils ranging in age from 50–230 years in four moraine sequences, Mellor (1984, 1986b) also found clay contents generally < 2%. Fine sand and coarse silt dominate in the majority of these soils, which do not exhibit age- or depth-related trends in the clay fraction.

The slow rate of clay production by weathering processes is also supported by Mellor's (1986b) comparison of textural differences between soil profiles in the oldest moraines (230 years) and those beyond the glacier foreland boundaries. At the latter group of sites, all probably deglaciated about 9000 ^{14}C years B.P., slight increases in clay content occur in the upper horizons but clay fractions still do not exceed 3%. However, the interpretation of this evidence as indicating an absence of chemical weathering is weakened by the possibilities for the translocation of weathering products within the profiles.

SEM observations of the number of weathered grains (expressed as a percentage of 50 counted grains) provide further clarification and more positive evidence of rates of weathering (Mellor, 1986b). In profiles from the moraines (230 years), < 8% of the grains possess weathering features, with none at the alpine sites (Storbreen and Vestre Memurubreen, Jotunheimen). Beyond the glacier foreland boundaries at the sub-alpine sites (Austerdalsbreen and Haugabreen), between 34 and 50% of grains are weathered in E horizons, 18 to 30% in B horizons and only 6–8% in C horizons. In the corresponding profiles at Storbreen and Vestre Memurubreen the proportion of weathered grains is less, reaching 12–24% in B horizons and 0–6% in C horizons. Separate results for the silt (2–60 μm) and fine sand (60–200 μm) fractions were in excellent agreement. Unweathered grains show the fresh, sharp and angular features commonly associated with glacial comminution (Whalley & Krinsley, 1974; Whalley, 1982). Weathering takes the form of etching, which appears to be the result of dissolution controlled by crystallographic structures, cleavage traces and imperfections (Wilson, 1975; Wilson & McHardy, 1980). Such dissolution probably releases exchangeable bases, together with iron and aluminum (Mellor, 1986b) and has been attributed to the presence of organic acids (Dearman & Baynes, 1979).

In the glacier foreland soils, a lack of depth-related trends, both in the SEM results and in terms of micromorphological evidence (Mellor, 1986a) may reflect the random incorporation of preglacially-weathered grains during till formation (Mellor, 1986b) or even the small number of sample points. However, micromorphological analysis did detect a statistically

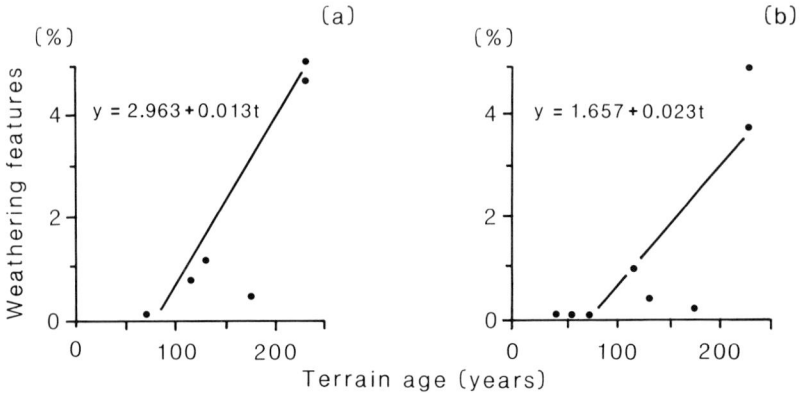

Fig. 4.13. Micromorphology point-count data chronofunctions for weathering features within (a) Ae and (b) B horizons, Austerdalsbreen, southern Norway. Fitted regression lines are statistically significant (p < 0.05) (from Mellor, 1986a).

significant increase through time in the areas of weathering features in the Ae and B horizons at Austerdalsbreen (Fig. 4.13). These features are largely represented by cleavage expansion of mica grains, possibly caused by hydration (Mellor, 1985; Ellis, 1983).

Darmody, Thorn & Rissing (1987) provide further important evidence for chemical weathering from their mineralogical analyses of the < 6 μm (< 0.006 mm) fraction in the uppermost 20 cm of the soils at Storbreen. The primary minerals in the samples – mica, feldspar and amphibolite, in decreasing abundance – reflect the local pyroxene-granulite gneiss bedrock. Secondary minerals include vermiculite and interstratified minerals, including some regularly interstratified mica-vermiculite (hydrobiotite). The results show an increase in the amount of secondary minerals with age together with a corresponding decrease in the mica, both on moraine crests (Fig. 4.14) and proximal slopes. There are no secondary minerals in the most recently-deglaciated samples (9 years), which exhibit uniformly high values of mica with depth. On the oldest terrain (230 years) depth gradients are exhibited in the quantities of both mica and the secondary minerals. These patterns are fully consistent with the weathering of mica to hydrobiotite and vermiculite. Bearing in mind that the interpretation is complicated by clay and silt translocation, low clay contents, heterogeneity of till, and possible aeolian infall, the patterns strongly suggest that the weathering of mica increases towards the soil surface, where the secondary minerals become more abundant through time. The fact that Mellor (1986c) has independently found upward increases in hydrobiotite and reciprocal

Moraine date (A.D.)

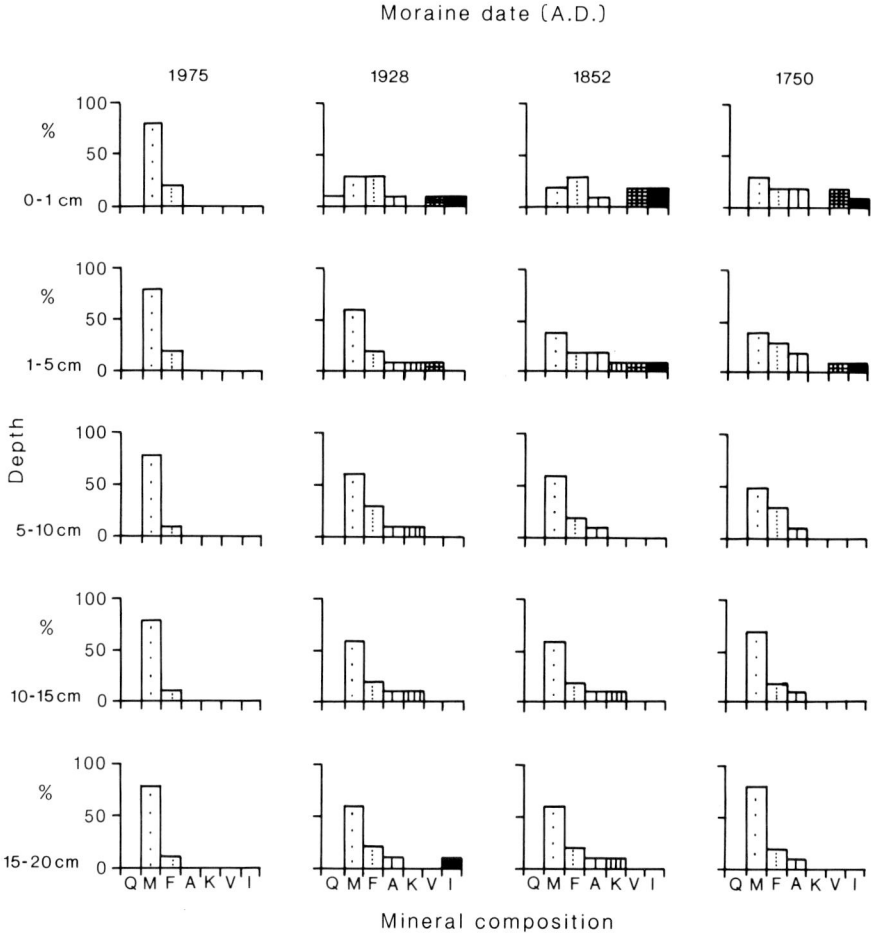

Fig. 4.14. Relative abundance of various minerals in soils (< 6 μm fraction, uppermost 20 cm) from moraine crests of three ages (and from a till plain deglaciated in 1975), Storbreen, Jotunheimen. Mineralogy: quartz (Q), mica (M), feldspar (F), amphibole (A), kaolinite (K), vermiculite (V), interstratified (I) (from data in Darmody, Thorn & Rissing, 1987).

decreases in biotite at both Austerdalsbreen and Haugabreen considerably strengthens this conclusion.

Overall, with the possible exception of calcareous substrates, 230 years would appear insufficient time for weathering trends to be reflected in clay:silt ratios. The short timescale under consideration on most glacier forelands (relative to the restricted range and rate of operation of chemical weathering processes) suggests that chemical weathering is of relatively

(a)   (b)

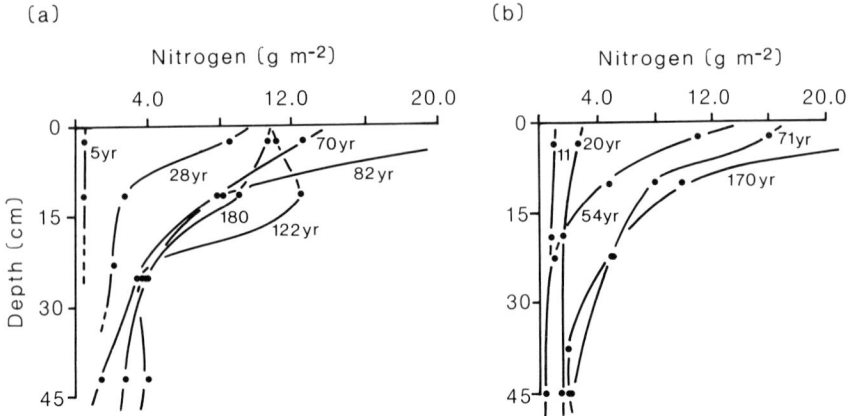

Fig. 4.15. Depth functions for total nitrogen (per 2.5 cm column) with increasing terrain age from Alaskan glacier forelands: (a) Glacier Bay, (b) Mendenhall Glacier (from Crocker & Major, 1955; Crocker & Dickson, 1957).

minor importance in most glacier foreland soils. However, the susceptibility to weathering of some minerals, particularly mica, means that pedogenic weathering should not be ruled out in consideration of the chemistry of soil development.

4.2.7   Nitrogen

Crocker & Major's (1955) well-known diagram from Glacier Bay (Fig. 4.4d) shows a distinct maximum in the total nitrogen content of the soil system (forest floor and mineral soil). During the phase of vigorous alder growth, nitrogen accumulates at an annual rate of about 4.9 g m^{-2} yr^{-1}, attaining values > 250 g m^{-2} (per 45.7 cm soil column) within about 55 years. After about 100 years, when alder thickets have been replaced by spruce forest, concentrations decline to about 200 g m^{-2}. In the mineral soil, the annual rate of increase in the level of nitrogen is considerably lower (1.5 g m^{-2} yr^{-1}) and here too there may also be a decline from maximum values, but at a later date. Depth functions (Fig. 4.15a) show that most of the nitrogen in the mineral soil is located within the upper 20 cm, where major changes in its distribution occur during the spruce forest stage. Particularly prominent is the fall in the amount of nitrogen within the uppermost 10 cm after about 100 years. Overall, the accumulation and subsequent depletion of nitrogen seems to be closely associated with the role of alder in nitrogen fixation, which is fully discussed in section 4.3.3.

Whereas Ugolini (1966, 1968) and Bormann & Sidle (1990) have confirmed the general pattern of change in nitrogen at Glacier Bay, most studies elsewhere have not detected a significant decline in nitrogen in the later stages of soil development. At the Herbert and Mendenhall Glaciers, Crocker & Dickson (1957) suggest that nitrogen in the soil system as a whole approaches an apparent steady state after about 110 years, with values of about 300 g m^{-2} or more. The rate of accumulation decreases after about 60 years but there is no evidence in their data for an absolute decline in the nitrogen content of either the soil system or the upper layer of the mineral soil (Fig. 4.15b). Nevertheless, the variability and the small number of data points is such that the data are not incompatible with a slight loss of nitrogen from the system (Crocker & Dickson, 1957: 176).

A similar ambiguity exists in the paper of Sondheim & Standish (1983) relating to terrain deglaciated by the Robson Glacier. After application of a rigorous sampling design and curve-fitting procedure, they conclude that *mean* nitrogen content reaches a temporary steady state of about 0.04% in less than two centuries within the uppermost 15 cm (Fig. 4.16). However, they also state (p. 514) that there is a small but definite decrease in the nitrogen content between their two oldest soils, despite the contradictory evidence of confidence intervals that exhibit considerable overlap (at one standard error of the mean). Clearly, interpretation is rendered difficult by the small numerical differences involved in relation to likely sampling error.

On the moraines of the Klutlan Glacier, relatively high concentrations of nitrogen (1.0–2.0%) are reached within the surface organic-rich horizon during the first 15 years (Jacobson & Birks, 1980). Thereafter, levels remain high but appear to decline slightly to levels of 1.0–1.5% after 200 years. The same chronosequence exhibits a much slower increase in nitrogen in the uppermost mineral horizon, apparently declining from about 0.7% to 0.4% between 200 years and > 268 years (the estimated age of the oldest surface).

Most other studies are at least consistent with increasing nitrogen values through time, usually with a relatively rapid rise in the early stages. At Bödalsbreen, southern Norway, concentrations of nitrogen in the mineral soil reach about 0.06% on the youngest moraine examined (50–70 years old), rising to about 0.09% on the oldest moraine (about 230 years old) (Vetaas, 1986). However, levels of about 1.0% in the organic horizon on the oldest moraine are considerably higher than on the next youngest (165 years). In the similar podzolic soils at Engabreen, a major increase in nitrogen content of all horizons accompanies the major rise in the carbon content between about 70 and 100 years (sites 3 and 4) (Alexander, 1982). From site 4 onwards a steady state is hypothesised for carbon and nitrogen

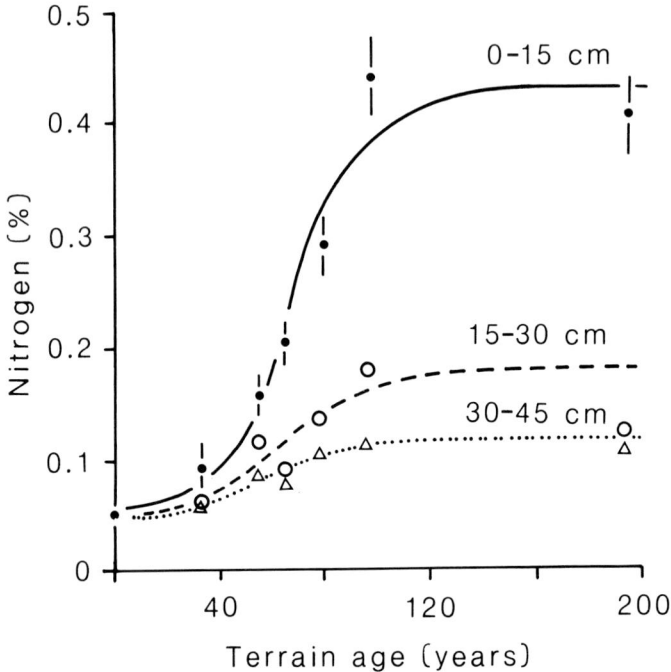

Fig. 4.16. Chronofunctions for total nitrogen at three depths on moraines of increasing age, Robson Glacier, British Columbia, Canada. Mean values are shown with the standard error of the mean (bars). Best-fit relationships, with an exponential depth term and a logistic age term, are shown (from Sondheim & Standish, 1983).

contents in the F horizon, nitrogen value reaching about 0.8%. In the mineral horizons, similar trends are apparent until about 255 years (site 8), after which there is a fall in the levels of both carbon and nitrogen in the E horizon, from a level of about 0.15%. However, as the estimated age of the oldest site in the chronosequence (site 9) is > 2500 years, there is uncertainty as to the temporal significance of this difference.

Although Fitter & Parsons (1987) report no trend with time in water-soluble nitrogen in front of the Athabasca Glacier, Canada, Stork (1961), Jenny (1965, 1980), Viereck (1966) and Smith (1984) indicate increasing nitrogen values through time in chronosequences above the tree line from northern Sweden, the Alps, Alaska, and South Georgia in the sub-Antarctic, respectively. Only Viereck's data, from outwash sequences in front of the Muldrow Glacier, suggest a major reduction in the rate of accumulation after the pioneer stage (25–30 years), attaining levels of about

0.17% after 200 years at about 5 cm depth. The detailed pattern shown by Stork at Storglaciär, in the mid-alpine belt of Swedish Lappland, is unclear because of a possibly anomalous data point. Both of Smith's chronosequences from South Georgia are related only to a relative timescale. Data from the Rhonegletscher chronosequence, which involve low-alpine soils, is presented by Jenny at the level of the ecotessera (soil system + vegetation system). Nitrogen accumulates to the extent of about 300 g m^{-2} after 300 years, an initial rate of accumulation of 0.98 g m^{-2} yr^{-1}, falling to 0.69 g m^{-2} yr^{-1} at 300 years.

4.2.8 Phosphorus

There have been few investigations of the phosphorus content of glacier foreland soils and these are hardly comparable because different fractions have been analysed using different extraction methods. Fitter & Parsons (1987) state that there are no significant trends with time in the water-soluble phosphorus content of soils in front of the Athabasca Glacier. Stork (1961) reports a rapid decline in the phosphorus content of the surface layer of soils within 50 years of deglaciation at Storglaciär in northern Sweden. This decline, from about 15 mg 100 g^{-1} to < 5 mg 100 g^{-1} is attributed to leaching of minerals released from the parent rock during the formation of the till. However, Stork provides no information on the extraction procedure.

At the Klutlan Glacier, Jacobson & Birks (1980) measured the NH_4F-extractable fraction, which Walker & Syers (1976) term 'nonoccluded P'. According to Walker & Syers, this is an inorganic fraction that represents ions adsorbed at the surface of iron and aluminium oxides, hydrous oxides, and calcium carbonate. However, Jacobson & Birks interpret their data as providing an estimate of phosphorus available to plants. In the light of the relatively high values of this fraction in the organic horizon, this would appear to be a useful interpretation in this case. Within the uppermost mineral horizon, Jacobson & Birks' results show a very small but statistically significant linear increase in values through time from about 2 to 4 mg kg^{-1} after about 250 years. In addition, Fitter & Parsons (1987) demonstrate a weak linear relationship using Jacobson & Birks' data from the uppermost layer (O1) of the surface organic horizon, the regression line rising from about 6 mg kg^{-1} to about 11 mg kg^{-1} over the same time interval.

Bormann & Sidle (1990) found a large decrease in the NH_4F-extractable phosphorus in the surface organic horizon at Glacier Bay, from about 220 μg g^{-1} in the alder stage (terrain age 64 years) to 60–70 μg g^{-1} in

Table 4.1. *Total phosphorus* (P_T), *acid-extractable phosphorus* (P_{Ca}) *and organic phosphorus* (P_o) *(ppm) in the* < 2 mm *fraction of soils on outwash deposits in front of Franz Josef Glacier, New Zealand (after Walker & Syers, 1976).*

Horizon	Age								
	0 years			100 years			1000 years		
	P_T	P_{Ca}	P_o	P_T	P_{Ca}	P_o	P_T	P_{Ca}	P_o
1	766	719	12	652	368	197	867	25	549
2				715	640	33	693	93	409
3				733	681	14	620	178	304
4							698	483	131
5							794	679	47

subsequent stages under spruce (terrain age > 110 years). This decrease appears to have been due to the uptake of phosphorus by spruce, foliar phosphorus content in the spruce stage being much higher than in the alder stage.

In the most comprehensive investigation to date, Walker & Syers (1976) present information from a chronosequence of soils on glacio-fluvial deposits near the Franz Josef Glacier, New Zealand, based partly on the work of P. R. Stevens (Stevens, 1968; Stevens & Walker, 1970). Data are given for total phosphorus (P_T), acid-extractable phosphorus (P_{Ca}) and organic phosphorus (P_o). The main trends over the first 1000 years (Table 4.1) appear to be a reduction in acid extractable P_{Ca} and a corresponding rise in the organic P_o. A similar pattern has been reported by Archer (1973) in relation to soils on tills and glacio-fluvial outwash in front of cirque glaciers in the Ben Ohau Range, also in New Zealand. Both trends are most evident in the surface horizon but are also detectable in lower horizons. Over a much longer timescale, involving surfaces up to 22 000 years old, all fractions decline, with eventually > 90% of the original phosphorus being lost (Walker & Syers, 1976; see also Burrows, 1990).

Taken in its totality, the rather meagre information available points to the existence of a variety of patterns and a number of different processes. An explanation of changing phosphorus values through time clearly requires a consideration of the initial conditions (the mineral composition of the parent material), physico-chemical processes and biological mineral cycling.

4.3   Environmental controls on pedogenesis

The chronosequences described in section 4.2 indicate that a variety of processes contribute to the observed patterns of pedogenesis in front of retreating glaciers. On each glacier foreland, the precise pattern is controlled by a particular combination of soil-forming factors. Attention is given in this section to the nature of this control and to the ways in which such factors influence the characteristics and rate of pedogenesis. As pointed out in section 4.1.1, soil-forming factors may be effective either in terms of the initial conditions of the site or as influx variables during soil formation.

4.3.1   Parent material

Jenny (1980) defined parent material exclusively as part of the initial state of the soil system. However, this is an oversimplification, as parent materials may change more-or-less continually due to gains (e.g. the deposition of loess or air-borne salts) and losses (e.g. pervection or erosion) (Crocker, 1952). Parent material influences many soil properties to varying degrees. Its influence tends to be greatest in the early stages of soil development and in drier regions (Birkeland, 1974). There tends to be a progressive diminution of parent material effects through time (cf. Chesworth, 1973, 1976a,b). In the context of glacier forelands, slow rates of soil development and relatively short timescales, combined with a dynamic physical environment, ensure that parent material effects can rarely be ignored.

Parent material control of soil development on recently deglaciated terrain is evident in both physical and chemical soil properties. The texture of different parent materials has a major effect on the rate of pedogenesis. For example, Gellatly (1987) describes soil development on moraines of the Classen Glacier, New Zealand, where the parent materials are derived from a fine- to medium-grained greywacke. In the case of fine-textured till deposits (composed of fine gravel supported by a silt and sand matrix), soil development follows soon after deposition and site stabilization, whereas moraines comprising coarse, blocky material experience a considerable delay in the onset of soil development. The deposition of loess, which is trapped by a developing plant cover, quickly helps to build up the A horizon on the fine-textured till. On the blocky substrate, fines tend to be washed down between the boulders to accumulate at depth. Even after a period of about 580 years, the cover of vegetation and soil on the blocky substrate is extremely sparse and restricted to pockets of loess and weathered rock material which may have accumulated between the boulders. Once such pockets of fines begin to accumulate, rates of soil development may proceed quite rapidly as precipitation input and leaching processes are concentrated between the boulders (Howell & Harris, 1978). However, as Gellatly points out, it may take over 1000 years for this blocky parent material to reach a similar point of 'inception' as that reached soon after deposition of a matrix-supported till. Thus, order-of-magnitude differences in the rate of soil development can be caused by variations in the texture of till parent material.

Differences may be found between tills and parent materials of other origins. Within the ranges found on glacier forelands, finer grain sizes and poorer sorting are generally more favourable to rapid soil development than coarser and well-sorted materials. Sorted deposits of glacio-fluvial origin in outwash plains and river terraces tend therefore to be character-ized by slower rates of soil development than unsorted tills in moraine slopes and till plains. Archer, Simpson & Macmillan (1973) attribute differences in the soils from two sites on a lateral moraine of the Tasman Glacier, New Zealand, to differences between parent materials of colluvial and frost weathering origins; the colluvium being relatively rich in silt and the frost weathered material being relatively rich in sand.

Initially, soil chemical properties are highly dependent on the chemical composition of the parent material, and subsequent rates of change in soil chemistry may also be significantly affected. The importance of this has already been discussed in relation to pH and base status in section 4.2.4. A good example is provided by Stork (1961) from glacier forelands in the Kebnekajse massif, Swedish Lappland. Amphibolite, alternating with micaschist and gneiss constitute the main rock types of the massif. In front of Storglaciär, where the parent material is derived from amphibolite, a basic rock, initial pH values of about 8.1 fall to about 5.5 on terrain deglaciated for 50 years. The acidity of the till in front of Isfallsglaciär, also derived from amphibolite, is much the same. In front of Tarfalaglaciär, however, where the till is derived from micaschist, initial pH values are much more acid at about 5.9.

The relationship between initial pH and rate of pH decline with increasing age has been examined for 18 chronosequences in southern Norway by Messer (1984). Correlation coefficients between initial pH value and pH decline in successive time intervals indicate that initial pH has a significant influence during the first 200 years of soil development in this region (Fig. 4.17). After the first 25 years, the correlations are highly significant ($p < 0.05$); the fact that the relationship for the first 25-year period is only marginally significant ($p < 10.0\%$) probably reflects greater

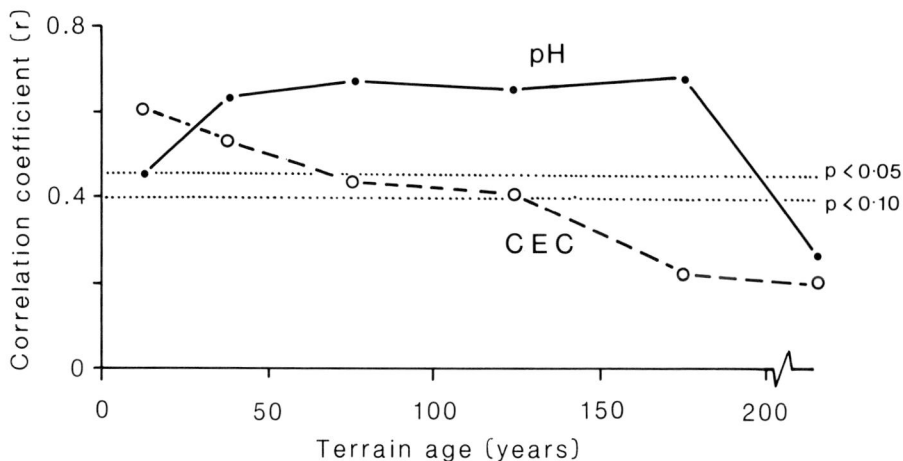

Fig. 4.17. The correlation between initial soil property (pH or CEC) and rate of change in the property with increasing terrain age, based on data from 18 glacier forelands, Jotunheimen, southern Norway: (a) pH (solid line); and (b) CEC (broken line). Two levels of statistical significance are shown (from Messer, 1984).

variability in the initial pH values. As the rate of decline in pH falls towards the end of the 230-year chronosequences, and acidity levels tend to converge (Messer, 1988), the correlation coefficient ceases to reach a statistically significant level.

Messer (1984) has also examined the relationship between initial CEC values and the rate of increase in CEC in a similar way (Fig. 4.17). In the case of CEC, the correlation coefficients are statistically significant ($p < 2.0$) only during the first 50 years of soil development, becoming marginally significant between 50 and 150 years ($p < 10.0\%$) and insignificant thereafter. It would appear, therefore, that factors other than the initial characteristics of the parent material, such as climate (Messer, 1988) and vegetation, increase in their relative importance as the rate of change in CEC accelerates after the first 50 years.

A different approach is to compare rates of soil development with C horizon characteristics as surrogates for initial conditions (Bockheim, 1980). After reviewing 15 soil properties in a wide range of chronosequences, many of which do not relate to glacier forelands and involve very long timescales, Bockheim (1980) concluded that rates of increase in the clay content of the B horizon and solum thickness are the only soil property changes significantly affected by the parent material. Both properties were found to be positively correlated with the clay content of the C horizon.

Bockheim also concluded that the decline through time of both pH and percent base saturation was unrelated to parent material. Messer's (1984) results can be reconciled with those of Bockheim if the precision of her results and the rapidity of pH change are taken into account together with the much longer timescales involved in the latter's chronosequences.

4.3.2   Topography

In view of the importance of catenary variation in soils generally (Birkeland, 1974; Gerrard, 1981; Hall, 1983), topographic variables (including site aspect, slope angle and slope position) would be expected to exert significant control on soil development in front of retreating glaciers. As in the case of parent material, topography is most commonly regarded as part of the initial state of the soil system. However, this is again an oversimplification as topography and its associated geomorphological and pedological processes do not necessarily remain unchanged through time. Altitudinal controls will be considered elsewhere in relation to climate (see section 4.3.4).

Although the importance of topography has been recognized in most studies of glacier foreland soils, there have been few detailed investigations of topographic effects. On the contrary, a prime objective of almost all such studies has been to eliminate the influence of topography by sampling soils from sites within an acceptably narrow range of slope angles (usually near-flat sites) or sites deemed 'optimal' for soil development. Thus, the information available is far from providing an understanding of the processes responsible for catenary differentiation.

Apart from investigations involving relatively old moraines and hence much longer timescales (e.g. Berry, 1987; Birkeland & Burke, 1988), the only studies that have specifically examined topographic controls at the scale of the soil catena and in the glacier foreland context have been carried out in Norway (Messer, 1984; Parkinson & Roberts, 1985; Vetaas, 1986) and in the Pyrenees (Parkinson & Gellatly, *in prep.*). Messer (1984) employed a sampling design stratified according to age and 'site type'. She recognized four site types: (1) moraine crests; (2) moraine slopes; (3) intermoraine depressions; and (4) intermoraine optimum sites. At six of her 18 glacier forelands, the best developed soils were sampled separately within each site type.

Moraine crests are almost invariably characterized by the poorest soil development in terms of soil depth. This is attributed by Messer (1984) primarily to site exposure. In winter, moraine crests are characterized by a thin snow cover and are often exposed to severe climatic conditions when free of snow. In the summer months the same sites are likely to be most affected by strong winds and soil drought. Exposed and freely-draining moraine crests therefore tend to develop a relatively poor vegetation and soil cover, which is in turn susceptible to disturbance and erosion.

Messer's best developed soil profiles are frequently found on the lower segments of proximal slopes where slope angles tend to be less steep. This general pattern is probably caused by a variety of geomorphic and pedogenic interactions (cf. Furley, 1968, 1971; Gerrard, 1981). Steep slope angles are not only more susceptible to mass movement but also to overland flow and slope wash. Messer (1984) observed that on steep slopes, heavy rain and snow meltwater drains primarily by overland flow with limited percolation. Thus, better developed soil profiles on the lower slopes may be attributed in this case to a combination of less exposed conditions, greater down-profile translocation of the products of organic matter decomposition, and possibly the down-slope translocation of pedogenic products from farther up slope. According to Messer, distal slopes at her sites rarely attain the same degree of soil development as proximal slopes because they are steeper.

Intermoraine areas exhibit very varied soil profiles. Some, the intermoraine optimum sites, are at least as well developed as those of the proximal slopes; others, where not prone to erosion by meltwater streams, tend to develop distinctive snowbed or peaty soils, particularly in depressions or hollows (Messer, 1984). Intermoraine optimum sites are characterized by low-angle slopes and good drainage at elevations beyond the influence of the erosional activity of glacio-fluvial streams.

Systematic measurements of a wide range of soil properties on the crests, proximal slopes and distal slopes of moraines of four different ages in front of Bödalsbreen have been made by Vetaas (1986). His data indicate that whilst the development of a podzol profile is slowest on moraine crests, it is more rapid on distal slopes than on proximal slopes (Fig. 4.18). Traces of a leached (E) horizon are found in the distal slope of moraine D (terrain age < 60 years). A similar E horizon, absent from the proximal slope of moraine D is found in the proximal slope of moraines C and B (at least 100 years older). After about 230 years the podzol profile is better developed on the distal slope with thicker Ah and E horizons. The differences are attributed by Vetaas (1986) to the effect of a strong glacier wind, which reduces the accumulation of litter and the formation of humus on moraine crests (and to a lesser extent proximal slopes), and from which north-facing distal slopes receive greatest shelter.

In Okstindan, northern Norway, Parkinson & Roberts (1985) found

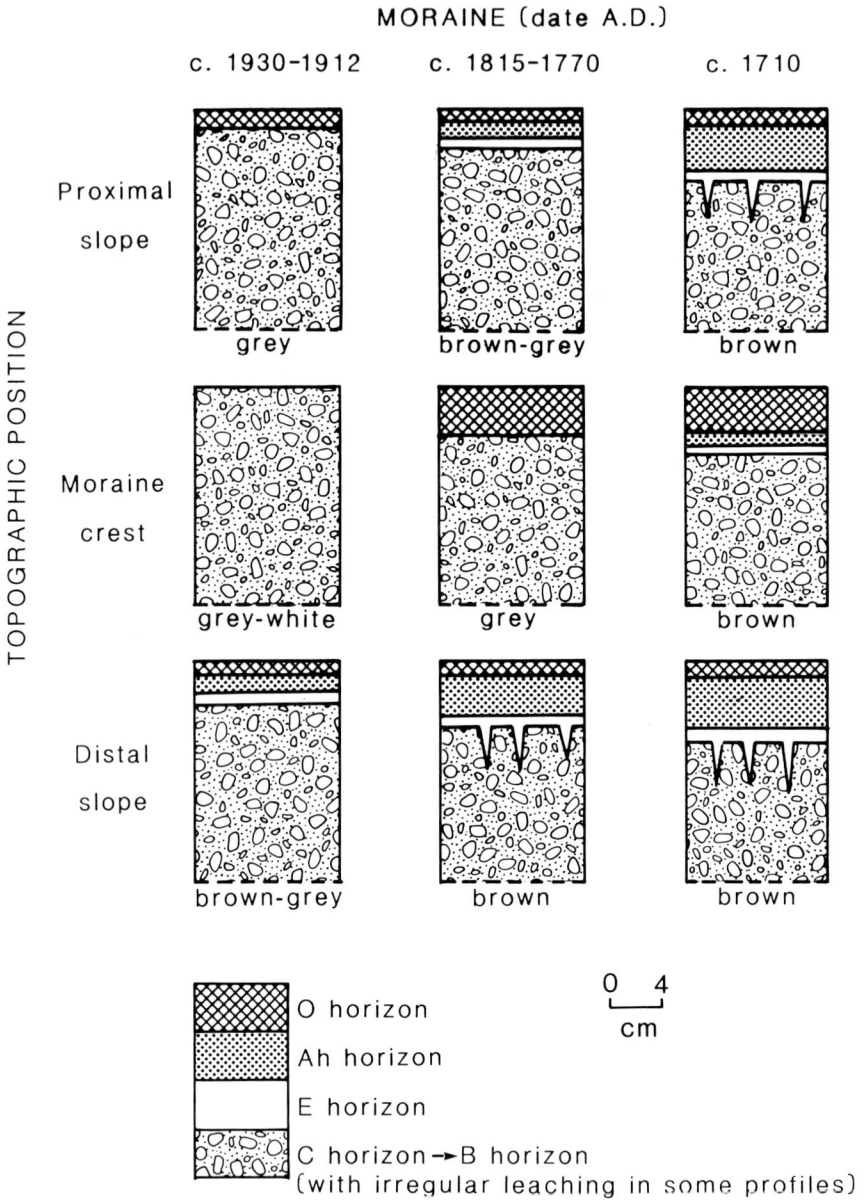

Fig. 4.18. Variation in the nature and rate of podzolic soil development on moraine crests and on the proximal and distal slopes of moraines A–D, Bödalsbreen, southern Norway (from Vetaas, 1986).

that pH, organic matter and texture do not vary in a consistent pattern between top, middle and basal slope positions on two Neoglacial moraines in front of Austre Okstindbreen and Corneliussensbreen. However, exchangeable cations exhibit patterns of preferential leaching with the highest Ca:Mg, Ca:K and Ca:Na ratios at the top of each slope. Similar microtopographic patterns in cation ratios, which reflect the relative ion replaceability ranking of Bohn, McNeal & O'Conner (1979) (Na > K > Mg > Ca, with Ca the least mobile), have been detected on Neoglacial moraines of the Glacier de la Munia and other small glaciers in the Cirque de Troumouse, French Pyrenees (Parkinson & Gellatly, *in prep.*).

4.3.3   Biota

The recognition of an independent biotic factor (defined as the available biota) is not entirely satisfactory because the dependent and independent aspects of the biota are especially difficult to separate (Crocker, 1952, 1959; Jenny, 1958). It is nevertheless clear that the biota, and particularly plants, play an important if not dominant role in soil development generally. Indeed, Jacks (1965) defines soil formation as the process whereby organisms make and maintain a habitat, and Ugolini & Edmonds (1983) state that it is indisputable that soil formation proceeds many times faster in the presence of biota than without. Here, attention is confined to the effects of plants on soils; reciprocal relationships will be considered in chapter 6.

Control of soil development by plants is most obvious in those soil properties most directly dependent on the supply of organic matter to the soil, such as the nature and thickness of surface organic horizons and the organic carbon content of mineral horizons. From many glacier foreland chronosequences there is evidence that soil development is slow until vegetation becomes well established. In some cases inferences have been made about the effects of particular species or plant communities on both physical and chemical soil properties. A good example is provided by Persson (1964) from the glacier foreland of Skaftafellsjökull, Iceland, where the mosses *Rhacomitrium canescens* and *R. lanuginosum* sometimes seem to be impregnated with fine sand carried by the wind. The moss shoots successively grow up over the sand while their lower parts are decomposed into humus which gives the surface layer its dark colouration. A rare example from a glacier foreland in the tropics is given by Hope (1976) from in front of the Meren Glacier, Irian Jaya, New Guinea. The pH of ground water in the highly calcareous till is 9.5 to 10.5. At a site deglaciated in A.D.

1920–30, exposed till remains at about pH 10, whereas the 2 cm thick moss mat and the surface of the underlying mineral soil have pH values of 7 and 9, respectively. Further acidification is attributed by Hope to the thickening of the humus layer.

There are numerous other examples of changes, sometimes relatively abrupt, in the rate of soil development that are closely synchronous with vegetation change (see sections 4.1.2 and 4.2) and where it is assumed that the latter affects the former. However, some soil changes have been found to be out of phase with vegetation changes. For example, Persson (1964) suggests that much of the early reduction in pH of the surface soil layer in front of Skaftafellsjökull occurs due to leaching alone, prior to any major vegetation development; only a later, further fall in pH is attributed to the effect of plants. Similarly, in an Afroalpine chronosequence on Mount Kenya where the abrasion pH of most of the minerals in the parent material (alkali feldspars) is 10–11, the greatest drop in pH occurs in the first 100 years (Mahaney, 1990) when there are only scattered individual plants with essentially no cover (Spence, 1989). Again, on the foreland of Nooksack Glacier in the North Cascades National Park, Washington, supraglacial debris and newly-deposited till is distinctly more acid (pH 4.2–5.5) than the abrasion pH of the granodiorite bedrock (pH 7.3–8.6), before colonization by vegetation (Zasoski & Bardo, 1979; Bardo, 1980).

Some of the best evidence for the role of plants is provided from Glacier Bay by Crocker & Major (1955) who demonstrate the remarkable acidifying effect of *Alnus crispa* in two ways. First, pH levels under alder are compared with those under other species and also with those associated with areas of bare ground which have apparently always been devoid of macroscopic plants (Fig. 4.19). The ages of *Alnus*, *Populus trichocarpa* and *Salix* spp. were determined by dendrochronology of the shrub specimen in question, whereas the age of *Dryas drummondii* was estimated. Although all the shrubs appear to influence the acidity of the soil, only the alder has a major effect. This has been confirmed by Ugolini (1968) who states that for contemporaneous surfaces colonized by different types of vegetation, the pH values of the uppermost 3 cm of soil are 6.5 (under alder), 6.8 (under *Dryas*) and 7.0 (under mosses).

A major effect of alder is also revealed by Crocker & Major's second approach, which involved the measurement of pH levels along a transect from the oldest (initial) individual of an expanding alder thicket to the adjacent bare ground beyond the thicket. The sequence of samples, taken from a 29-year-old surface, yielded the following pH values associated with increasingly old individuals over a horizontal distance of approximately

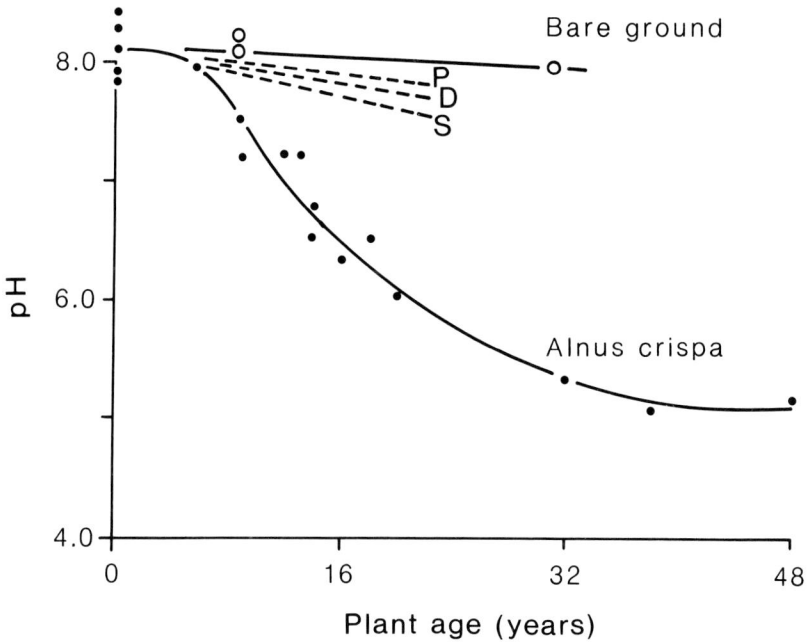

Fig. 4.19. The pH of surface soils (uppermost 5 cm) beneath different plant species with increasing plant age, Glacier Bay, Alaska. Open circles represent samples from bare ground; filled circles represent samples beneath *Alnus crispa*; broken lines represent samples beneath *Populus balsamifera* ssp. *trichocarpa* (P), *Dryas* spp. (D) and *Salix barclayi* (S) (from Crocker & Major, 1955).

9 m: 7.9 (bare ground), 7.2 (9-year-old alder), 7.1 (12-year-old alder), and 6.5 (18-year-old alder).

The litter under alder has a reaction of pH 5.6 to 6.1, tending to become more acid with time and reaching 4.2 to 4.6 in the final stages of alder dominance (Crocker & Major, 1955). The forest floor is consistently more acid than the soil (Fig. 4.4b) and the fermentation–humus (FH) layer is more acid than the undecomposed litter (L) layer (Crocker & Dickson, 1957). Spruce (*Picea sitchensis*) litter is no more acid than the late alder litter with the consequence that there is little change in the reaction of either the forest floor or the mineral soil during the period of spruce dominance.

All the above data suggest acidification originating from alder litter, which would also account for the enhanced leaching of calcium carbonate beneath alder. Under 18-year-old alder on a surface with an estimated age of 31 years, calcium carbonate levels had fallen from initial values of around 5% to about 0.3% (Crocker & Major, 1955). Adjacent ground of

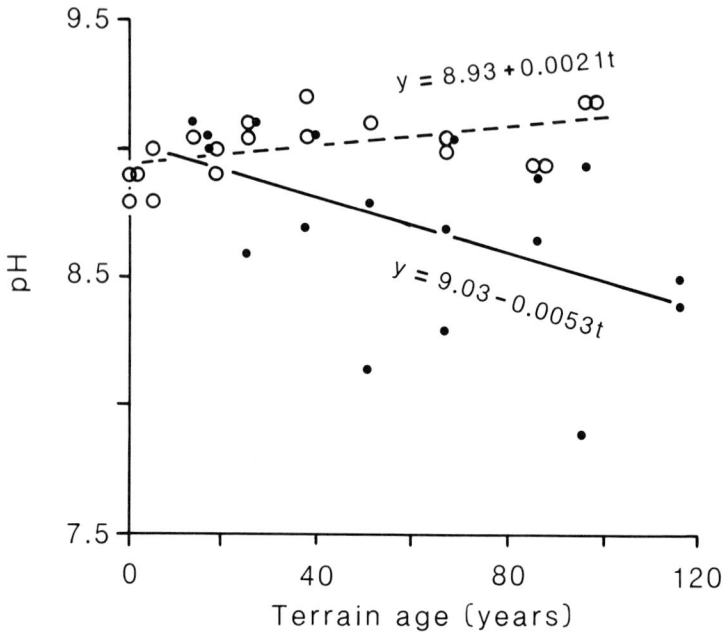

Fig. 4.20. The pH of surface soils (uppermost 5 cm) beneath *Dryas drummondii* and in unvegetated areas, with increasing terrain age, Athabasca Glacier, Alberta, Canada. Both regression lines represent statistically significant relationships (p <0.05) (from Fitter & Parsons, 1987).

similar age but devoid of macroscopic vegetation contains about 3.8% calcium carbonate in the equivalent surface horizon. In another area, with a terrain age of about 20 years, the percentage calcium carbonate in the fine earth fraction (<2 mm) had been reduced from 7–9% to 6, 5 and 3–4% under *Salix*, *Populus* and *Dryas*, respectively. Crocker & Major's supposition that small 'loci' of calcium carbonate may not be completely dissolved in the descending acid solutions, and hence that values for calcium carbonate may lag a little behind the pH values, is supported by Ugolini's (1968) analyses, which show that the calcite is leached more readily than the dolomite (see section 4.2.4).

At a higher altitude (2000 m) on the recessional moraines of the Athabasca Glacier, and in the absence of alder, but with moderate to high levels of calcium carbonate (15–85%), Fitter & Parsons (1987) have investigated the effect of *Dryas drummondii* in a similar way. In the top 5 cm of the soil, pH declines from 9.0 to 8.5 under *Dryas* but not in bare areas <0.5 m away (Fig. 4.20). The rate of acidification (0.5 units per 100 years) is much slower than at Glacier Bay (0.3 units per 30 years). This difference is presumed by Fitter & Parsons to be caused by much slower growth rates of *Dryas* at higher altitudes.

Whilst the evidence described above is strongly suggestive of causal relationships between particular species and changes in soil pH, the findings are not conclusive. A comparison of a soil characteristic between two areas in the field is by its very nature, a confounded experiment; that is, more than one alternative hypothesis cannot be ruled out (cf. Hurlbert, 1984; Underwood, 1986). For example, the differences between the bare soil and the soil beneath the plant may be due to reasons other than the presence of the plant. In this case, one alternative possibility is that the bare areas are disturbed by cryoturbation, maintaining a high pH near the soil surface. This could account for the statistically significant increase in pH through time in the bare areas in front of the Athabasca Glacier (Fig. 4.20).

Crocker & Major (1955) have used the same approach in their investigations at Glacier Bay of the rapid increase in soil nitrogen associated with the presence of *Alnus crispa* and *Dryas drummondii*. Their data are summarized in Table 4.2, which shows the total nitrogen in the soil beneath several plant species and in bare ground on terrain of equivalent age. It should be noted that the data for *Dryas* relate to three different locations and that the *Dryas* mat refers to both the living *Dryas* and any organic residues upon the surface of the mineral soil. Crocker & Major conclude that although the accumulation of nitrogen by *Dryas* is considerable when compared with the values from the bare till, alder is much more effective. Not only is the quantity of litter at their one alder site greater than the weight of the *Dryas*-mat, but its percentage nitrogen is twice as great. A small effect of *Dryas* was also detected in a similar experiment near the Athabasca Glacier, where values of water-soluble nitrogen are significantly higher beneath *Dryas* ($3.8 \mu g\ g^{-1}$) than in adjacent bare areas ($2.4 \mu g\ g^{-1}$), the mean difference of $1.6 \mu g\ g^{-1}$ being greater than the 95% confidence limit of $1.3 \mu g\ g^{-1}$.

Both *Dryas drummondii* and alder are endowed with coralloid root nodules in which atmospheric nitrogen is fixed symbiotically (Lawrence, 1958, 1979; Lawrence *et al*, 1967). *Dryas* plants transplanted from the field into a nitrogen-free rooting medium exhibited foliar % nitrogen levels twice as high in nodulated plants as in plants without nodules. Tests by Bond with ^{15}N on detached nodules and also on an intact plant confirm that the nodules are the site of active nitrogen fixation, and that the fixed nitrogen is exported from the nodules to other parts of the plant (Lawrence *et al.*, 1967).

Table 4.2. *Total nitrogen and organic carbon in soils beneath different plant covers on terrain of equivalent age at Glacier Bay (after Crocker & Major, 1955).*

Nature of the site	Surface age (years)	Horizon depth (inches)	Nitrogen %	g m²	Organic carbon %	g m²
Bare[a]	24	0–6	0.0075	3.3	0.02	43
Algal crust[b]	24	0–0.4	0.093	—	1.89	—
Sparse moss[b]	31	0–2	0.015	7.7	0.17	87
		3–6	0.006	5.4	0.04	40
Dryas[b1]	24–25	Mat	1.313	12.20	32.3	300
		0–2	0.019	9.46	0.32	160
Dryas[b2]	24–25	Mat	1.330	18.6	33.8	473
		0–2	0.21	12.3	0.37	217
Dryas[c1]	—	Mat	1.036	20.50	32.5	640
		0–1	0.025	5.96	0.54	132
		1–3	0.0065	3.39	0.15	81
		3–6	0.002	2.29	0.08	79
Dryas[c2]	—	Mat	1.075	20.3	30.6	570
		0–1	0.030	7.3	0.60	146
		1–3	0.010	6.8	0.16	115
		3–6	0.007	7.4	0.20	209
Populus[b]	24–25	0–2	0.018	12.4	0.34	234
(19 years old)		3–6	0.007	7.0	0.09	97
Salix barclayi[b]	24–25	0–2	0.027	19.3	0.49	345
(20 years old)		3–6	0.004	4.0	0.09	96
Alnus[c]	—	Litter	2.076	46.0	28.4	630
(19–20 years old)		0–2	0.063	28.4	1.08	489
		3–6	0.005	5.4	0.08	83

[a] = Anchorage Cove, [b] = Goose Cove, [c] = Hugh Miller Inlet

Knowledge of fixation in association with alder goes back further (e.g. Bond, 1955; Virtanen *et al.*, 1955). The symbiont is the actinomycete *Frankia* which, with the exception of the legume family and one other species (where the symbionts are from the bacterial genus *Rhizobium*), is the symbiont in all known nodulating species (Bond, 1976; Fitter & Hay, 1987). Only small amounts of nitrogen compounds leak from the nodules into the soil directly (Virtanen, 1957); mainly they move up the stems as free amino acids (Virtanen & Miettinen, 1952), later to be incorporated into proteins in the leaves. There, nitrogen accumulates by the latter part of the growing season in mid-August to a concentration of almost 3% of the dry weight of the leaf, a nitrogen level comparable to that developed in alfalfa and other legumes long used in agriculture for adding nitrogen to the soil (Lawrence, 1958).

Nitrogen fixation is known or suspected in association with a variety of other species of higher plant present during soil development on glacier forelands. At Glacier Bay, the soapberry (*Shepherdia canadensis*) and two members of the legume family (*Astragalus alpinus* and *Hedysarum alpinum*) possess root nodules and fix nitrogen. It is also possible that *Rubus spectabilis* may be a nitrogen-fixer (D. B. Lawrence, personal communication; cf. Becking, 1979, 1984). Elsewhere in North America, *Dryas octopetala*, *Astragalus nutzotinensis*, *Lupinus latifolius*, *Oxytropis campestris*, *O. deflexa* and *Hedysarum mackenzii* have also been named as nitrogen fixers (Tisdale, Fosberg & Poulton, 1966; Viereck, 1966; Fraser, 1970; Scott, 1974b; Jacobson & Birks, 1980; Sondheim & Standish, 1983), as have species of *Lotus* and *Trifolium* in Europe (Ellenberg, 1988) and *Carmichaelia*, *Coriaria* and *Gunnera* in New Zealand (Wardle, 1980; Burrows, 1990).

Apart from the work of Lawrence *et al.* (1967) on *Dryas drummondii* at Glacier Bay, however, only one other study has involved the direct examination of biological nitrogen fixation by vascular plants on recently-deglaciated terrain. Using the acetylene-reduction technique, Blundon & Dale (1990) sampled three stages of plant community development in a 200-year chronosequence in front of the Robson Glacier, British Columbia. *Hedysarum boreale* var. *mackenzii*, which is the major nitrogen fixing agent, accounts for 72% of the nitrogen input in the pioneer herb stage, 79% in the intermediate *Dryas* stage, and 88% in the *Picea engelmannii* forest stage. The estimated annual input of nitrogen from all sources decreases from 57.2 m mol C_2H_4 m^{-2} yr^{-1} in the pioneer stage to 41.6 m mol C_2H_4 m^{-2} yr^{-1} in the intermediate stage and 7.3 m mol C_2H_4 m^{-2} yr^{-1} on the oldest moraine.

In areas free of vascular plants, there may be a slow increase in the amount of nitrogen present in the surface soil horizon (Lawrence *et al.*, 1967). Small quantities of nitrogen may be added to the surface in precipitation or by aeolian deposition, which may include leaf litter fragments (Lawrence *et al.*, 1967; Bormann *et al.*, 1989) and dead animals. One remarkable example of the latter is provided by large quantities of grasshoppers embedded within Grasshopper Glacier, Knife Point Glacier and other glaciers in the Wind River Range, Wyoming, U.S.A. (Lockwood *et al.*, 1991). Another small supply may originate from the ammonium chloride present in some igneous rocks (Ingols & Navarre, 1952). Lawrence (1979) argues that the soil at Glacier Bay is too cold for effective fixation of nitrogen by free-living bacteria, even by the photosynthetic ones known to have that capability. He maintains that the main source is likely to be biological fixation by blue-green algae, which thrive in the mat-like 'black crust' that develops on the soil surface within a few years of deglaciation. Although they may be particularly important in the Arctic and Antarctic, blue-green algae appear to be less important in front of Alpine glaciers (Moiroud & Gonnet, 1977). In front of Robson Glacier, soil micro-organisms (e.g. *Gloecapsa*, which grows as a black crust on the mineral substrate surface) account respectively for 26%, 20% and 8% of the nitrogen input in the three successional stages referred to previously.

The black crust phenomenon at Glacier Bay has been subdivided into three types by Worley (1973): (1) *Lophozia* liverwort mat; (2) Cyanophyta blue-green algae mat; and (3) lichen mat. According to Lawrence (1979) nitrogen fixing blue-green algae live both independently in the Cyanophyta mat and in symbiosis within the bodies of *Lempholemma radiatum* in the lichen mat. It is also possible that nitrogen fixers are associated with the liverwort mat (cf. Griggs, 1933; Griggs & Ready, 1934; Burdon & Dale, 1990). Crittenden (1975) has investigated nitrogen fixation on the glacier foreland of Sólheimajökull, Iceland. Five lichen species exhibit appreciable rates of acetylene reduction, including species of *Peltigera* and *Stereocaulon*, which are amongst the most abundant macrolichens in developing *Rhacomitrium* heaths on the more recent moraines. He concludes that lichens are probably the major contributors to the fixation of nitrogen in the chronosequence.

The initiation of biological mineral cycling on recently-deglaciated terrain is well illustrated by soil nitrogen which accumulates as a direct result of biological fixation. At Glacier Bay, where alder is particularly important for the rapid accumulation of nitrogen in the soil profile, the later fall in the amount of nitrogen (and phosphorus) in the upper part of the profile (Figs. 4.4d & 4.15a) also appears to be biologically controlled as it corresponds with the early spruce dominance (see section 4.2.7). In contrast to total nitrogen, the organic carbon content of the soil remains high (Fig. 4.4c). This is reflected in the carbon:nitrogen ratios. While the ratios for both the litter and the uppermost horizon of the mineral soil at the alder thicket stage are no more than about 15, with the development of the spruce forest, the ratios increase to 30 or more (Crocker & Major, 1955). Uptake of nitrogen and other nutrients may be assisted by mycorrhiza. On moraines of the Athabasca Glacier, for example, *Dryas drummondii* exhibits clearly visible ectomycorrhizal sheaths, even on the youngest surfaces (deglaciated for 15 years). On the oldest material (deglaciated for 90 and for 115 years, respectively) there are small amounts ($<5\%$ of total root length) of vesicular-arbuscular endomycorrhizal infection (Fitter & Parsons, 1987). Further evidence for interaction between biota and soils, including relationships between foliar nutrient content and soil nutrients, will be considered in section 6.1.7.

4.3.4 Climatic controls

Many pedologists consider climate to be the most important soil-forming factor, a viewpoint enshrined in the concept of a zonal soil (Bunting, 1965; Cruickshank, 1972; Duchaufour, 1982). Zonal soils are mature soils that reflect the influence of zonal (regional) climate over a long period of time. Traditionally, climate has been regarded as the dominant factor unless the effects of parent material or topography are overriding (intrazonal soils) or where soils have yet to reach maturity (azonal soils).

The influence of climate is both direct and indirect, through its effects on the biota and the geomorphic processes that contribute to soil formation. Moisture and temperature are generally considered to be particularly important controls on most physical, chemical and biological pedogenic processes. These processes require the presence of water and their rates are highly dependent on temperature (Birkeland, 1974). In mature landscapes zonal soils tend therefore to be characteristic of freely draining sites and mesic environmental conditions generally.

Even though the most common types of parent material tend to be freely-draining tills and glacio-fluvial deposits, glacier foreland soils do not fit readily into existing soil classifications which are based largely on mature soils of zonal status. By definition, glacier foreland soils are azonal, and they often do not possess the diagnostic criteria of zonal types. Nevertheless, they may be usefully considered as developing towards typical zonal soils.

The podzols characteristic of the glacier forelands of coastal Alaska, which have been described in some detail (see especially sections 4.1.2 & 4.3.3), are highly dependent on the oceanic climate. At Glacier Bay, the mean annual temperature is about 5°C and annual precipitation values are about 1900 mm (Noble, Lawrence & Streveler, 1984). These conditions permit the development of recognizable podzols after about 150 years during the spruce forest stage (Ugolini, 1966). In front of the Robson Glacier, just west of the continental divide in the Canadian Rockies, at an altitude of about 1680 m, a cooler, drier and a generally more continental climate leads to brunisolic soils (Sondheim & Standish, 1983). Assuming that the mean annual soil temperature is above 0 °C, orthic eutric brunisols are characteristic of the oldest moraines. After about 200 years of development, these soils do not approach the morphological characteristics of podzols possessed by much older forested surfaces nearby.

In the even more continental climate of the Klutlan Glacier, Yukon Territory, at an altitude of about 1220 m, with a mean annual temperature of about − 8 °C and a mean annual precipitation of about 400 mm (Driscoll, 1980), the soils also appear to be brunisolic. In relation to the comparable vegetation stages at Glacier Bay, the changes in soil chemical properties on the Klutlan moraines are generally slower by some 50–100 years (Jacobson & Birks, 1980). This is true despite the fact that the species in the two vegetation successions are substantially different and the similarities in the vegetation stages between the two areas are at the physiognomic rather than specific level.

Patterns in the nature and rate of soil development in relation to altitude, continentality and climate have been examined in greater detail in southern Norway by Mellor (1985, 1987) and Messer (1988). Mellor determined a large number of physical and chemical soil properties from moraine sequences on four glacier forelands, whilst Messer analysed four soil properties from 18 forelands. An important feature of these studies is the level of standardization achieved in both field and laboratory methods. The chronofunctions derived from these studies are truly comparable and enable a quantitative assessment of the extent of regional variation in and the degree of climatic control on soil development.

Mellor's (1987) soil chemical properties at the four glacier forelands are summarized in Table 4.3. Values of most properties tend to decrease with depth and increase through time, the main exception being pH, which exhibits the opposite trend. Exchangeable sodium contents are especially high at the two western sites (Haugabreen and Austerdalsbreen), which is consistent with a marine origin, at least in part. There are also some differences in parent materials, the tills at the western sites being relatively poor in iron-bearing minerals. However, the main differences between these sites and those in Jotunheimen (Storbreen and Vestre Memurubreen) can be attributed to climate. The glacier forelands of Haugabreen, Austerdalsbreen, Storbreen and Vestre Memurubreen lie at altitudes of about 650, 400, 1225 and 1500 m, respectively. These altitudes respectively correspond with mean annual temperatures of 3, 5, − 1 and − 2.5 °C, and mean annual precipitation values of the order of 3000, 2250, 1500 and 800 mm (Mellor, 1987). The western sites lie within the sub-alpine birch woodland zone, whereas the Jotunheimen sites are above the tree line in the alpine zone.

Similarities and differences in the nature and rate of soil development on

Table 4.3. *Range of values displayed by soil chemical properties on four glacier forelands in southern Norway (after Mellor, 1987).*

Soil property	Glacier foreland			
	Haugabreen	Austerdalsbreen	Storbreen	V. Memurubreen
pH	3.60–5.50	3.60–5.50	4.10–6.50	4.00–6.50
organic C (%)	0.10–36.10	0.10–24.60	0.10–18.67	0.10–4.53
$CEC^{\dagger}$	4.00–51.00	1.00–43.00	1.00–23.00	1.00–10.00
$Ca^{\dagger}$	0.18–7.13	0.07–4.90	0.13–6.40	0.32–1.44
$Mg^{\dagger}$	0.02–4.96	0.01–2.60	0.03–2.89	0.04–1.00
$K^{\dagger}$	0.02–1.41	0.03–2.60	0.03–0.78	0.01–0.21
$Na^{\dagger}$	0.30–13.90	0.21–8.99	0.10–0.20	0.07–0.15
Fe_p (%)	0.01–0.22	0.01–0.37	0.01–0.18	0.01–0.13
Al_p (%)	0.01–0.21	0.01–0.19	0.01–0.23	0.01–0.10
$Fe_d - Fe_p$ (%)	0.08–0.78	0.01–0.48	0.03–0.37	0.91–1.42
$Al_d - Al_p$ (%)	0.01–0.14	0.01–0.03	0.01–0.07	0.01–0.47

† = meq $100g^{-1}$

the four forelands are clearly seen in the patterns of horizon differentiation through time (Fig. 4.21). The results are broadly similar for the two western sites and for the two Jotunheimen sites. The main differences between the soils from the two areas involve the nature of the surface horizon and the presence of a bleached (E) horizon at the western sites. At the latter sites, the surface horizon is almost entirely organic (LF) whereas in Jotunheimen it possesses a relatively large mineral content (A horizon). Application of separate principal components analyses (PCA) to the chemical data for each horizon at each foreland led Mellor (1987) to identify two dominant groups of pedogenic processes common to all sites. The first group of processes involves organic matter accumulation, plant nutrient cycling and decomposition; the second group involves the translocation of organic material, iron, aluminium and exchangeable cations. According to Mellor, the combination of the two process groups identified by PCA indicates that podzolisation is operating within the soils at all four sites. However, podzolisation is more intense at the western sites where soils on the older moraines may be classified as orthic humo-ferric podzols according to the Canadian system (Canada Soil Survey Committee, 1978; Mellor, 1987). In Jotunheimen, the corresponding soils are orthic dystric brunisols (arctic-alpine brown soils). It should also be noted that the particularly slow rates of profile development at Vestre Memurubreen reflect the severity of climatic conditions within the mid-alpine belt.

(a)

(b)

(c)

(d)

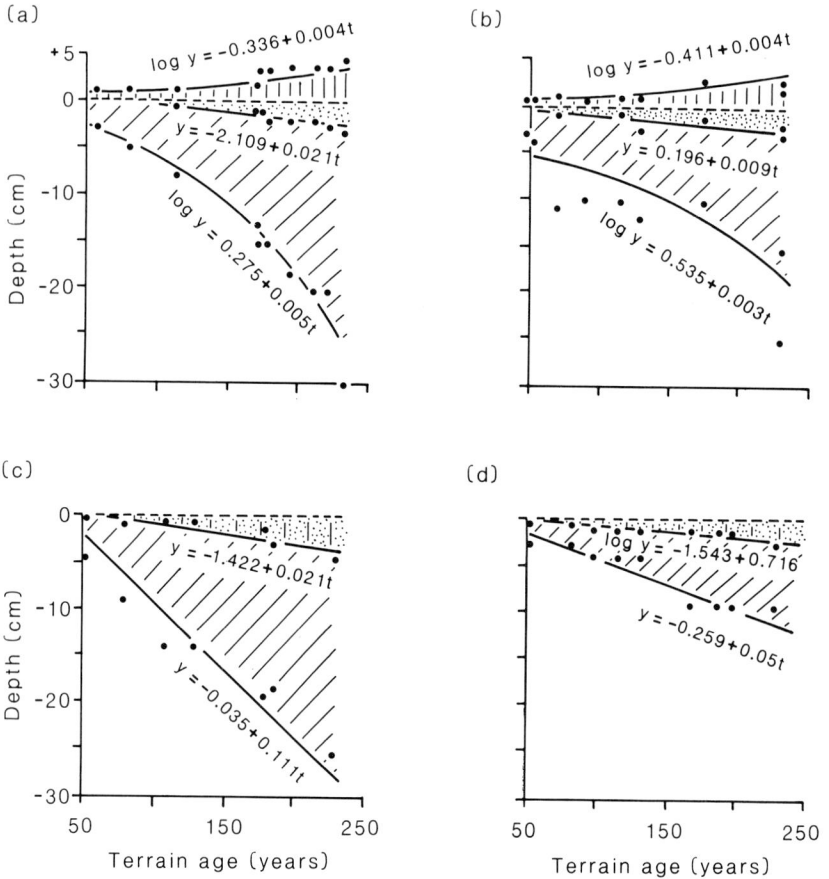

Fig. 4.21. Horizon thickness chronofunctions for four southern Norwegian glaciers: (a) Haugabreen, (b) Austerdalsbreen, (c) Storbreen, (d) Vestre Memurubreen. All regression lines represent statistically significant relationships (p
<0.05). Shading represents horizons: vertical lines, organic (LF) horizon; stipple
bleached (E) horizon; vertical lines and stipple, organic-mineral (A) horizon;
diagonal lines, visual B horizon (from Mellor, 1987).

Results presented by Messer (1988) substantiate and extend these
conclusions (Fig. 4.22). The data for pH, cation exchange capacity (CEC)
and loss-on-ignition were derived from the uppermost 10 mm of the mineral
soil and the best-fit transformation was assessed for each soil property
separately (Messer, 1984). Of the 72 (transformed) relationships between
soil property and terrain age, 57 reach the 5.0% level of statistical
significance. The altitudinal range of these glacier forelands extends from

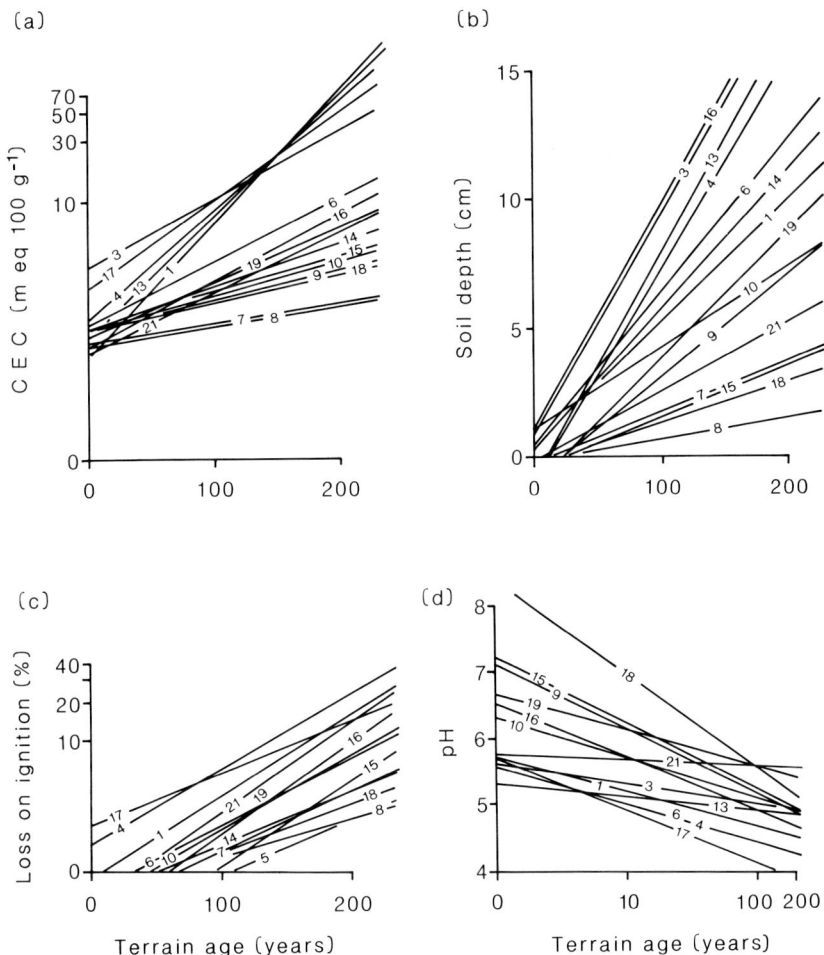

Fig. 4.22. Regional variation in chronofunctions from 21 glacier forelands in southern Norway: (a) CEC, (b) soil depth, (c) loss-on-ignition, (d) pH. Only statistically significant relationships are shown (p < 0.05). Glacier names: (1) Bergsetbreen, (2) Besshöbreen, (3) Bödalsbreen, (4) Böyabreen, (5) Gråsubreen, (6) Haugabreen, (7) Högvaglbreen, (8) Koldedalsbreen, (9) Austre Memuru-breen, (10) Vestre Memurubreen, (11) Austre Nautgardsbreen, (12) Vestre Naut-gardsbreen, (13) Nigardsbreen, (14) Slettmarkbreen, (15) Storbreen, (16) Stygge-dalsbreen, (17) Supphellebreen, (18) Svartdalsbreen, (19) Svellnosbreen, (20) Veobreen, (21) Visbreen (from Messer, 1988).

Table 4.4. *Spearman's rank correlation coefficients (r_s) between rates of change in soil properties and climate based on 18 glacier foreland chronosequences from southern Norway (after Messer, 1984, 1988).*

Climatic variable	Soil property			
	pH	LOI	CEC	Depth (cm)
Altitude (m)	−0.004	−0.313	−0.763***	−0.640***
MAT (°C)	−0.111	−0.239	+0.861***	+0.410*
MAP (mm)	−0.062	−0.067	+0.783***	+0.510**

LOI = loss on ignition (%); CEC = cation exchange capacity (meq $100g^{-1}$);
MAT = mean annual temperature; MAP = mean annual precipitation
Statistical significance levels: ***($p < 0.01$); **($p < 0.05$); *($p < 0.10$)

80 m (Suphellebreen) to 1900 m (Gråsubreen), which corresponds with mean annual temperatures of about 5 to −5.5 °C. Messer's (1987) precipitation values, which range from 1290 to 525 mm, are probably underestimates as she calculated these from the west-east gradient in precipitation exhibited by data from meteorological stations, without any correction for altitude effects. Spearman's rank correlation coefficients (r_s) between rates of change in each soil property and glacier snout altitude, mean annual temperature and mean annual precipitation, respectively, are given in Table 4.4.

Strong relationships to climatic parameters are suggested by high and statistically significant correlations involving the rate of increase in CEC through time. The rates differ by almost an order of magnitude between the low-altitude western forelands and the high altitude eastern ones. In Fig. 4.22a, glaciers 1, 4, 13 and 17 (Bergsetbreen, Bøyabreen, Nigardsbreen and Suphellebreen, respectively) possess distinctly higher rates than all the others. Although Haugabreen (glacier 6) lies farther west, the glacier snout lies at a higher altitude (935 m) than the other western glaciers; hence a rate of increase in CEC that is little higher than those found on the glacier forelands in Jotunheimen. The highest altitude foreland of all in eastern Jotunheimen (glacier 5, Gråsubreen) is characterized by the lowest values of CEC and the lowest rates of increase.

Linear increases in soil depth through time also exhibit wide regional variations in rates that are significantly correlated with climate (Fig. 4.22b & Table 4.4). Again, the western sites, which tend to be located at lower altitudes in the sub-alpine birch woodland zone with milder and wetter climates, exhibit faster rates of soil development. Both CEC and soil depth here reflect the organic phase of pedogenesis. The lack of any significant correlation between loss on ignition and climate (Fig. 4.22c and Table 4.4) relates not only to soil organic matter content being the result of a balance between organic inputs to the soil and their rates of decomposition and removal, as suggested by Messer (1987), but also to the fact that soil samples were taken from the uppermost layer of the mineral soil (thus avoiding the surface organic horizon most characteristic of the low altitude sites). Regional variations in the rate of decline in pH with time (Fig. 4.22d) seems to be related to factors other than climate, particularly the initial pH of the parent material (see sections 4.2.4 & 4.3.1).

At the highest altitudes investigated by Messer (1987) the brown soils that are dominant in front of most Jotunheimen glaciers are commonly replaced by humic regosols or regosols (cf. Ellis, 1979, 1980a), which are characterized by an organic horizon overlying apparently unaltered parent material. Here, where organic matter production and decomposition are at a very low level beneath sparse bryophyte and lichen communities, rates of soil development are relatively slow.

Birkeland, Burke & Benedict (1989) compare accumulation indices for iron and aluminium and a depletion index for phosphorus in chronosequences from Baffin Island (Canadian Arctic), and alpine locations in western North America (Sierra Nevada and Wind River Range), the Nepal Himalaya and the Southern Alps of New Zealand. Although only partially based on glacier foreland chronosequences, they propose the following ranking of rates of soil development and hence pedogenic development: Southern Alps > Khumbu Glacier area, Nepal $\simeq$ Wind River Range > Sierra Nevada > coastal Baffin Island > inland Baffin Island. The ranking is in broad agreement with regional climate, with accumulation and depletion greatest in the warmest and wettest environments, and least in the coldest and driest environments.

Even slower rates of pedogenesis characterize glacier forelands in the Arctic and Antarctic, particularly in the relatively dry and continental areas. Campbell & Claridge (1987) consider such regions as being characterized by soils in which the effectiveness of the biotic factor approaches but not necessarily reaches zero. Here, algae, lichens and mosses perform biological weathering (Ugolini & Edmonds, 1983). Under the most extreme conditions, the development of soils is confined to physico-chemical soil forming processes and may be termed 'abiotic' or 'ahumic' (Tedrow & Ugolini, 1966). Bockheim (1990) considers the dominant processes in the cold desert soils of the Transantarctic Mountains to be salinization, rubification (reddening due to oxidation of iron-bearing minerals) and desert pavement formation.

At high latitudes and in high-alpine environments, permafrost may be an additional climatically-controlled influence on soil development, resulting in impeded drainage and affecting the soil disturbance regime. Except under extreme polar conditions, where thin glaciers are frozen to their beds, permafrost is normally absent beneath glaciers because of the insulating effect of the ice above (Liestøl, 1978). After deglaciation, permafrost requires some time to form (cf. Payette, Gauthier & Grenier, 1986). Although it is present at a depth of 20–30 cm in much older 'climax' soils, it is not found (at least to a depth of 75 cm) on outwash of the Alaskan Muldrow Glacier deposited 200 to 300 years ago (Viereck, 1966). According to Viereck, permafrost development would be expected as vegetation sucession proceeds with continued thickening of moss or humus layers, which insulate the ground more effectively in summer than in winter and hence produce lower soil temperatures (Benninghoff, 1952).

4.4   Soil formation in time and space

Soil formation may be considered as consisting of two overlapping steps: the accumulation of parent material and horizon differentiation (Simonson, 1959, 1978). The processes involved in the former were considered largely in chapter 3, and those involved in the latter have provided the basis of the present chapter. However, any attempt to define where physical environmental processes end and where pedogenic ones begin is in many ways artificial, as both operate together in the landscape.

A related problem is the definition of soil horizons *per se*. This is rather arbitrary where soils are in the early stages of development. Standard horizon nomenclature tends, therefore, to be abandoned in the study of glacier foreland soils (e.g. Viereck, 1966). Indeed, many researchers have adopted particular depths rather than horizons as the sampling unit, an approach advocated strongly by Messer (1988, 1989). This not only avoids the difficulties of identifying horizons but also appears more appropriate for embryonic soils, where the locus of pedogenesis lies near the soil surface in a heterogeneous landscape. It should also be pointed out that results from the sampling of soils at particular depths are likely to be of more use than horizons where it is intended that soil data be used for interpreting vegetation succession.

4.4.1 Soil development and equilibrium concepts

Simonson (1978) recognizes four groups of pedogenic processes *sensu stricto*, which are responsible for horizon differentiation: additions, losses, transfers and translocations. Mellor (1984, 1985), with specific reference to recently deglaciated terrain, recognizes three: organic, translocatory and weathering processes. Most such schemes view soil formation as a complex of processes under the control of the soil forming factors, emphasizing the interactions of climate and biota working through time on parent materials in particular topographic situations (Arnold, 1983).

Organic processes are extremely important on glacier forelands, where the accumulation and decomposition of organic matter are often decisive in determining the nature and rate of horizon differentiation. Normally, the translocation of organic matter, sesquioxides, cations and fine mineral particles are also significant. Physical and chemical weathering appear to be relatively ineffective, with most textural properties in particular being inherited from the parent material rather than developed *in situ* by weathering processes. As soils develop, the declining influence of parent material is evident in several soil properties. Control is increasingly exerted by regional climate, both directly and indirectly through the biota. Topographic factors are effective at a variety of scales through modification of the local climate and geomorphological processes.

Soil formation on glacier forelands has both autogenic (endogenous) and allogenic (exogenous) components. As pointed out by Crocker (1952) there are parallels with autogenic and allogenic succession (see sections 5.4.2 & 6.3). Whereas autogenic change is self-driven or driven by biotic factors and might be considered a development *sensu stricto*, allogenic change is driven by external environmental factors. The latter includes the continued accumulation of parent material, seen by Simonson (1979) as overlapping with horizon differentiation. Thus, the steady accumulation of loess in front of Classen Glacier, New Zealand, produces a loessic A horizon of exaggerated thickness and a 'diluted' soil profile (Gellatly, 1986). This process may continue indefinitely. Rodbell (1990) concludes that the deposition of loess has been more-or-less continuous throughout the Holocene on moraines of the Ashburton Glacier, New Zealand, commonly producing loess thicknesses of 50–100 cm in 10 000 years. Many other soil-forming processes are essentially allogenic, ranging from those associated with the changing geomorphological environment of the paraglacial landscape (see sections 3.2 & 3.4) to the changes in local or regional climate that accompany glacier retreat (see section 3.3). Such processes make a major and sometimes dominant contribution to the nature and trajectory of soil development *sensu lato*.

Although it is normally assumed that glacier foreland soil chrono-sequences represent the formation of monogenetic soils (i.e. soils formed in a single interval of horizon differentiation and in an unchanging environment), polygenetic soils are also represented. For example, in front of Classen Glacier, if the rate of loess deposition exceeds the rate of horizon differentiation, the A horizon becomes buried. Successive intervals of loess deposition produce successive buried (fAh) horizons (Gellatly, 1986). The progress of soil development may be similarly interrupted by other kinds of depositional event, by disturbances (such as cryoturbation or solifluction, which destroy distinct horizons) or by erosion, which can truncate or completely remove profiles. Johnson, Keller & Rockwell (1990) suggest that all soils are polygenetic and that the older they are, the more polygenetic they become. These authors also recognize regressive as well as progressive pedogenesis, and hence they refer to soil evolution rather than soil development (see also, Johnson & Watson-Stegner, 1987).

Some soil chronosequences are nevertheless suggestive of a monogenetic development tending towards a steady state equilibrium (Chorley & Kennedy, 1971; Smeck, Runge & Mackintosh, 1983). Birkeland (1974) has argued that in general organic matter probably reaches a steady state condition more rapidly than any other soil attribute, possibly within 200 years. This appears to be the case at Glacier Bay (Crocker & Major, 1955; Ugolini, 1968; Bormann & Sidle, 1990), at Engabreen (Alexander, 1982) in the podzolic soils of the Swiss Alps (Fitze, 1982) and in the mountain meadow soils of the Caucasus (Gennadiyev, 1979). However, in other cases, particularly under the influence of less favourable climates and biota, a longer time period is required. In the Jotunheimen-Jostedalsbreen region of southern Norway, for example, the work of Mellor (1984, 1985, 1986a, 1987) and Messer (1984, 1988, 1989) suggests that a steady state has not been reached after 230 years. Messer (1984) calculated the time necessary for attainment of a steady state in the uppermost 10 mm of her soils by extrapolating chronofunctions to the values characteristic of soils beyond the respective glacier foreland boundaries. The latter sites were deglaciated about 9000 B.P. and are assumed to be in a steady state condition. Based on chronofunctions for loss-on-ignition data, the mean 'estimated time of attainment' is 260 years, with a 95% confidence interval of 51 years, and individual estimates ranging from 177 to 409 years (Messer, 1984).

Other soil properties related to organic processes would also be expected to attain a steady state condition in a relatively short period of time, if uninterrupted. Fig. 4.16 provides an example of the attainment of an apparent steady state in soil nitrogen in front of the Robson Glacier within about 100 years, although this may be a transient or temporary condition (Sondheim & Standish, 1983) (see section 4.2.7). In southern Norway, 'estimated times of attainment' for cation exchange capacity in the upper-most 10 mm of the soil range from 170 to 970 years (with a mean value of 412 ± 101 years). In this case, the calculated values are significantly correlated with mean June temperatures ($r = +0.55$; $p < 0.02$) (Messer, 1984).

Data for most other soil properties or for whole soil profiles, and from soils in cold, dry environments, are only compatible with the attainment of a steady state after a much longer period of time than is available on most glacier forelands. An example of such a theoretical steady state is provided by Messer (1984) in relation to chronofunctions of soil depth. 'Estimated times of attainment' in the Jotunheimen-Jostedalsbreen region range from 418 to 6291 years with a mean value of 1840 ± 968 years. Even though horizons may be recognizable after a few hundred years or less, full horizon differentiation normally requires a much longer period of soil development. At Engabreen, a distinct A horizon develops after an interval of 40–45 years, and a B_w horizon within about 120 years, closely followed by an albic E horizon within 200 years. After about 250 years sufficient mobile iron and aluminium has accumulated to meet the diagnostic criteria for a B_s (spodic) horizon. Thus, a morphologically and chemically distinctive cryorthod (orthic podzol) has developed within 255 years (Alexander, 1982) but, given more time, major enlargement of horizon thicknesses and further horizon differentiation is likely.

The slow rate of change in the later stages of soil development may give a false impression of a steady state. This appears to be the case in front of Franz Josef Glacier, New Zealand (Stevens, 1963, 1968; Burrows, 1990). On sandar surfaces deglaciated for 10 000 years, the soil is a well-developed gley-podzol, with a highly-weathered yellow-brown B horizon. Further changes are detectable on older surfaces (> 13 000 years old). Similarly, in an Afroalpine chronosequence on Mount Kenya, the data of Mahaney (1990) suggest an asymptotic rise of both soil organic carbon and soil nitrogen (to about 25% and 2%, respectively) for some 6000 years, only to be followed by a reduction to about half of these values in older soils (after *ca.* 12 000 years). On such long timescales, however, there is less control over possible controlling factors. In the extreme case of Antarctic cold desert soils, it has been hypothesized (Bockheim, 1990) that many soil properties require > 250 ka to equilibrate.

For some soil properties the concept of a steady state appears to be inappropriate, even as a theoretical ideal. Instead of a balance of input and output of materials as can be envisaged for a steady state equilibrium, these properties are better considered as tending towards a terminal state because inputs and outputs do not balance. Yaalon (1971, 1983) designates the processes involved as 'self-terminating' (see also, Torrent & Nettleton, 1978). Such a concept seems applicable to the leaching of carbonates, to the depletion of phosphorus and to the accumulation of iron and aluminium in soils; it may also be applied to the rapid removal of fines in the early stages of soil formation by pervection (see section 3.2.3), to the early reduction in soil pH by leaching, and to chemical weathering (both in terms of the texture and mineralogy of soils). Fitze (1980, 1982) recognizes the non-steady state nature of iron and aluminium accumulation in podzols on moraines in the Swiss Alps, where the ratio $Al_d:Fe_d$ is a good indicator of soil age. Energy dispersive x-ray analysis of reddish-brown grain coatings in spodic horizons shows higher ratios in the outer parts of coatings than in the interior parts. Most other properties 'seem to attain a steady state within a few hundred years' (Fitze, 1982: 266).

4.4.2   Spatial variation in soil chronosequences

The results and conclusions outlined above have been reached by viewing the soils on each glacier foreland as representing a chronosequence and by deriving chronofunctions. However, the assumption that it is possible to reconstruct a temporal sequence from the relationship of soils to terrain age is difficult to test. The fact that statistically significant chrono-functions are increasingly being established is improving the basis for inference but does not provide a critical test. A statistically significant correlation with age is not necessarily related in a simple way to a time sequence at a single site; it may be related to an environmental gradient that is confounded with terrain age.

A rare opportunity to test the predictions of a soil chronosequence has occurred at Glacier Bay, where Bormann & Sidle (1990) resampled some of Crocker & Major's (1955) sites in 1985. The directly measured changes in the plots are broadly consistent with the changes inferred from the chronosequence. For example, the early rise in nitrogen content of the O horizon documented by Crocker & Major in the late alder stage, and the subsequent decline under spruce attributed to vegetative uptake, are

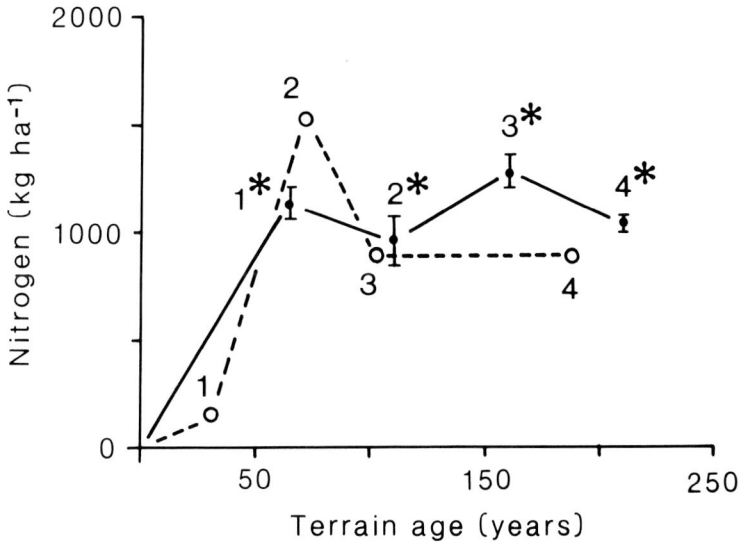

Fig. 4.23. Nitrogen content of O horizons at Glacier Bay, Alaska: a test of the soil chronosequence concept. Open symbols from Crocker & Major (1955); filled symbols (with standard error bars) from Bormann & Sidle (1990). Numbered sites: (1, 1*) Forest Creek; (2, 2*) Muir Point; (3, 3*) Beartrack Cove; (4) Bartlett Cove; (4*) Lester Island (from Bormann & Sidle, 1990).

supported by the direct measurements (Fig. 4.23). At their Forest Creek site, continued development of alder between the two samplings was accompanied by a more than ten-fold increase in the O horizon nitrogen content; at Muir Point, a 40% decrease occurred over the same time interval, during which alder had been replaced by spruce (see sections 5.1.2 & 5.1.3 for tests of vegetational chronosequences).

The traditional approach to soil chronosequences attempts to circumvent the problem of confounding by holding as constant as possible environmental factors other than time (see section 4.1.1). This usually takes the form of soil sampling only at sites falling within specified environmental limits of, for example, slope angle. However, as few environmental factors tend to be measured in this approach, there is great uncertainty as to the validity of the inferred temporal sequence. A modified approach has been adopted by Mellor (1984, 1986a, 1987) and Messer (1984, 1988, 1989) who use the best developed soil within each age zone. Here, as the best developed soil is found in the optimum environment, it can be argued that the resulting chronofunction should represent a temporal sequence characteristic of that environment. Nevertheless, this approach breaks down if the optimum environment is not identical on terrain of different ages.

Table 4.5. *Analysis of variance of soil properties on moraines of Robson Glacier, British Columbia (after Sondheim & Standish, 1983).*

Soil property	No. of samples	% variance accounted for					
		σ_E^2	σ_P^2	σ_S^2	σ_{MD}^2	σ_D^2	σ_M^2
ln(C)	180	7	2*	3*	8**	44**	37**
ln(N)	180	9	2*	3*	9**	47**	30**
C:N	180	62	7	1	3	7*	19**
pH	36	13	14	—	5	28**	40*
CaCO$_3$	36	53	40	—	5	1	0
Fe$_d$	36	25	8	—	10	13*	44*
Al$_d$	36	37	35	—	0	28**	0
(Fe$_d$ + Al$_d$)	36	23	13	—	10	19*	36*
Fe$_p$	36	65	31	—	5	0	0
Al$_p$	36	18	66*	—	16	0	0
(Fe$_p$ + Al$_p$)	36	53	7	—	28	0	12
Sand	36	71	0	—	15	0	15
Silt	36	50	2	—	36	0	12
Clay	36	52	0	—	4	22*	22

Variance components (σ^2): E = error term; P = pits; S = segments; D = depths; M = moraine age; MD = moraine-depth interaction term
Stars indicate statistically significant values.

An important limitation of soil chronosequence studies has been recognized by Sondheim & Standish (1983) who point to the lack of replication or attention to soil variability in most work. These authors use a rigorous sampling scheme, with replicates, to assess the variability of soils on moraine crests near the Robson Glacier. Each of six moraines were subdivided into five segments within each of which two soil pits were randomly located and soil samples were taken at three different depths. Mean values of particular soil properties could then be examined, together with error bars (e.g. Fig. 4.16). They also employ analysis of variance to partition the total variance of each soil property into a number of components and to measure the relative importance of the variance attributable to moraine age, depth, interaction of moraine age and depth, segments of moraines, and pits within segments (Table 4.5).

Approximately 90% of the variability in organic carbon is explained by the moraine age, depth, and moraine-depth interaction terms. Variability between segments within moraines accounts for only 3% and variability between pits within segments for 2%. Thus, age- and depth-related variation is clearly much larger than the locational differences between pits and segments. Other soil properties with a high degree of the variance explained by the moraine age and depth terms are: soil nitrogen, pH, dithionite-extractable iron and dithionite-extractable (iron + aluminium). For all the remaining variables, the error, pit and segment terms explain more than half of the total variance. Thus, although these results indicate that for some soil properties age-related trends are dominant, other properties appear to be largely unrelated to terrain age. This suggests that environmental factors other than time are significant even where, as in this case, the microtopography is relatively smooth.

Greater meso-scale variability has been demonstrated on much older (Preboreal) moraine ridges in Norway, where small topographic differences between sites may be sufficient to produce different soil types (Alexander, 1986; McCarroll & Ware, 1989). Bormann & Sidle (1990) have described considerable within-site variation in physical and chemical properties at Glacier Bay, but they did not sample more than one site per surface age. Their sites ranged in area from 0.02 to 0.08 ha and they attribute the within-site variability to 'different rates of succession, seed sources or other random factors'.

The environmental plexus of Whittaker (1987, 1989) provides a further basis for assessing the importance of terrain age relative to other factors. This was derived from a representative sample of sites from the full range of environmental conditions present on the Storbreen glacier foreland within a restricted range of altitudes. His plexus diagram (Fig. 3.16) clearly shows soil variables (soil depth, litter depth and rooting depth) as part of the *terrain-age factor complex* (see section 3.4). Nevertheless, even soil depth, the soil variable closest to terrain age in the plexus, exhibits statistically significant correlations with environmental variables (particularly frost churning and disturbance) and has a correlation coefficient (τ) with terrain age of only $+0.49$ ($p < 0.001$) (Whittaker, 1987). Environmental factors independent of time probably account for much of the variability left statistically unexplained by this correlation coefficient.

At a much smaller scale, a significant degree of soil heterogeneity has been established by Messer (1984) in apparently uniform plots of inter-moraine till plain on the Storbreen glacier foreland (Table 4.6). At least some of this heterogeneity can be accounted for by small-scale spatial variation in the environment. The uppermost 10 mm of the soil was sampled for pH and loss-on-ignition from grids of points within areas of 16 m^2. Results indicate both soil property- and age-dependent heterogeneity. The coefficients of variation indicate relatively high heterogeneity levels in relation to loss-on-ignition values, which may in turn be attributed to non-uniform colonization by plants. Heterogeneity is particularly high at first as organic matter is added in the earliest stages of soil development.

Table 4.6. *Soil heterogeneity in apparently uniform 16 m² plots on the Storbreen glacier foreland (after Messer, 1984).*

Terrain age (years)	Soil property							
	Loss-on-ignition (%)				pH			
	mean	S.D.	C.I.	C.V.	mean	S.D.	C.I.	C.V.
8	0.08	0.07	0.04	87.5	7.02	0.11	0.06	1.57
40	1.00	0.32	0.17	32.0	5.60	0.22	0.12	3.93
200	3.03	1.55	0.85	51.2	5.29	0.13	0.07	2.46

S.D. = standard deviation; C.I. = 95% confidence interval; C.V. = coefficient of variation

In relation to pH, initially low values of heterogeneity are presumably related to the nature of the till parent material, increasing slightly as pH values fall due to leaching. Subsequently, the degree of heterogeneity appears to fall as pH stabilizes at values close to 5.

Variation between glacier foreland chronosequences has been considered quantitatively at the regional scale by Messer (1988), while analyses at the global scale have been attempted by Bockheim (1980) and by Birkeland, Burke & Benedict (1989). Such comparisons between chronosequences, and others made in this chapter, indicate that environmental factors, particularly climate, have a major impact on the nature and rate of soil development at the larger scales (see section 4.3.4). At present, there is little further that can be gleaned from the scattered literature because, as pointed out by Messer (1988), most studies are not directly comparable due to the lack of standardization in field and laboratory methods.

The soils of recently-deglaciated terrain and their development are clearly more complex than has generally been appreciated in the past. As in the case of the physical environment (see chapter 3, particularly section 3.4), glacier foreland soils are characterized by variability and dynamism. Variability is manifest at many different spatial scales; dynamism is present in a variety of forms. Indeed, it could be argued that spatial variation is ubiquitous and that dynamism is inherent even in the absence of environmental change (cf. Muhs, 1984). A comprehensive understanding of glacier foreland soils requires an integrated approach to spatial variation and temporal change. It also necessitates a detailed knowledge of the interactions between soils and their physical and biological environments, and an appreciation of the position of soils in the landscape as a whole.

5

Plant succession: patterns and environmental factors

Following deglaciation, development of the vegetation is initiated as plants colonize relatively stable land surfaces and a succession of communities begins. The next two chapters are devoted to the patterns and processes that constitute the vegetational component of landscape development. In this chapter, a survey is made of the patterns of plant colonization and community succession that have been inferred by adopting a chronosequence approach. It is concluded by a discussion of the environmental factors that exert major controls on these patterns.

5.1 Vegetational chronosequences: methodological considerations

5.1.1 Concept and limitations

Most ecological investigations on glacier forelands have interpreted the vegetation patterns as representing chronosequences. The justification for this is found in the functional factorial methodology which has been outlined in some detail in chapter 4 (see section 4.1.1). As in the case of soils, *ideal* vegetational chronosequences probably do not exist and *actual* vegetational chronosequences may be represented as:

$$v = f(t, cl, o, r, p \ldots) \tag{10}$$
$$\text{or} \quad v = f(t, S_o, I) \tag{11}$$

where, in relation to vegetation or a particular property of vegetation (v), terrain age is the *dominant* factor (t), and the other environmental factors are *subordinate*: climate (cl), organisms (o), relief (r), parent material (p); the initial state of the system (S_o) and the influx variables (I) (Jenny, 1980).

Vegetational chronosequences are characterized by similar limitations to soil chronosequences. Most important of these are the existence of differences between sites, both in terms of their initial environmental

145

conditions and their environmental histories (see section 4.1.1). Neverthe-less, judicious use of vegetational chronosequences can provide approxi-mations to vegetation changes at particular sites. This has been demon-strated in a small number of studies where changes inferred from chronosequences have been tested against independent evidence of change (see section 5.1.2).

Studies at Glacier Bay, Alaska, provide one of the best known examples of a vegetational chronosequence. This parallels the soil chronosequence already described in some detail in chapter 4 (see especially section 4.1.2). The present outline is based largely on Decker (1966), who has elaborated the schemes of Cooper (1923b, 1939), Crocker & Major (1955) and Lawrence (1958) to recognize eight intergrading successional stages (see also Reiners, Worley & Lawrence, 1971).

Along the eastern arm of Glacier Bay, on gravelly till deglaciated for up to about five years, the early pioneer stage (I) consists largely of seedlings of mountain avens (*Dryas drummondii*) and prostrate willows (e.g. *Salix arctica*) together with mosses (especially *Rhacomitrium canescens* and *R. lanuginosum*) and herbs, such as broad-leaved willow-herb (*Epilobium latifolium*) and the variegated horsetail (*Equisetum variegatum*). This stage passes into the *Dryas*-mat stage (II), which occupies terrain deglaciated for about five to ten years and is characterized by extensive disc-shaped mats of *Dryas* from 0.1 to 4 m or more in diameter, which may coalesce to cover up to 90% of the landscape. The mat stage grades into the late pioneer stage (III) in which young Sitka alder (*Alnus crispa*) and poplar or black cottonwood (*Populus balsamifera ssp. trichocarpa*), generally less than 2 m tall, are well-established. Thus, terrain deglaciated for between 15 to 20 years has the appearance of scattered shrubs amongst a carpet of *Dryas*.

Extensive clumps of *Alnus*, 2 m or more in height, dominate the landscape after 20 to 25 years. This open thicket stage (IV) passes rapidly into the closed thicket stage (V) in which alder with thick decumbent basal branches forms an almost impenetrable dense cover. Thus, after 25 to 30 years, *Dryas* has disappeared except in rare open areas, and only a sparse understorey remains consisting of a few ferns and herbs. Scattered through the alder thickets are individual erect-growing specimens of poplar, shrubby willows (e.g. *Salix sitchensis* and *S. alaxensis*) and Sitka spruce (*Picea sitchensis*), which became established before or at the same time as the alder. After 30 to 40 years, poplars reach a height of 4 to 5 m and hence emerge above the alder canopy to form a clearly discernible line on the horizon, which is used to define the poplar line (stage VI).

The spruce forest stage (VII) develops as alder is replaced by spruce while poplar maintains its position. Although scattered spruce trees can be observed above the alder–poplar–willow canopy on younger terrain, it takes at least 75 to 90 years for the spruce to become truly dominant. Under the mature spruce forest canopy, few alders remain and the forest floor is densely covered by a moss blanket, commonly with *Rhytidiadelphus loreus* and *Hylocomium splendens* as the most important species.

Spruce forest was regarded by Cooper (1923b) as a sub-climax, the final stage arising with the successful invasion of western hemlock (*Tsuga heterophylla*) to form the spruce–hemlock forest stage (VIII). However, developments beyond the spruce forest stage are rather speculative; even after 200 years the forest is usually dominated by spruce, sometimes with an understorey of hemlock. Based largely on evidence from elsewhere in south-eastern Alaska, Lawrence (1958), Ugolini & Mann (1979), and Noble, Lawrence & Steveler (1984) have suggested that further stages, involving the spread of *Sphagnum*, forest deterioration and muskeg development, occur on much longer timescales of hundreds if not thousands of years (see sections 5.4.2 & 6.1.7).

Methodological limitations have been considered, at least implicitly, by all those who have worked on the Glacier Bay chronosequence. In practice, however, as in most other studies on the ecology of recently-deglaciated terrain, it has been assumed that the chronosequence is representative of changes through time at a particular site. Fastie (1990) has explicitly pointed out that such an inference is only valid if all stages in the chronosequence are developing along similar successional pathways. F. R. Stephens, in an unpublished manuscript, has suggested that researchers in south-eastern Alaska (including Glacier Bay) have failed to recognize that well-drained moraines exhibit a successional sequence to spruce without a preceding alder stage (but see Cooper, 1939: 139). Stephens suggests, therefore, that the 200-year-old surface at Glacier Bay is in effect part of a different chronosequence.

Most ecological studies of glacier foreland chronosequences are based on observed *correlations* between vegetation zonations and terrain age. Such correlations have been demonstrated in a variety of ways, ranging from simple descriptions to complex ordination procedures. Convincing correlation coefficients have been calculated between terrain age and the most important axes of vegetational variability at Storbreen, Jotunheimen. Using detrended correspondence analysis (DCA), for example, Whittaker (1989) has shown terrain age and the first DCA axis to be correlated at $\tau = +0.76$ (P < 0.001). This correlation coefficient is, moreover, higher than that between the vegetation axis and any of the large number of environmental variables measured.

By use of ordination diagrams it has also been possible to order

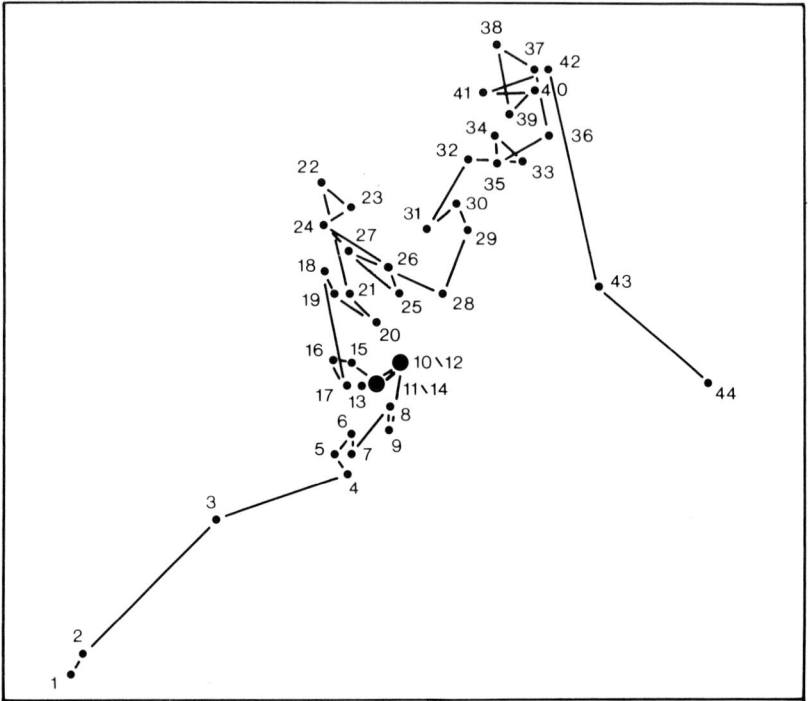

Fig. 5.1. Non-metric multidimensional scaling (NMDS) ordination of 44 vegetation sites at Styggedalsbreen, Jotunheimen. Sites are numbered and linked in order of increasing terrain age. Kruskal's stress statistic is 19.2% (from R. W. Bickerton, unpublished).

individual sites from a chronosequence according to their vegetational similarities. This approach was used by Bickerton (unpublished) at Styggedalsbreen, Jotunheimen. He used non-metric multidimensional scaling (NMDS) to order 44 sites of increasing terrain age. The order of the chronosequence is more-or-less represented in the two-dimensional ordination diagram in which the sites are numbered in terms of the terrain-age sequence (Fig. 5.1). Furthermore, with the exception of the two oldest sites, which were located on a relatively exposed moraine ridge and may therefore be regarded as anomalous, the near-linear arrangement suggests the species display approximate unimodality in time (Prentice, 1977). Comparable results were produced by Prentice (1977) in a similar analysis of data from the Klutlan glacier foreland, Yukon Territory, Canada. However, the fact that the sequence in the ordination reproduces the sequence of the chronosequence indicates no more than good correlation between vegetational variability and terrain age.

Such correlations are not necessarily indicative of causal relationships. Even under uniform environmental conditions, successional pathways at different sites may vary. Different vegetational histories may also result from different initial environmental conditions, or from differences in their subsequent environmental histories. That is, a particular correlation between vegetation and site age could arise from multiple successional pathways (Fastie, 1990).

Attempts to hold constant environmental variables other than time, by restricting the degree of environmental variation between the sites employed in describing a chronosequence, can only be a partial solution. Similar present-day environmental conditions do not necessarily reflect common initial environmental conditions, common environmental histories or common vegetational histories.

Replication of sites in a chronosequence does not provide a satisfactory solution. Even though high vegetational variability amongst replicates on terrain of equal age can indicate multiple successional pathways, low variability does not necessarily indicate a single pathway; all plots of the same age could have developed atypically in response to an atypical seed rain, substrate or climate (Fastie, 1990). Replication in this context is similar to pseudoreplication in ecological experiments, whereby possible causes are confounded (Hurlbert, 1984).

Indications of alternative successional pathways may be obtained by taking into account the vegetational variation present on terrain of various ages (cf. Matthews, 1978, 1979a, b; see section 5.3.1), and by analysing the interrelationships between vegetation, terrain age and environmental variables (cf. Whittaker, 1987, 1989; see section 5.4.3). Additional indications can be obtained from stand ages, if these are estimated independently of terrain age. This approach was employed by Birks (1980a) who, on the basis of ring counts, demonstrated little overlap in the ages of the woody plants between the various vegetation types ('noda') comprising a chronosequence on moraines of the Klutlan Glacier, Yukon Territory. Although these approaches improve the basis for inference, they do not provide critical tests of successional pathways; they indicate possible pathways and suggest unlikely ones.

Critical testing requires independent evidence of succession at particular sites with which to compare successions inferred from chronosequences. Fastie (1990) has suggested two potentially useful approaches: (1) comparison against other chronosequences; and (2) use of non-chronosequence methods. However, the former approach is flawed by the same problems as within-chronosequence replication: similarities between chronosequences may not have arisen via the same successional pathways. Two types of non-chronosequence method may be recognized and have been applied to the critical testing of a small number of chronosequences. The first involves the direct observation of succession; the second has been termed retrospective analysis.

5.1.2   Tests of chronosequences: observed successions

Directly observed successions are relatively rare in the context of glacier forelands. Photographic records may give a general picture of the nature and direction of succession but are highly dependent on the existence of early photographs, which were often made for no ecological purpose. Impressive examples of photographic evidence of succession can nevertheless be cited. Fægri (1933) provides three photographs of the same area of the glacier foreland of Nigardsbreen, southern Norway, taken in 1874, 1896 and 1931, respectively. These photographs are shown in Fig. 5.2. Over the time interval of 57 years, copses of *Alnus-Betula* developed on one side of the foreland. On the other side of the foreland, subjected to grazing by goats, there is no marked change. Since then, > 100 years after the earliest photograph, establishment of birch copse has greatly accelerated due to the discontinuation of goat grazing in the area (Fægri, 1986). Over an interval of 49 years at the Robson Glacier, British Columbia, Tisdale, Fosberg & Poulton (1956) describe marked changes in the composition and general aspect of the vegetation on all but the two oldest moraines. Much less change had occurred in the depressions between moraines and least change had occurred on sandar surfaces. Blundon & Russell (in prep.) have since extended the Robson photographic record to 77 years.

The use of photographs is enhanced if combined with field observation and recording of vegetation change. Cooper (1931) utilized this approach to investigate the development of patches of *Dryas* at Glacier Bay. Photographs taken in 1916 and 1929 show change from a mainly bare area with numerous small patches of *Dryas* to an almost complete covering within 13 years. A species count in the field in 1916 revealed 21 species. By 1935, 46 species were noted in the same area, including all but four of the species recorded in 1916. Although the species lists were incomplete, Cooper (1939) considers that it is safe to conclude that the flora approximately doubled over nineteen years. Cooper (1939) provides similar evidence of *Alnus* thicket development at Glacier Bay. Photographs taken in 1929 and 1935 show a considerable increase in the size of the *Alnus* thicket, of which there was no sign in 1916. Over an interval of 13 years, the thicket had increased in area from 0.20 ha to 1.21 ha and the height of individual plants had increased from 3.7 m to 4.6 m.

At Robson Glacier, the dominant species are known from 1914, 1953, 1963 and 1984 (Blundon, 1989; Blundon & Russell, in prep.). These are listed in Table 5.1, which permits comparisons within and between moraines and demonstrates that the same species have been colonizing newly-deglaciated terrain for the last 77 years, a conclusion supported also by photographs. For example, the vegetation on moraine 2 in 1914 was similar to that on moraine 4 in 1953, moraine 5 in 1963 and moraine 8 in 1984. These surfaces were 50, 46, 51 and 51 years old, respectively, when they were sampled. Again, the vegetation on moraine 3 in 1963 was similar to that on moraine 5 in 1984, when these surfaces were approximately 73 and 72 years old, respectively. Thus Blundon & Russell (in prep.) conclude that the successional sequence on the younger moraines was probably similar to that on the older moraines.

At four outlets of the Jostedalsbreen ice cap, southern Norway – Nigardsbreen, Bergsetbreen, Böyabreen and Åbrekkebreen (Brenndalsbreen) – H. J. B. Birks, H. H. Birks, H. F. Lamb and H. E. Wright Jr re-recorded in 1987 (Nigardsbreen) or 1990 the flora of moraines first recorded by Fægri in 1930 (Fægri, 1933; Birks *et al.*, in prep.). The vascular plant data from each foreland have been analysed by canonical correspondence analysis (CCA) using moraine age and time since original survey as 'external' predictor variables. Preliminary results from all four forelands indicate significant vegetation change over the last 60 years. However, the recorded changes are not entirely the same as successional pathways predicted from the chronosequence, the actual changes apparently involving processes such as geomorphic change, land-use effects and soil changes, which may have differed over the last 60 years.

Direct changes have been established by Spence (1989) in a different way. In front of Tyndall and Lewis Glaciers on Mount Kenya, changes were observed in the position of colonizing species relative to the glacier snout. The distances from each glacier terminus at which each species was first encountered had been measured 26 years earlier by Coe (1964, 1967). On re-recording the distances, and taking into account the rate of glacier retreat, it was shown that most of the species first recorded by Coe in 1958 had advanced in the direction of the glacier.

Re-recording of permanent plots provides the most detailed and useful approach to the direct observation of succession. However, few researchers have had the foresight to establish such plots on glacier forelands. In some cases, the results of only short periods of observation have been published (e.g. Jochimsen, 1963; Einarsson, 1971; Turmanina & Volodina, 1978, 1979; Elven & Aarhus, 1984). Although such records tend to be dominated

Fig. 5.2. Directly-observed succession: photographs of the glacier foreland of Nigardsbreen, southern Norway, in (a) 1874, (b) 1900, (c) 1931, (d) 1987 (a & b taken by K. Knudsen (University of Bergen Library, Catalogue Nos. KK1038/ KK96921; c by K. Fægri; d by H. J. B. Birks).

Table 5.1. *Dominant species on moraines (M1–5 & M8) of Robson Glacier, British Columbia, Canada, in 1914 (Cooper, 1916), 1953 (Heusser, 1956), 1963 (Tisdale, Fosberg & Poulton, 1966) and 1984 (after Blundon & Russell, in prep.).*

Moraine date (A.D.)	Year of description			
	1914	1953	1963	1984
MO (1783)	*Picea-Salix-Betula*	*Picea-Arctostaphylos-Salix*	—	*Picea-Arctostaphylos-Shepherdia*
M1	—	*Salix-Betula-Arctostaphylos*	*Salix-Arctostaphylos-Picea*	*Picea-Arctostaphylos Salix*
M2 (1864)	*Salix-Betula*	*Salix-Dryas-Arctostaphylos*	—	*Picea-Arctostaphylos-Salix*
M3 (1891)	*Dryas-Arctostaphylos-*	*Hedysarum-Salix-Dryas*	*Hedysarum-Salix-Dryas*	*Picea-Hedysarum-Salix*
M4 (1907)	Sparse	*Hedysarum-Salix*	—	*Hedysarum-Salix-Arctostaphylos*
M5 (1912)	Bare/Ice	*Hedysarum-Salix-Dryas*	*Hedysarum-Salix-Dryas*	*Hedysarum-Dryas-Salix*
M8 (1939)	Ice	Sparse	—	*Hedysarum-Dryas-Salix*

by short-term fluctuations in species' populations, successional trends have sometimes been detected.

Cooper initiated studies of permanent quadrats at Glacier Bay in 1916. He established three 1 m² quadrats at each of three sites of known age (17, 24 and 37 years, respectively) and recorded eight of the quadrats four times (in 1916, 1921, 1929 and 1935) (Cooper, 1923c, 1931, 1939). They have since been recorded on many additional occasions by D. R. Lawrence, although he has published a description of the vegetation changes from one quadrat only (Lawrence, 1979 & pers. comm.). Cooper (1939) summarized the changes that he observed as follows; (1) up to 1916, dominance by perennial herbs; (2) 1916–21, firm establishment of individual willows; (3) 1921–9, increase of mat plants, especially *Dryas*; and (4) 1929–35, maintenance of dominance by mat plants and the first noteworthy appearance of upright willow shoots.

Subsequent changes in Cooper's quadrat No. 1 have been described by Lawrence (1979). *Rhacomitrium* moss covered 30% of this quadrat and *Equisetum variegatum* was also abundant when the quadrat was first recorded by Cooper. Willow shrubs became increasingly established over the next 19 years, although much bare gravel was still evident in 1941, when Lawrence began his observations (Fig. 5.3). *Alnus* was not observed until 1955. In that year, one alder in its third or fourth year of growth and numerous alder seedlings were present, and a soapberry shrub (*Shepherdia canadensis*), rooted outside the quadrat, covered about half of the plot. By 1967, the alder canopy over the plot was 4.6 m tall and provided a 98% cover, all the *Rhacomitrium*, the willows and the soapberry had died but the *Equisetum* was thriving, and numerous individuals of *Pyrola asarifolia* had become established. At Lawrence's visit to the plot in 1972, alder leaf-fall had accumulated a rich organic layer 8.9 cm deep on top of the mineral soil, the *Pyrola* plants were markedly less frequent and moss growth was present on the larger fragments of alder twig litter and on the bases of the living stems.

These observed changes corroborate, to some extent, the early stages of succession inferred from the chronosequence at Glacier Bay (see section 5.1.1). However, the changes described by Lawrence (1979) from one quadrat may be misleading because only a few seedling *Dryas* plants became established and these did not survive for long. There is, moreover, no evidence from the permanent quadrats bearing on the establishment of spruce and the development of the spruce forest stage. Thus Cooper's quadrats, which probably represent the oldest permanent plots on terrain of known age following glacial recession, and have a history of observation stretching over 70 years, are as yet unable to provide a critical test of the whole chronosequence (but see also section 4.4.2).

Lüdi (1945), working on recently-deglaciated terrain in front of the Grand Glacier d'Aletsch, set up six large permanent plots between the glacier margin and the crest of a lateral moraine. The date of deglaciation for each plot varied from about 1860 to 1945 and terrain ages varied from three years to about 84 years. Each plot therefore corresponds with a different stage in the chronosequence. These plots, which vary in area from 300 to 600 m², were re-recorded in 1971 and the observed changes have

Fig. 5.3. A permanent quadrat: Cooper's quadrat No. 1, on terrain deglaciated by Grand Pacific Glacier, Glacier Bay, Alaska, in A.D. 1879. Set up by W. S. Cooper in 1916, the quadrat has since been photographed by him (up to 1935) and D. B. Lawrence (since 1941). Selected photographs: (a) 1921, (b) 1935, (c) 1949, (d) 1955 (first appearance of *Alnus*); (e) 1967; (f) 1982.

(d)

(e)

(f)

Table 5.2. *Five successional stages inferred from the chronosequence in front of the Grand Glacier d'Aletsch, Swiss Alps (after Richard, 1968, 1973).*

Stage No.	Terrain age (years)	Important species
1	5–10	*Oxyria digyna, Cerastium pedunculatum, Linaria alpina*
2	10–30	*Oxyria, Cerastium, Salix* spp., *Rhacomitrium canescens*
3	30–60	*Salix* spp., *Trifolium pallescens, Trifolium badium*
4	60–100	*Salix* spp., *Rhododendron ferrugineum, Trifolium* spp., *Empetrum hermaphroditum*
5	> 100	*Vaccinium myrtillus, Rhododendron, Larix decidua, Picea abies*

been analysed by Richard (1973). Results are presented in terms of changes in the number of species belonging to various phytosociological groups (Fig. 5.4) and the height growth of tree species (Fig. 5.5). These patterns can be compared with the successional trends inferred from the chronosequence (Table 5.2).

Although all plots are characterized by an increase in the number of species between 1944–8 and 1971, by far the greatest change was observed in the youngest (plot 1). Three times as many pioneer species were growing in plot 1 in 1971 as in 1948 (Fig. 5.4a). At the same time the number of species representative of all the other phytosociological groups also increased substantially. However, there are distinct differences in the changes observed between the phytosociological groups in the other plots. There was a modest increase in the number of pioneer species in plot 2, the age of which had increased from 24 to 51 years, and a modest decrease in the number of pioneer species in plot 3, which had increased in age from 44 to 71 years. These changes provide independent evidence of initial rapid colonization by pioneer species, a subsequent fall-off in the rate of addition, and the later replacement of pioneer species by other groups.

Whilst plots 2–5 exhibit negligible change in the relatively small number of species characteristic of calcicolous grassland (Fig. 5.4b), they all show an increase in the number of species found in acidic grassland (Fig. 5.4c). This pattern of change is consistent with later colonization by aciophilous species characteristic of later successional stages. Many forest species are also later colonizers, as demonstrated by the substantial increase in their numbers in plot 2 and moderate increases in plots 4 and 5 (Fig. 5.4d). Tree

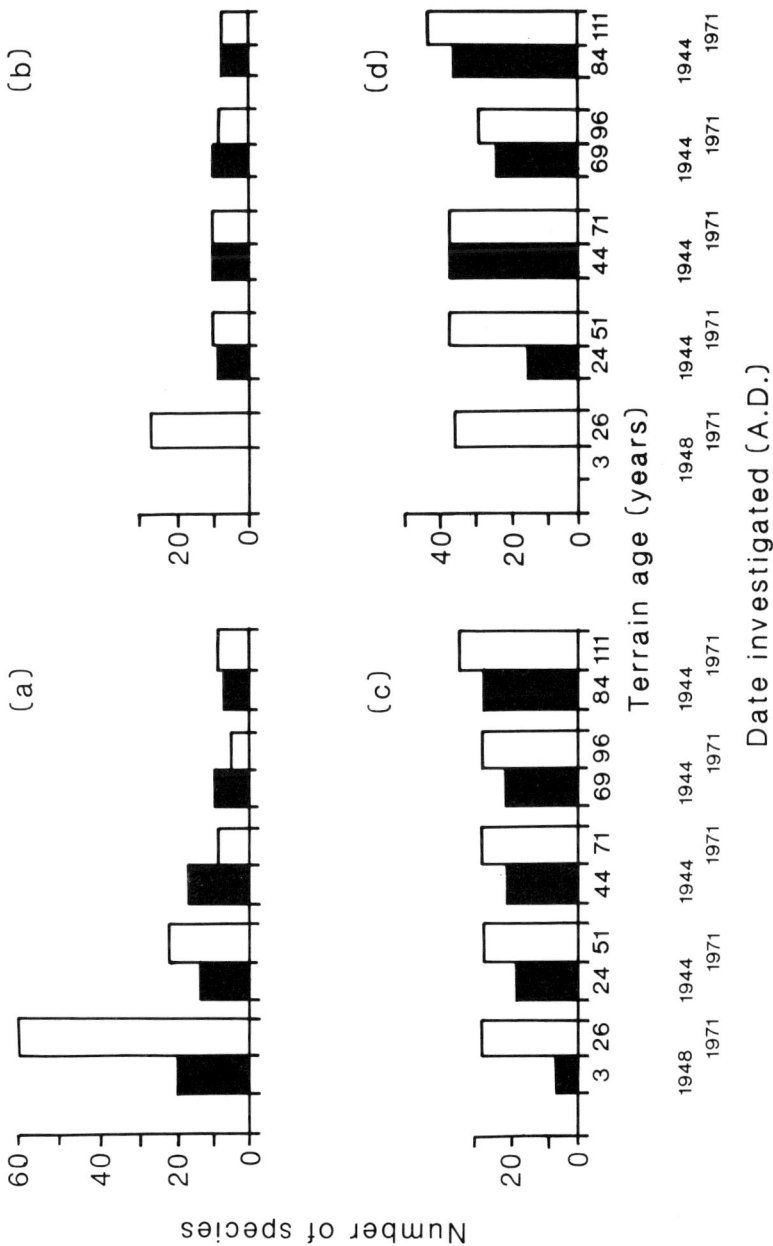

Fig. 5.4. Change in the species composition of vegetation over a 23–27 year period based on permanent plots on the foreland of the Aletschgletscher, Swiss Alps, 1944–71: (a) pioneer species; (b) species of calcareous grassland; (c) species of acid grassland; (d) forest species. Each pair of columns represents two sampling dates (from Richard, 1973).

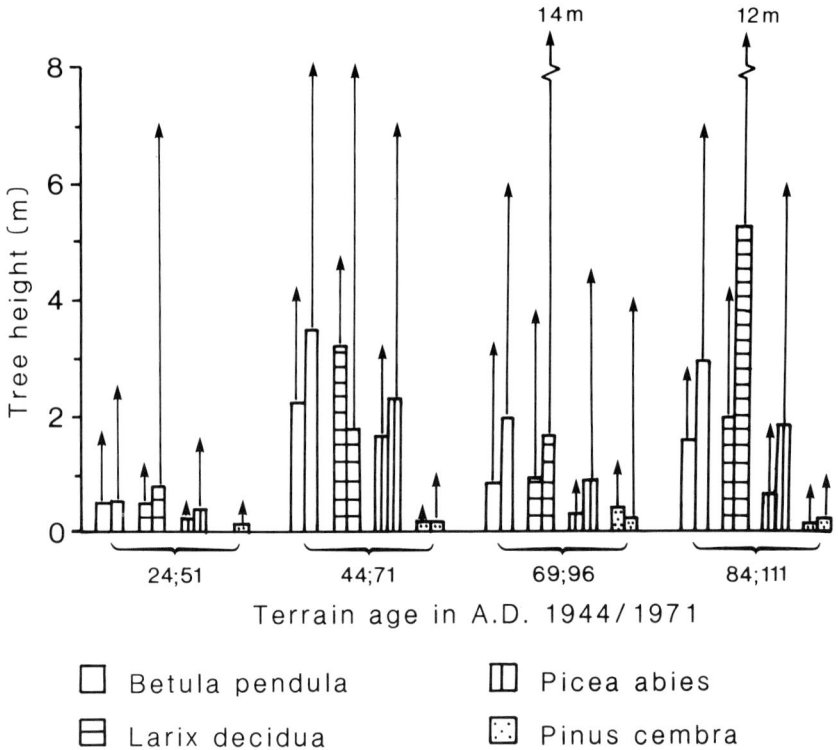

Fig. 5.5. Change in the height of trees over a 27-year period in permanent plots, near the Aletschgletscher, Swiss Alps, 1948–71. Each pair of columns represents the two sampling dates; column height is mean tree height; arrows indicate maximum tree height (from Richard, 1973).

growth in plots 2–5 (Fig. 5.5) suggests a slow net advance of birch (*Betula pendula*) and larch (*Larix decidua*), little change in pine (*Pinus cembra*) and an intermediate state for spruce (*Picea abies*), the growth of which is retarded by intense grazing by chamois (Richard, 1973).

Bearing in mind the limitations of a small number of sites and the short period of observation, the results from the Grand Glacier d'Aletsch are fully consistent with the succession inferred from the chronosequence. The former limitation has been largely overcome in a further long-term study involving permanent quadrats on the Storbreen glacier foreland, Jotunheimen, Norway (Matthews & Whittaker, 1987; Whittaker, 1991). Whittaker analysed changes in 105 of 638 permanent plots (each of area 16 m²) that had been established 12 years earlier by Matthews. These plots are located within an altitudinal range of 1150 to 1350 m on the southern side of the

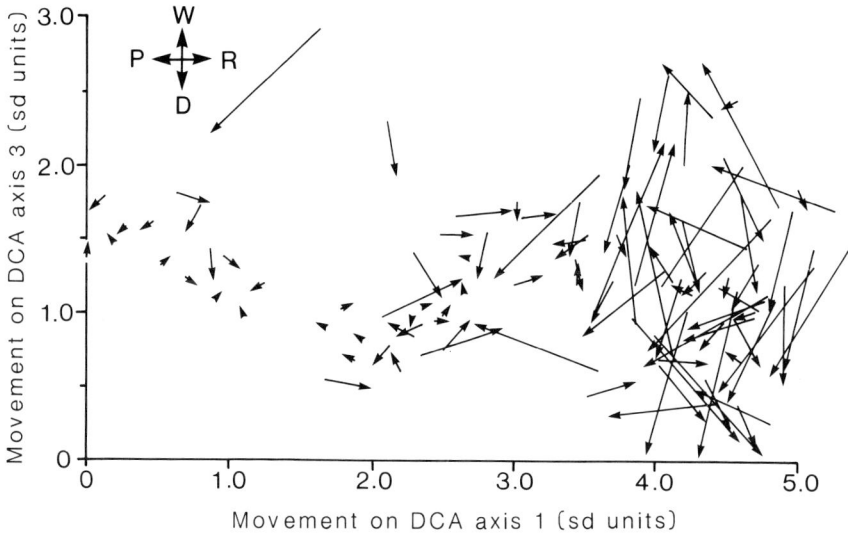

Fig. 5.6. Vegetation change over a 12-year period, Storbreen, Jotunheimen: detrended correspondence anlaysis (DCA) ordination of data from 105 permanent quadrats. Vectors indicate the magnitude and direction of change in relation to DCA axes 1 and 3. Units (sd's) are standard deviations of species turnover. Interpretation of movement along axes: P, progression; R, retrogression; W, wetting; D, drying (from Whittaker, 1991).

foreland. A detrended correspondence analysis (DCA) was carried out on the two data sets in combination so that changes in the position of the sites could be observed in a common ordination space. The observed changes are shown with respect to the first and third DCA axes in Figs, 5.6 & 5.7.

Site scores on DCA axis 1 are strongly (negatively) correlated with terrain age and hence this axis may be interpreted as representing a time axis (Whittaker, 1989, 1991). In vegetational terms, axis 1 represents a chronosequence from herbaceous pioneer communities to dwarf-shrub heaths, the study area lying well above the local tree line. Axis 3 is closely related to a moisture/exposure/snow melt environmental complex and scores on this axis are positively correlated with moisture conditions. Negative movements with respect to axis 1 (right to left in Fig. 5.6) therefore represent progressive successional changes as would be inferred from the chronosequence. Movement in relation to axis 3 is interpretable in terms of community response to changing moisture conditions (see below).

By grouping the sites and presenting the mean resultant movement according to terrain age class, Fig. 5.7 provides a simplification of the

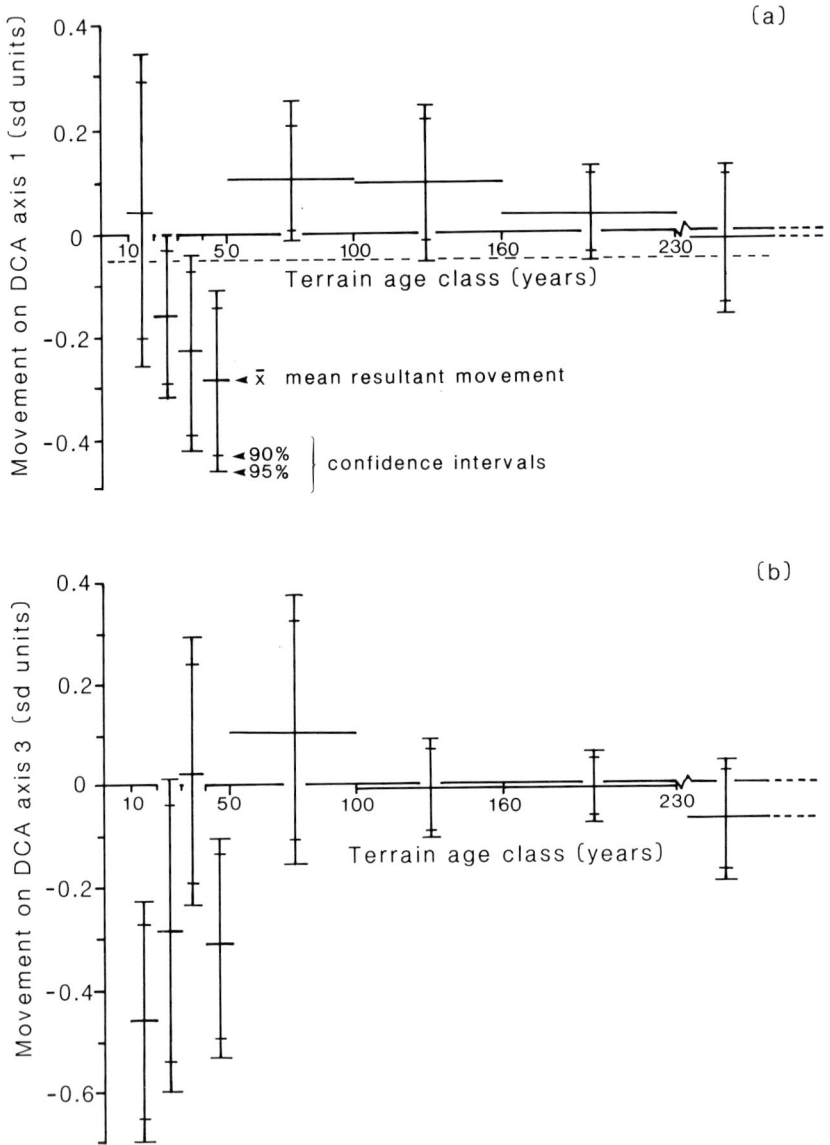

Fig. 5.7. Magnitude and direction of vegetation change over a 12-year period in relation to terrain age, Storbreen glacier foreland, Jotunheimen. Mean resultant movement (with 95% and 90% confidence intervals) is indicated for terrain-age classes with respect to (a) DCA axis 1, and (b) DCA axis 3 (from Whittaker, 1991). See also Fig. 5.6.

highly stochastic changes depicted at the level of the individual site in Fig. 5.6. Student's *t* statistic was used to test whether the changes in each age class were significantly different from zero, and to calculate confidence intervals (Whittaker, 1991). The overall trend in relation to axis 1 records progressive change but this is accounted for principally by the sites on the younger terrain. Although the mean *gross* movement (i.e. the absolute movement irrespective of sign) for the 10–19 years age class is the largest for any age class, the *net* change is a very small retrogressive change that is not statistically significant. Significant progressive change occurs in the three age classes involving terrain ages between 20 and 49 years. A weak retrogressive tendency characterizes the older terrain, reaching the 10% level of statistical significance only in the 50–99 years age class. The overall trend in relation to axis 3 is towards communities characteristic of drier site conditions and is produced by significant changes in the 10–19, 20–29 and 40–49 years age classes.

Thus, on the younger terrain (< 50 years), and despite the large stochastic element, these results strongly support the existence of progressive successional changes, which are similar to the successional changes inferred from the chronosequence. Independent of these progressive changes, but also significant at the younger sites, is a second type of change (that detected with respect to axis 3). The latter is probably caused by desiccation in response to continuing glacier retreat (Whittaker, 1991). On older terrain, a longer time interval would seem to be required in order to detect progressive changes and hence to test the validity of the older successional stages inferred from the chronosequence. At least over the 12 year time interval investigated so far, it appears that retrogressive change is at least as likely as progressive succession, and that both are conditioned by disturbance processes.

5.1.3   Tests of chronosequences: retrospective analysis

Chronosequences may also be tested by retrospective analysis, that is by the palaeoecological reconstruction of actual successional pathways at particular sites. This has been attempted on the moraines of the Klutlan Glacier, Yukon Territory, largely from the pollen record in lake sediments (Birks, 1980b), and at Glacier Bay from the tree-ring record of stand history (Fastie, 1990).

Birks used modern pollen assemblages in surface lake muds and moss polsters from each moraine in the Klutlan chronosequence to interpret the pollen stratigraphy of cores from three lakes – Triangle Lake, Cotton Pond and Heart Lake – on the oldest moraine (the Harris Creek Moraine, HCM).

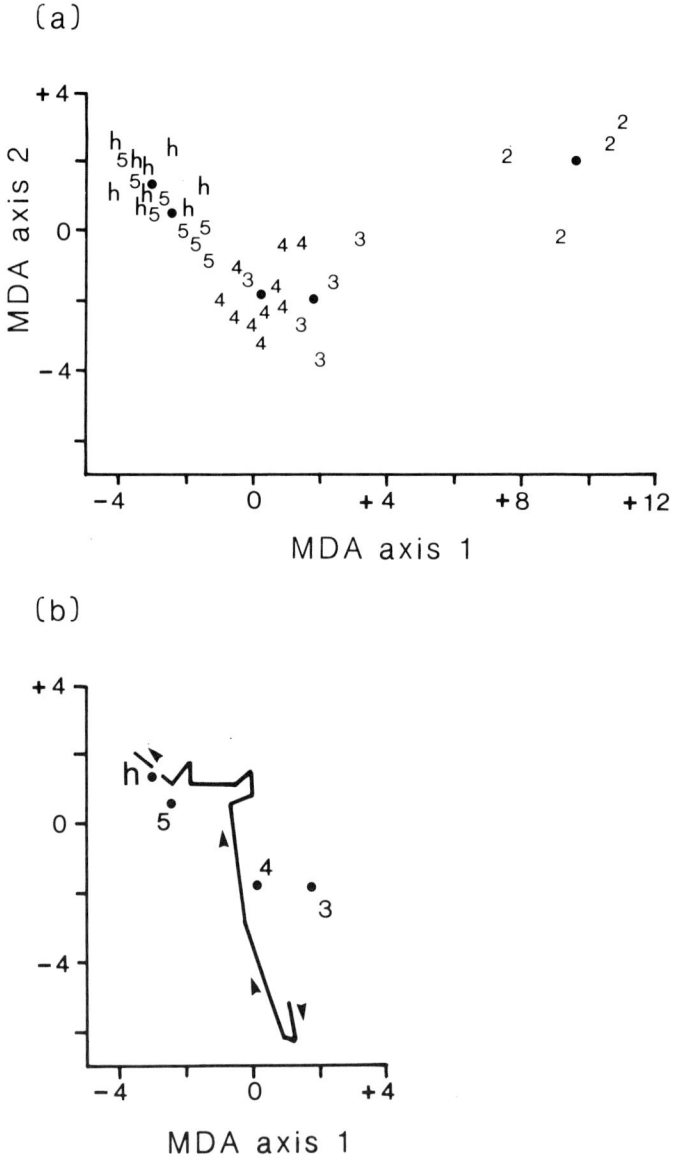

Fig. 5.8. Comparison between modern pollen from surface samples in a chrono-sequence and fossil samples from lakes on the Harris Creek Moraine, Klutlan Glacier, Yukon Territory, Canada: (a) surface pollen samples (labelled 2–5 and h) grouped from five moraines (K-II to K-V, and H) and their respective group centroids (filled circles); (b) fossil pollen samples from Triangle Lake joined in stratigraphic order and related to the group centroids from (a). All samples and centroids are positioned in relation to ordination axes 1 and 2 of a multiple discriminant analysis (MDA) (from Birks, 1980b).

In addition, pollen stratigraphy was examined from Gull Lake on the adjacent upland, in order to distinguish regional vegetation changes from successional vegetation change. The data from Gull Lake cores suggest that there has been little regional vegetation change, at least since about 800 years ago.

Modern pollen spectra from the various moraines were summarized by multiple discriminant analysis (MDA, also known as canonical variates analysis). This technique provides a graphical representation of the similarities and differences between sites from various stages of the chronosequence (Fig. 5.8a) to which the position of fossil assemblages from the lake cores can be added (Fig. 5.8b). The positions of the centroids (mean site scores) of the five moraines (K-II, K-III, K-IV, K-V and HCM) suggest that the pollen signatures of the respective moraines are generally distinct, although there is some overlap between individual spectra from older moraines (Fig. 5.8a). High *Hedysarum* and *Salix* pollen values characterize spectra from the youngest terrain (K-II). High pollen values of *Sheperdia canadensis*, *Betula* and *Salix* distinguish K-III and K-IV, and dwarf shrub pollen values tend to be high on K-IV. On K-V and HCM, *Picea* values exceed 30% but there are few other differences that are consistent between the two oldest moraines.

The sequence of modern pollen assemblages from progressively older moraines in Fig. 5.8 clearly represents all but the earliest stages of the vegetation chronosequence that has been described in some detail by Birks (1980a). The later stages of the chronosequence agree well with the stratigraphic sequence from Triangle Lake (Fig. 5.8b). The basal pollen spectra at Triangle Lake are similar to modern assemblages on Moraines K-III and K-IV, representing a *Shepherdia*-dominated vegetation. Through time (up-core), this basal assemblage is replaced by a *Picea-Betula-Salix* assemblage similar to those on K-IV and K-V today. In turn, this is replaced by a *Picea*-dominated assemblage that occurs today on HCM. Unfortunately, no fossil assemblages were found comparable to the modern pollen assemblages of K-II with high *Hedysarum* pollen values. This is attributed by Birks (1980b) to rapid accumulation of minerogenic sediments from massive erosion, the melting of ice-cored moraines, and the inwash of material from unstable lake slopes, all of which militate against a clear pollen stratigraphic record of the earliest stages of vegetation development. Further support for the succession inferred from the chronosequence is provided by Birks (1980a) from macrofossils preserved in the humus layers of soils on moraines K-IV and K-V. Leaves of *Dryas drummondii* (particularly characteristic of younger moraines), *Salix glauca* and *Arctos-*

taphylos uva-ursi are overlain by *Picea* needles and remains of the mosses *Tomenthypnum nitens* and *Rhytidium rugosum*.

Stratigraphic records from Cotton Pond and Heart Lake indicate a different kind of succession. Melting of ice underlying HCM occurred comparatively recently and a 'trash-layer' (Florin & Wright, 1969), rich in fungal remains, coarse detritus and wood, formed from supraglacial forest, soil and litter was deposited prior to lake sedimentation. The pollen statigraphy closely matches the modern pollen spectra from K-V and HCM without any progressive succession. Thus, although pollen stratigraphy and hence vegetation history at Triangle Lake corroborate the later stages of the vegetation succession inferred by Birks (1980a) from the chronosequence, the different results from Cotton Pond and Heart Lake demonstate that closely similar sites may have experienced different environmental histories associated with post-depositional disturbance.

Fastie (1990) used a second type of retrospective analysis to test the vegetation chronosequence at Glacier Bay. He examined the past growth rate of spruce trees from different sites within the chronosequence. Tree-ring widths were measured from a total of 400 increment cores from six sites. Preliminary analyses of annual growth increments (Fig. 5.9) indicate that spruce trees near the mouth of the bay at Bartlett Lake (terrain age about 210 years) exhibit a different pattern of growth to trees located 45 km to the north at Muir Point (terrain age about 100 years). In the early stages of growth, the growth rings are thicker at Bartlett Lake than at Muir Point; in the later stages of growth, the pattern is reversed.

According to Fastie (1990), such differences in the growth of spruce from comparable stages of its life cycle indicate differences in successional pathways between sites. In the early stages, spruce at Bartlett Lake appears to have grown without substantial competition from an alder shrub thicket, whereas the trees at Muir Point reflect suppressed growth, apparently by dense alder thicket. Many decades of dominance by alder at Muir Point may well have caused persistent differences between the young forests and the forests at Bartlett Lake. After a century of succession at Muir Point, where alder continues to be important, spruce forest is half as dense as it appears to have been in the forests at Bartlett Lake after its first century of succession (by which time closure of the spruce forest would have eliminated alder). Similarly, other trends along the chronosequence at Glacier Bay, such as the decline in soil nitrogen between mid-successional Muir Point and late-successional Bartlett Lake, may not be a true representation of the changes that actually occurred at the older sites in the past (cf. sections 4.2.7, 4.3.3 & 4.4).

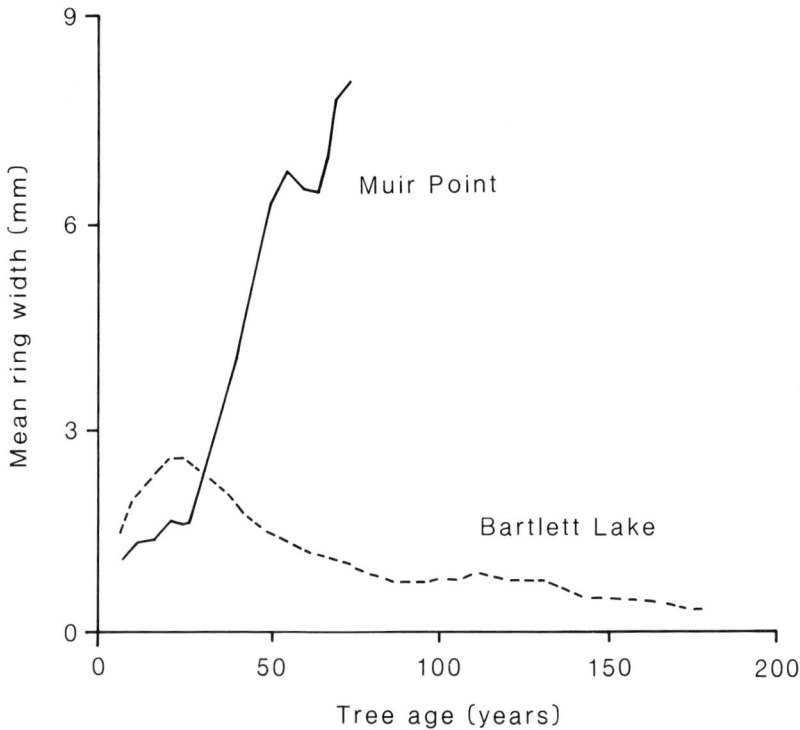

Fig. 5.9. 5-year mean tree-ring widths for dominant individuals of *Picea sitchensis* from randomly-sampled plots at Muir Point (terrain age *ca* 100 years; *n* = 40 trees) and Bartlett Lake (terrain age *ca* 210 years; *n* = 41 trees), Glacier Bay, Alaska (from Fastie, 1990).

The importance of these tests carried out by Fastie and others is that successional changes inferred from chronosequences must not be accepted uncritically. In most cases where tests have been made, successions inferred from chronosequences have been supported in a general way. However, such tests are time consuming and may be difficult to perform; they may not be possible in specific situations and they are often limited in scope. In particular, directly-measured changes are limited by short periods of observation and pollen analysis has major problems of resolution. In no case has there been an adequate test of a whole chronosequence.

5.2 Inferred successional trends

The aim of this section is to identify general trends in the patterns of successional change that have been inferred from vegetational chronosequences on recently-deglaciated terrain. Because of spatial variation in the plant cover both within and between glacier forelands, this is a more difficult task than would appear at first sight. Such variation militates against generalization. Deviations from these trends will therefore be developed further in subsequent sections.

5.2.1   Cover

Perhaps the most obvious successional trend on recently-deglaciated terrain is the increase in vegetation cover and the corresponding decrease in bare ground with increasing terrain age. Rates are extremely varied. Under favourable conditions, plant colonization begins almost immediately. *Epilobium fleischeri* and *Saxifraga aizoides*, respectively, have been found growing on terrain deglaciated for approximately one year in front of Steingletscher in the Swiss Alps (Ellenberg, 1988) and Rotmoosferner in the Austrian Alps (Palmer & Miller, 1961). According to Hope (1976), *Epilobium detznerianum* colonizes in front of the tropical Meren and Carstensz Glaciers of Irian Jaya (West New Guinea) within a few months.

The subsequent increase in cover tends to be sigmoidal in form (Stork, 1963; Zollitsch, 1969; Sondheim & Standish, 1983), although it is patchy and highly dependent on local environmental factors. Coarse-textured substrates, rock outcrops, disturbances and very severe climates may all delay or prevent the attainment of a complete vegetation cover. The pattern on the Pasterze glacier foreland in the Alps is typical (Fig. 5.10). The data were derived from areas of up to 100 m² selected using phytosociological criteria. An initial slow increase is followed by rapid change in moist, sheltered and nearly undisturbed sites to reach approximately 100% cover in < 100 years (curve C). In very exposed, very disturbed, and moist sites that later dry out, closure is achieved after a much longer period (curve A). Using a 2 m wide belt transect in front of Storglaciär in northern Sweden, Stork (1963) recognizes a slow phase in the first 10–15 years, followed by a rapid increase, especially involving mosses, to reach approximately 100% cover after about 50 years. At Glacier Bay, *Dryas drummondii* produces a virtually continuous mat over large areas within about 20 years. At this stage bare ground is, however, still very frequent and it is not until 30–40 years have passed that *Alnus crispa* finally produces an almost complete cover (Crocker & Major, 1955).

5.2.2   Spatial organization

Patchiness or heterogeneity in the development of the vegetation cover has already been referred to or implied at many spatial scales. A

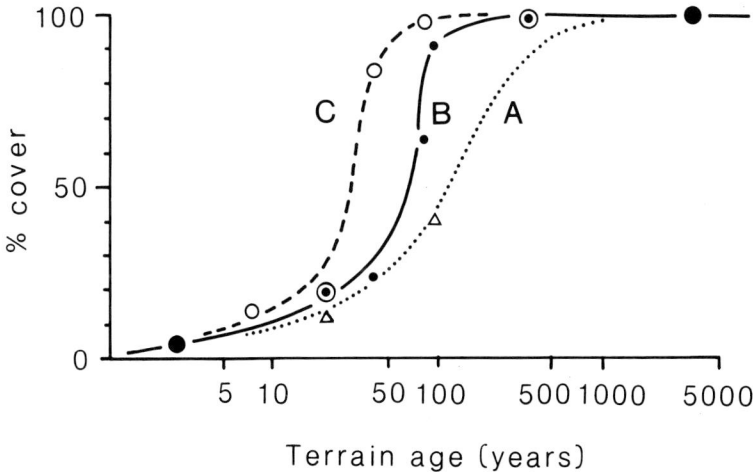

Fig. 5.10. Vegetation cover chronofunctions for three successional pathways (A, B and C, denoted by various symbols and lines) characteristic of respectively more favourable habitats, Pasterze Gletscher, Austrian Alps. Large filled circles represent all three pathways. Pathways and habitats are defined in Fig. 5.41 (from Zollitsch, 1969).

quantitative example is provided by Fægri (1933) from the glacier foreland of Nigardsbreen, southern Norway, where the areas occupied by various communities and by bare ground have been estimated on terrain of increasing age (Fig. 5.11). At each stage a mosaic of communities can be envisaged with appreciable areas still unoccupied by vegetation after 150 years of vegetation development at this site. A wide range of physical environmental factors combined with biological controls, such as patterns of immigration and species interaction, are responsible for such patchiness. Thus, although a complete vegetation cover is attained relatively rapidly in selected favourable areas, the development of the vegetation landscape as a whole is often a much slower process.

Apart from the larger-scale spatial patterns depicted on vegetation maps (section 5.3.1), there has been very little research done on glacier forelands relating to temporal trends in vegetation pattern or in the spatial organization of vegetation generally. Small-scale pattern analysis has been applied in only three studies. Anderson (1967) applied Greig-Smith's (1961a) analysis of variance technique to the distribution of *Salix herbacea* near Drangajökull, Iceland. He detected non-random heterogeneity with a single peak in variance at a block size of 40 cm. This corresponds well with the environmental heterogeneity at the site, plant growth being restricted to

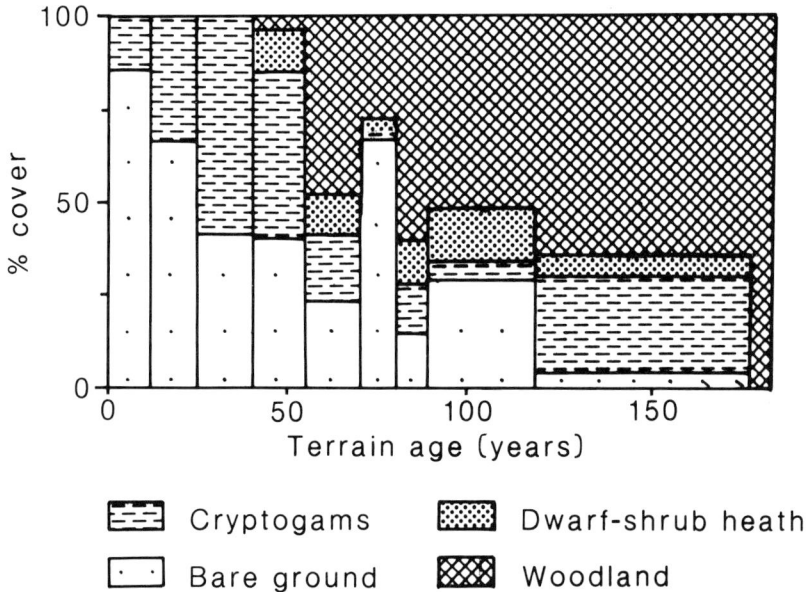

Fig. 5.11. Percentage cover of various plant communities on terrain of increasing age, Nigardsbreen, southern Norway (from Fægri, 1933).

the fine centres of sorted polygons of about 40 cm diameter. At larger block sizes, the pattern tends towards regularity which, according to Anderson, reflects topographic uniformity.

Dale & MacIsaac (1989) investigated both the scale and intensity of pattern exhibited by *Dryas drummondii* in two chronosequences on glacio-fluvial outwash adjacent to the Dome and Athabasca Glaciers, Canadian Rocky Mountains. At these sites *Dryas* made up 96% of the living plant cover in all quadrats analysed on terrain varying in age from 15 to 115 years. Their results (Table 5.3) show consistent scales of pattern at both sites, with some decrease on the older surfaces. Pattern intensity is more variable, but appears first to increase, as *Dryas* cover increases to about 50%, and then to fall on the older surfaces. Pattern scales appear to be controlled by strong environmental (habitat) pattern at scales of about 2 m (6–7 quadrats) and 6 to 8 m (19–25 quadrats), which respectively correspond with the average dimensions of abandoned river channel risers and abandoned river channels. Some decrease in the scale of pattern on the older surfaces is probably due to senescence and breakup of older *Dryas* mats whilst new clones continue to be recruited even on the oldest surfaces. These conclusions were supported by a new method for detecting the independent contributions of patch size and gap size.

Table 5.3. *Average pattern scale* (b*), and intensity* (d*) of* Dryas drummondii *for the two smallest scales of pattern on glacio-fluvial outwash at two sites in the Canadian Rocky Mountains (after Dale &* MacIsaac, 1989).

Surface code	Surface age (yrs)	Number of transects	Dryas (%)	b_1	d_1	b_2	d_2
Sunwapta River site							
A	18	3	3	7	13	22	6
B	30	4	45	6	51	25	39
C	47	2	75	7	39	20	28
D	115	1	79	4	37	14	31
Dome Glacier site							
A_1	37	4	27	6	33	18	26
B_1	56	3	80	6	30	19	21
C_1	94	3	89	5	29	18	11

Dale & Blundon (1990) used similar techniques to examine pattern development in six common species – *Dryas drummondii, Hedysarum boreale* var. *mackenzii, Salix vestita, S. glauca, S. barclayi* and *Picea engelmannii* – on five moraines of increasing age in front of Robson Glacier, British Columbia. Their results indicate little consistency of pattern scale either among transects within species or among species within transects, although many species show evidence of scales of pattern at a larger scale than those investigated. Neither is there convincing evidence for any change in the number of scales of pattern. Commonly, three to five scales of pattern are indicated, but patterns are irregular in part due to the low proportion of occupied quadrats. Pattern intensity of all species increases early in the chronosequence. Despite differences arising from the use of presence/absence or density data, *Dryas* and the *Salix* spp. attain peak pattern intensities on either the 1912 or the 1931 moraines, whilst peaks in *Picea* are characteristically later (on moraines dating from 1801 or 1891). Only *Dryas* and *S. barclayi* show clear early peaks followed by a consistent decrease through time.

It might be expected that the scale and intensity of plant morphological pattern would initally increase during succession as individuals and patches spread, followed by a decline in pattern intensity as they coalesce (Greig-Smith, 1964). Subsequent pattern scales and intensities probably vary greatly between species and communities depending on growth form, growth rates and the nature of species interactions. However, the nature of recently-deglaciated terrain in general, and quadrat covariance analysis of species' pairs on the moraines of Robson Glacier in particular, suggest that the mutual spatial arrangements of species are not simply a matter of plant-plant interactions but involve the spatial heterogeneity of the substrate (Dale & Blundon, 1991).

Several observations on whole communities suggest that loose spatial organization in the early stages is consolidated on older terrain to produce better defined communities. Thus, in front of Sesvenna Glacier in the Alps 'there appears at first a motley, but by no means accidental, mixture' (Braun-Blanquet, 1932: 308). At Skaftafellsjökull, Iceland, 'it is most probable that the first species appear in a more or less occasional mixture' (Persson, 1964: 350). On Mount Kenya, 'pioneer succession in front of the Tyndall and Lewis Glaciers appears to be haphazard' (Spence, 1989: 287). Fægri (1933) and Elven (1978b) respectively recognize increasingly distinct vegetation types during sucession in the sub-alpine and alpine zones of southern Norway. Turmanina & Volodina (1978) write of groups of associations becoming increasingly distinct with increasingly distinct community boundaries on terrain of increasing age on the forelands of the central Caucasus.

This pattern of poorly-defined, loosely-organized assemblages in the early stages of succession is reflected in the use of such terms as 'pre-communities' and 'half-communities' (Friedel, 1938), 'initials' (Elven, 1975, 1978c; Ammann, 1979), and 'simple groupings' (Solomina, 1989). In the later stages, on the other hand, these become 'full communities', 'associations' or 'complex groupings'. This progression towards greater organization in later successional stages is illustrated well by the increasing number of inter-specific associations detected by Elven (1978b) on terrain of increasing age in front of Hardangerjökulen, southern Norway. Based on grids of quadrats across moraines, positive associations between species become increasingly common as terrain age increases from 37 years, through 90 years, to 170–220 years (Fig. 5.12). On the youngest terrain only a small proportion of the species present are involved in statistically significant associations. As terrain age increases to about 90 years the number and proportion of positive associations increase. No clear cluster-ing of species is apparent, although the species can be interpreted as being disturbed along one axis corresponding to a snow-cover gradient. On the oldest terrain, the proportion of positive associations again increases and three clusters of species can be distinguished. These correspond with a visible zonation across the moraine ridges from a chionophobous, lichen-dominated and wind-eroded upper zone, via an *Empetrum hermaphroditum* heath, to a forb and moss-dominated snowbed zone at the base. The

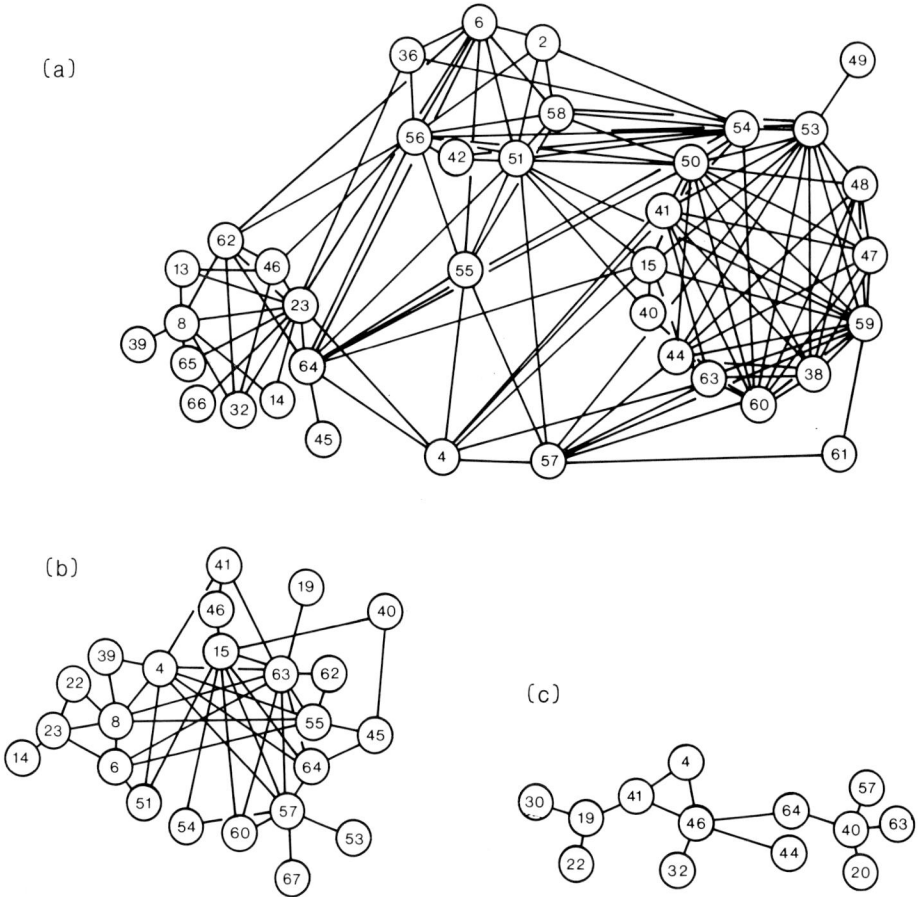

Fig. 5.12. Statistically significant positive associations (χ^2, p < 0.05) between species on terrain of varying age, Hardangerjökulen, southern Norway: (a) 170–220 years, (b) 90 years, (c) 37 years (from Elven, 1978b). Numbered species: 2, *Empetrum hermaphroditum*; 4, *Salix herbacea*; 6, *Antennaria alpina*; 8, *Gnaphalium supinum*; 13, *Taraxacum croceum* coll.; 14, *Veronica alpina*; 15, *Festuca vivipara*; 19, *Sagina saginoides*; 20, *Saxifraga oppositifolia*; 22, *Poa alpina* var. *vivipara*; 23, *Trisetum spicatum*; 30, *Deschampsia alpina*; 32, *Poa alpina* sterile; 36, *Hieracium alpinum* coll.; 38, *Gymnomitrion concinnatum*; 39, *Kiaeria starkei*; 40, *Pogonatum urnigerum*; 41, *Pohlia nutans*; 42, *Polytrichum hyperboreum*; 44, *P. piliferum*; 45, *Rhacomitrium canescens* 46, Hepaticae (indet.); 47, *Alectoria nigricans*; 48, *A. ochroleuca*; 49, *A. pubescens*; 50, *Cetraria cucullata*; 51, *C. ericetorum*; 53, *C. nivalis*; 54, *Cladonia gracilis* var. *gracilis*; 55, *C. macrophyllodes*; 56, *C. mitis*; 57, *C. coccifera* coll.; 58, *C. uncialis*; 59, *Cornicularia muricata*; 60, *Ochrolechia geminipara*; 61, *Pertusaria dactylina*; 62, *Psoroma hypnorum*; 63, *Solorina crocea*; 64, *Stereocaulon alpinum*; 65, Microlichenes (indet.); 66, *Cladonia ecmocyna*; 67, *Cladonia* spp. (indet.).

formation of clear vegetation zones is accompanied by the presence of statistically significant negative associations, which are absent from the two younger surfaces (Elven, 1978b).

5.2.3   Stratification and physiognomy

Horizontal and vertical structure tend to develop in parallel, an increase in cover commonly being accompanied by an increase in vegetation stratification. This is clearly shown in Fig. 5.13, which relates to data from the permanent plots in front of the Grand Glacier D'Aletsch, Swiss Alps (Lüdi, 1945) (see also, section 5.1.2). Each stratum in turn increases in cover, the development of ground and field layers tending to precede shrub and tree layers in those successions leading to woodland or forest. Total cover values exceed 100% on older terrain due to the existence of overlapping strata. Although isolated woody plants can be found amongst the pioneers on many forelands in the Alps (Lüdi, 1958), herbaceous perennials are generally followed by bushes and then trees. At Glacier Bay, low shrubs plus herbs, tall shrubs, then trees, appear in overlapping sequence (Reiners, Worley & Lawrence, 1971), whilst on the Klutlan Glacier moraines herbs tend to precede shrubs (Birks, 1980a). In general, the sequence from herbs to shrubs to trees seems to parallel the sequence of plant size, longevity and competitive powers (Grime, 1979). However, woody plants in general and trees in particular exhibit great variability in their rate of immigration and subsequent growth (see section 2.2.2).

The bryoid-thalloid stratum does not necessarily precede those strata comprising phanerogams. This is clearly the case at Glacier Bay, where the 'black crust' phenomenon seems not to be an essential precursor of later stages (Worley, 1973; Lawrence, 1979), and at the Klutlan Glacier. On the Klutlan moraines, there are no mosses or lichens in the pioneer *Crepis nana* nodum. Mosses are first found on terrain deglaciated for at least nine years, and a sequence from acrocarpous mosses to pleurocarpous mosses to lichens then parallels the sequence of phanerogams of increasing stature (Birks, 1980a). On the Pasterze glacier foreland, cryptogams are relatively unimportant and phanerogams predominate from the earliest stages (Fig. 5.14). The curves shown relate to communities characteristic of very wet to moist, nearly stable sites, which are only disturbed by running water. The increasing vegetation cover is largely accounted for by the increase in the size and number of species of flowering plants. At local sites with differing environmental conditions, there are even fewer species of moss and lichen, and these exhibit a more erratic pattern.

That phanerogams can precede or at least colonize at the same time as

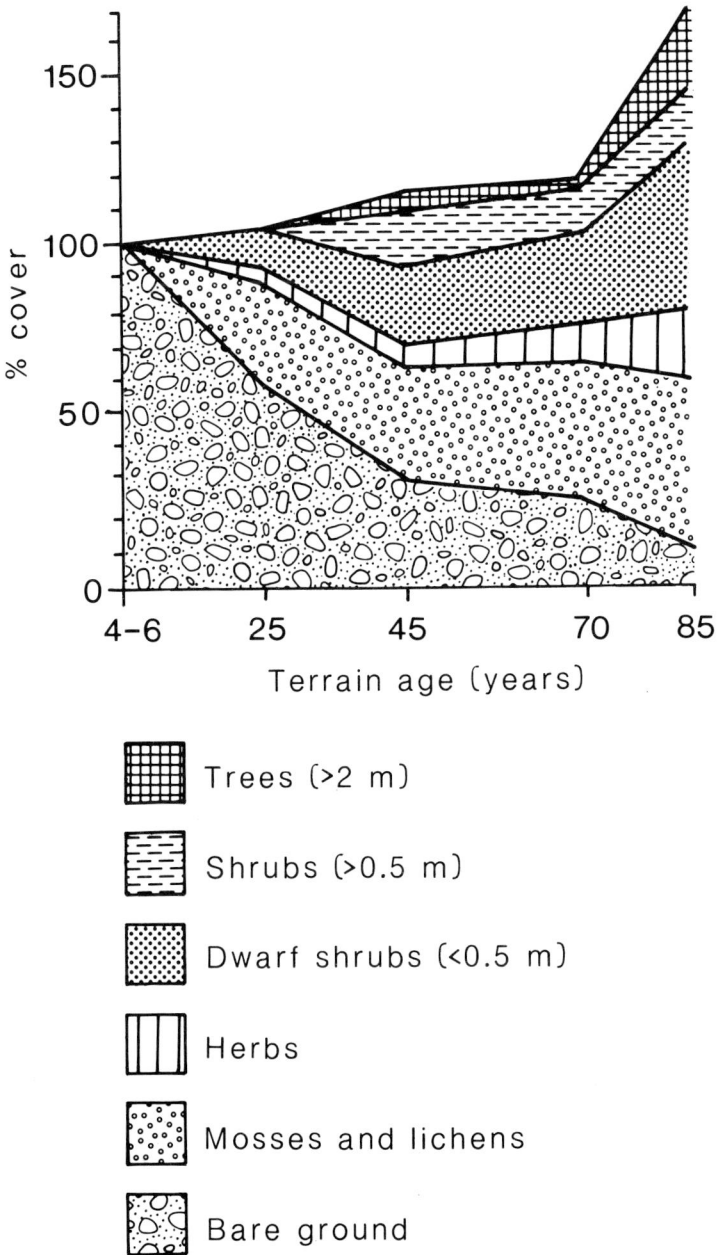

Fig. 5.13. Percentage cover of various vegetation strata on terrain of increasing age, Grand Glacier d'Aletsch, Swiss Alps (from Lüdi, 1945).

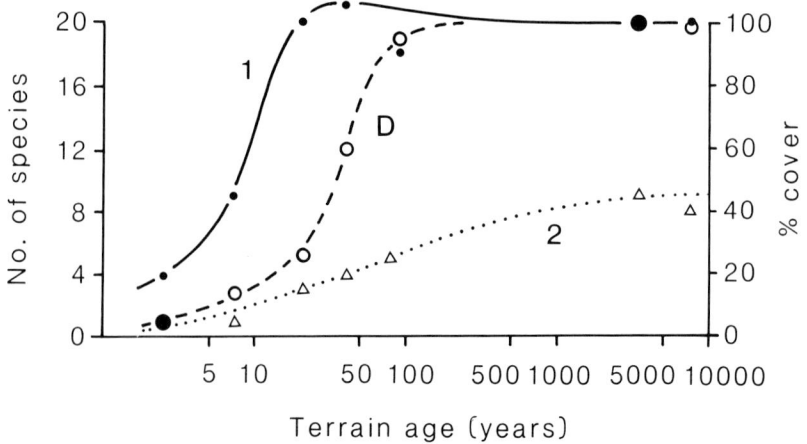

Fig. 5.14. Numbers of phanerogam (1, solid line filled circles) and cryptogam (2, dotted line, triangles) species in successional pathway D (very wet to moist, nearly stable sites), Pasterze Gletscher, Austrian Alps. Percentage vegetation cover (D, broken line, open circles) is also shown. Large filled circles represent two superimposed symbols (from Zollitsch, 1969).

cryptogams has been widely reported from the Alps (Braun-Blanquet & Jenny, 1921; Friedel, 1938; Lüdi, 1958; Palmer & Miller, 1961; Jochimsen, 1970; Richard, 1973; Ellenberg, 1988) and also from sites in Scandinavia (Elven, 1974, 1978b,c; Crouch, 1991), Greenland (Beschel & Weidick, 1973), the Tien Shan (Solomina, 1989), New Zealand (Somerville, Mark & Wilson, 1982; Burrows, 1990) and tropical mountains (Hope, 1976; Coe, 1967). *Senecio keniophyton* is clearly the first colonizer in front of the glaciers on Mount Kenya, ahead of mosses (Coe, 1964; Spence, 1989). On ten nunataks exposed over the last 50 years by downwasting of Omnsbreen, southern Norway, mosses and phanerogams colonize first, followed by lichens and hepatics (Elven, 1980). In front of the Franz Josef Glacier, New Zealand, the first colonists are herbs (e.g. *Raoulia* spp. and *Poa novae-zelandiae*) and mosses (e.g. *Rhacomitrium crispulum*); shrubs follow within 10 years of deglaciation (Burrows, 1990). Flowering plants are followed by mosses, then lichens on glacier forelands in the Tien Shan (Solomina, 1989).

Cryptogams appear to precede phanerogams or at least predominate at an early stage in many wetter and colder regions. In front of glaciers of the mid- and high-alpine zones of Swedish Lappland, vascular plants occur amongst the pioneers from a very early stage, but they do not attain high cover values until much later (Stork, 1963) (Fig. 5.15). Mosses colonize first and are predominant for the first 50 years. Lichens, with the exception of

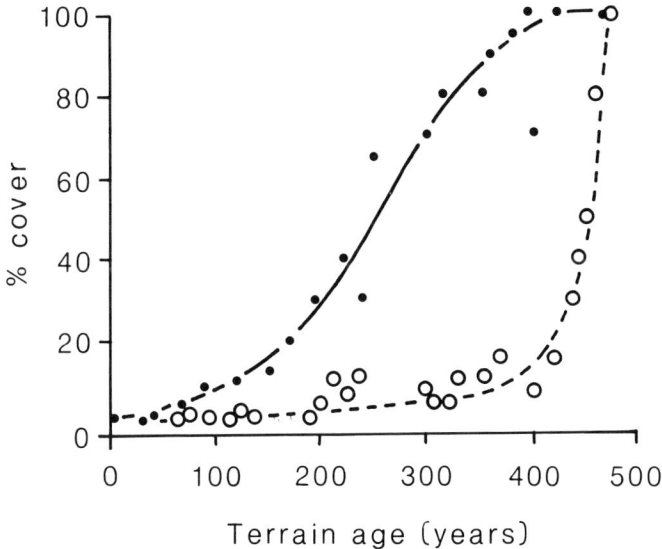

Fig. 5.15. (a) Percentage vegetation cover (solid line) and (b) percentage cover of the vascular plant component (broken line), Storglaciären, Swedish Lappland (from Stork, 1963).

Stereocaulon spp., and liverworts, with the exception of *Marchantia alpestris* (which grows under stones) are also important but tend to colonize later. Mosses tend to dominate in the early stages on the sub-alpine glacier forelands of the Jostedalsbreen outlet glaciers in southern Norway (Fægri, 1933; Vetaas, 1986), in Iceland (Persson, 1964), in Svalbard (Kuc, 1964), and in a chionophilous succession in New Zealand (Archer, 1973). However, it should be noted that substrate characteristics are important at a local scale. Phanerogams (especially forbs and woody perennials) often occur as pioneers on gravelly or stony till, and alongside cryptogams (and sometimes grasses) on sandy or silty glacio-fluvial deposits (cf. Grubb, 1986, 1987; Ellenberg, 1988).

A somewhat similar pattern is described from in front of a receding high-altitude glacier on the sub-Antarctic island of South Georgia by Smith (1984) (Fig. 5.16). A small number of moss and grass species establish on terrain deglaciated for 5 to 15 years. On terrain deglaciated from 15 to 30 years, there is a significant increase in the number and cover of moss species. The first occurrences of both forbs and terricolous lichens are found on terrain deglaciated for > 20 years. However, elsewhere, where the surface is stable and not influenced by cryoturbation, crusts of terricolous lichens

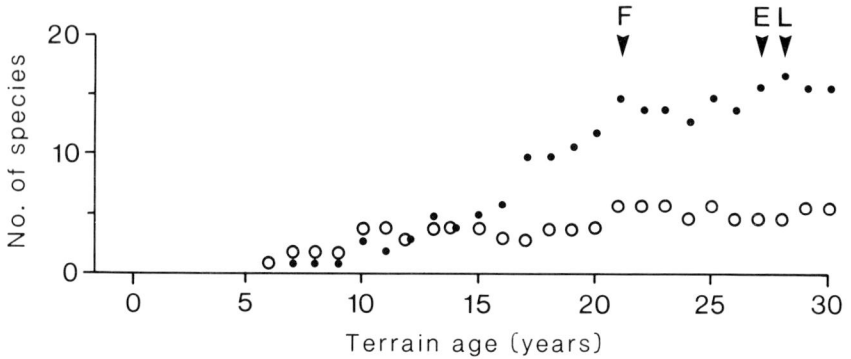

Fig. 5.16. Number of species of flowering plants (open circles) and mosses (filled circles) in front of a high-altitude glacier on the sub-Antarctic island of South Georgia. Other first occurences are indicated by arrows: E, epipetric moss; F, forb; L, lichen (from Smith, 1984).

sometimes create an organic substrate on which develop small moss mounds which are later colonized by liverworts around the periphery and vascular plants within the moss.

In the maritime Antarctic, simple cryptogamic successions have been described from moraine systems in the South Shetland Islands (Lindsay, 1971) and on Signy Island (Collins, 1976; Smith, 1972), and from the deglaciated headland of Bonaparte Point on Anvers Island (Smith, 1982). There, *Deschampsia antarctica*, one of only two Antarctic vascular plants, is sometimes a colonist on relatively old terrain, but mosses predominate throughout the successions. Antarctic fellfield communities on deglaciated soils appear to exhibit a pattern of colonization by successive associations of micro-organisms, micro- or rarely macro-algae, bryophytes and/or lichens, with the local development of closed moss stands in favourable habitats (Smith, 1984). The possibility of preliminary colonization by micro-organisms is discussed more fully elsewhere (see section 6.1.4).

Although they often become important later in successional sequences, terricolous lichens are dominant pioneers only in certain extreme, very exposed polar environments. Saxicolous lichens, on the other hand, are the normal pioneers on boulders and rock outcrops, irrespective of environment (Fægri, 1933; Stork, 1961; Orwin, 1970, 1972; Magomedova, 1979; Longton, 1988). Rock surfaces tend to exhibit a sequence of crustose to foliose to fruticose forms, possibly curtailed or followed by mosses and sometimes, eventually, phanerogams. However, such lithoseres may be truncated by weathering processes and they differ so much in terms of duration and kind from those on unconsolidated substrates that they are regarded here as almost separate phenomena.

There have been comparatively few investigations of glacier foreland fungi (Bridge Cooke & Lawrence, 1959; Sprague & Lawrence, 1959a, b, 1960; Horak, 1960; Baxter & Middleton, 1961). There is nevertheless some evidence for the existence of fungal succession. Sprague & Lawrence (1960) recognize a successional sequence in south-eastern Alaska from saprophytic and parasitic soil-invading filamentous forms to a complex of forest mushrooms. In the same region, Baxter & Middleton (1961) describe geofungi (obtained by preparing cultures from soil washed from the roots of arborescent plants) which indicate an increase in the number of species and species groups with terrain age.

5.2.4   Plant biomass

Few measurements of biomass have been made on glacier forelands. In general, biomass increases during succession as plant size, cover and stratification increase, although complications are introduced by the likelihood of large seasonal and annual variations, and an important below-ground component. Bormann & Sidle (1990) provide estimates of the above-ground biomass of the main tree species in the Glacier Bay chronosequence. The distribution of above-ground biomass follows the well-documented succession from *Alnus crispa* to *Picea sitchensis* (Fig. 5.17). *Alnus* is replaced by *Picea* as the dominant component of the biomass in the transitional early-spruce-forest stage, the other tree species together never attaining > 8% of the total biomass. Above-ground tree biomass therefore follows a sigmoidal pattern, increasing most rapidly in the transitional stage and reaching an asymptote of about 300 tonnes ha^{-1}. It is remarkable that this primary succession has nearly attained the average biomass for temperate coniferous forests of 307 tonnes ha^{-1} (Cole & Rapp, 1981) in a period of about 110 years (Bormann & Sidle, 1990).

Much lower values for above-ground biomass are characteristic of glacier forelands in the alpine zone. Some data from terrain deglaciated by the Glacier de Saint-Sorlin in the French Alps has been presented in a diagram by Moiroud & Gonnet (1977: 69). There, a pioneer vegetation dominated by *Leucanthemum alpinum* forms the first stage of a succession to Alpine meadow dominated by *Carex curvula*. On terrain deglaciated for 25 years, above-ground biomass is about 35 g m^{-2}. After 100 years, values increase to 220 g m^{-2}, which is still only 50% of the presumed asymptote of about 420 g m^{-2} characteristic of the mature Alpine meadow. Preliminary results indicate higher values from seven glacier forelands in Jotunheimen,

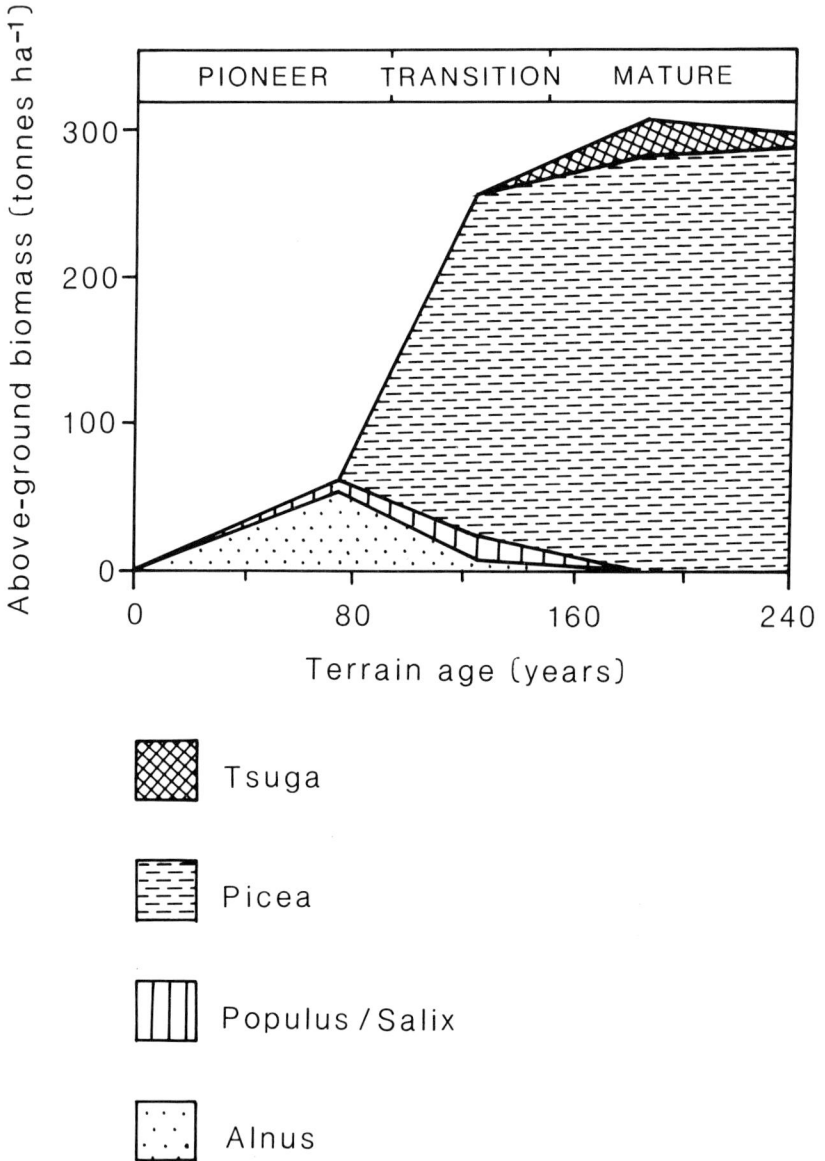

Fig. 5.17. Above-ground biomass of tree species with increasing terrain age, Glacier Bay, Alaska (from Bormann & Sidle, 1990).

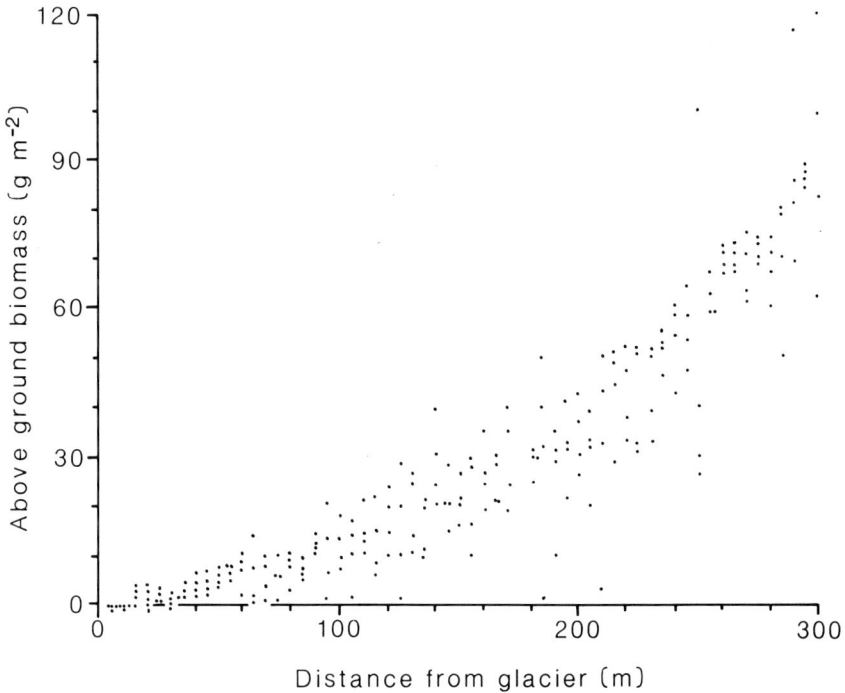

Fig. 5.18. Above-ground biomass with increasing terrain age, Storbreen, Jotun-heimen. Each point represents a sample plot of 0.25 m² (from P. J. Beckett & J. A. Matthews, unpublished).

southern Norway (P. J. Beckett & J. A. Matthews, unpublished). In the pioneer zones, on terrain deglaciated for 14 years, above-ground biomass ranges from 40 g m^{-2} at Storbreen to about 120 g m^{-2} at Leirbreen. The steady increase in above-ground biomass with increasing distance from the Storbreen glacier snout, shown in Fig. 5.18, is made up largely of mosses, grasses and forbs.

The importance of below-ground biomass is highlighted by the work of Turmanina & Volodina (1978, 1979) near the Djunkuat Glacier in the central Caucasus. Annual measurements were made over a five-year period of three stages in a succession to *Festuca varia* meadow. In pioneer groups the annual average ratio of below-ground : above-ground biomass is 2.2 (range 0.5 to 4.0). At an intermediate stage, in various communities of grasses and legumes, it rises to 5.4 (range 1.4 to 15.0), whilst in the *Festuca* meadow it reaches 6.2 (range 3.1 to 17.0). Annual variations for the pioneer and mid-successional stages are correlated with July mean temperatures, the above-ground biomass increasing disproportionately in warm summers. In the case of the *Festuca* meadows, the habitats of which tend to be relatively dry, this relationship breaks down, and annual biomass is greatest is cool, damp summers. Data from pioneer zones in Jotunheimen (based on a single year) indicate below-ground : above-ground ratios of about 3.0 at both Storbreen and Leirbreen, although difficulties in separating all the roots from the substrate suggest that these are minimum estimates (P. J. Beckett & J. A. Matthews, unpublished data).

5.2.5   Species diversity

Almost all studies indicate an increase in the number of species (richness) with increasing terrain age, at least on younger ground. Precise patterns vary considerably from site to site. In severe climates, such as the mid- and high-alpine belts of Scandinavia (Stork, 1961; Elven, 1980), Greenland (Gjaerevoll & Ryvarden, 1977) and the sub-Antarctic (Smith, 1984; Fig. 5.16), this increase in richness is relatively slow. More rapid increases, and in some cases higher richness values on older terrain, tend to characterize glacier forelands and nunataks in the low- and sub-alpine zones of Scandinavia and Iceland (Einarsson, 1970; Elven, 1980; Vetaas, 1986), the Alps (Zollitsch, 1969) and New Zealand (Sommerville, Mark & Wilson, 1982). Large differences may be found in the same region, as shown by Solomina (1989) for glacier forelands in the Tien Shan, where the characteristic number of flowering plant species usually ranges from 5 to 15 species on 20–25-year-old moraines, from 15 to 20 species on 50–100-year-old moraines, and from 20 to 30 species on 150-year-old moraines (Fig. 5.19).

Several investigators have detected a peak in richness early in succession followed by a decline. On the foreland of the Pasterze Gletscher in the Alps, an early peak in richness (all species) between 40 and 80 years after deglaciation is most marked (24 to 26 species) in moist, sheltered and nearly undisturbed sites (Fig. 5.20, curve C) (Zollitsch, 1969). A smaller and less distinct peak is characteristic of exposed and dry sites that are disturbed at the surface (curve B), whereas at very exposed, very disturbed and moist sites that later dry out (curve A) richness steadily increases with no evidence of an early peak (cf. Fig. 5.14). More species appear able to colonize where physical environmental conditions are most favourable and when vegetation cover is incomplete (cf. Fig. 5.10). All three habitat types appear to support about 20 species in later succession stages.

A similar pattern is indicated for the number of phanerogams in plots of 3 m² in favourable microtopographic sites in front of Hardangerjökulen, southern Norway (Elven & Ryvarden, 1975). On proximal moraine slopes

(a)

(b)

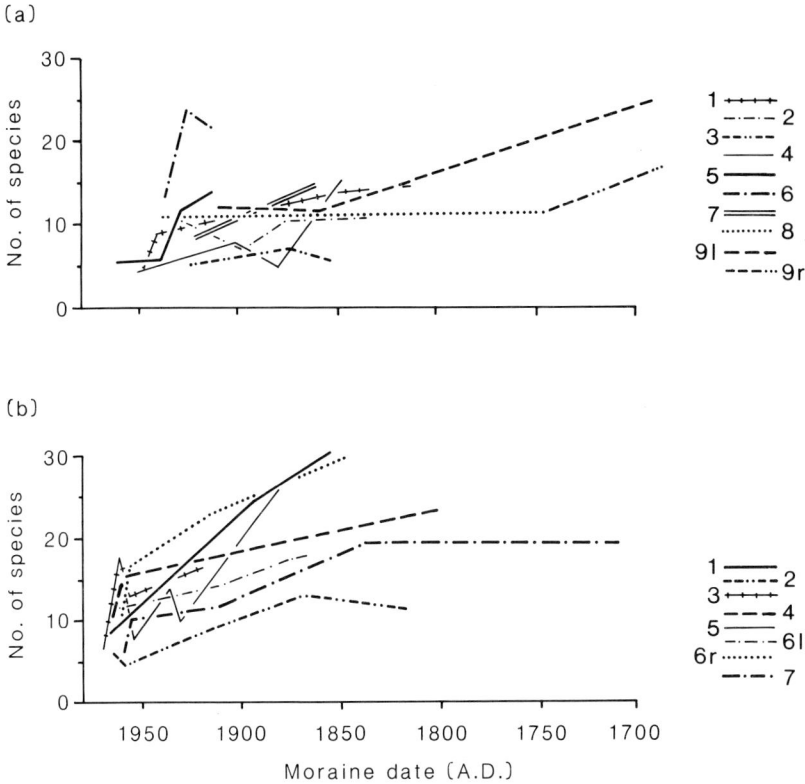

Fig. 5.19. Species richness (flowering plants only) on moraines of increasing age, Tien Shan, U.S.S.R.: (a) glaciers in the Chon-Kemin valley (1–7) and the Chon-Kyzylsu valley (8–9) (l, r = left and right lateral moraines, respectively); (b) glaciers in the Akshiyrak massif (1–7) (from Solomina, 1989).

(with a south-facing aspect) a steady increase in the number of phanerogam species occurs, reaching a peak of about 27 species on terrain deglaciated for 30 to 40 years (Fig. 5.21). On older moraines, the subsequent decline in the number of species appears to be more gradual than in front of the Pasterze Gletscher, falling to about 15 species after about 100 years. Inter-moraine areas exhibit a rather similar pattern, possibly with a later and less well-marked early peak. There is little evidence for an early peak on distal slopes, whilst moraine crests are characterized by a remarkably constant and comparatively low number of species (maximum of about 10 species) after the initial rise in the first 30 years.

Other evidence for an early peak in richness is presented by Persson (1964), Matthews (1978a), Sommerville, Mark & Wilson (1982), and

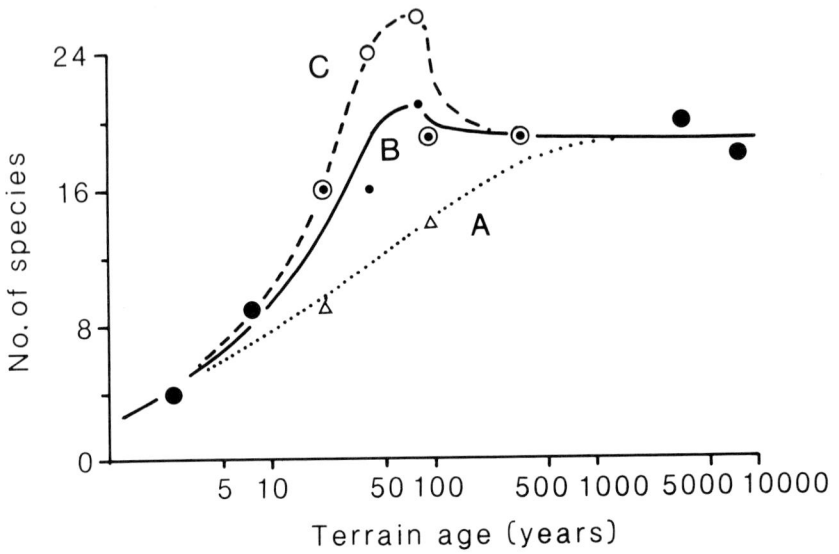

Fig. 5.20. Species richness (all species) in three successional pathways (A, B and C) characteristic of respectively more favourable habitats, Pasterze Gletscher, Austrian Alps. Large filled circles represent all three habitats. Pathways and habitats are defined in Fig. 5.41 (from Zollitsch, 1969).

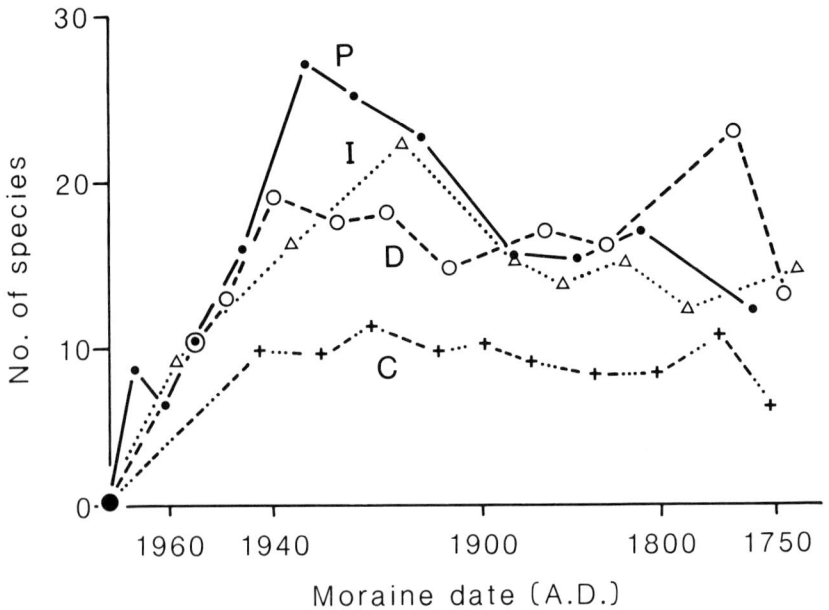

Fig. 5.21. Species richness (phanerogams only) with increasing terrain age and in relation to four microtopographic site types (C = moraine crest, P = proximal slope, D = distal slope, I = inter-moraine area), Hardangerjökulen, southern Norway (from Elven & Ryvarden, 1975).

Vetaas (1986). In front of Skaftafellsjökull, Iceland, the number of vascular plant species per square metre increases from 4.8 to 12.7 on ground deglaciated for 4–9 years and > 60 years, respectively (Persson, 1964). However, the total number of vascular plant species recorded reaches a peak of 56 species on terrain deglaciated for > 12 years, subsequently declining to about half this number after > 60 years. Maps of the number of vascular plant species in 16 m² quadrats on the Storbreen glacier foreland, Jotunheimen, indicate a patchy distribution with an initial peak richness per site on terrain deglaciated for 25 to 35 years but a variety of possible trends in later successional stages (Fig. 5.22).

Early peaks in richness appear to reflect the ability of a relatively large number of species to colonize open spaces in the early stages of succession provided environmental conditions are favourable. This may also reflect an overlap in the distribution of pioneer and later colonizers along the time gradient. A subsequent decline in richness is most popularly accounted for by increasing competition in later successional stages, particularly as vegetation cover approaches 100% (cf. Figs. 5.14 & 5.20). In severe physical environments (e.g. exposed, dry, cold or highly disturbed locations) the absence of an early richness peak seems to reflect the existence of relatively uniform conditions for plant growth at all successional stages, combined with a smaller number of colonizing species and slower rates of change.

Where environmental conditions are most severe, there may therefore be little or no trend in richness during succession. In a regional study of moraines and related Neoglacial deposits in the Teton Range, Wyoming, Spence (1985) found no significant relationship between the number of species per deposit and terrain age. Instead, species richness on these species-poor deposits is correlated with the diversity of surrounding vegetation, topographic diversity of the site and site aspect. The absence of any relationship with terrain age is likely to result from the prevalence of steep slopes and hence extensive mass wasting and disturbance. Paradoxically, lack of an early peak in richness at Glacier Bay after an intial steep rise (Reiners, Worley & Lawrence, 1971) may reflect the exceptionally rapid succession under a maritime climatic regime. It is likely that the increase in vegetation cover, particularly of *Alnus*, introduces an unfavourable biological environment for many of the competing colonizers at an early stage.

Aspects of diversity other than richness, namely equitability (evenness) and diversity *sensu stricto* (combining both richness and equitability components) have been investigated in a few studies (cf. Pielou, 1969; Magurran, 1988). Detailed results have been published for the Glacier Bay chronosequence by Reiners, Worley & Lawrence (1971) who have investigated whole communities and different vegetation strata. Diversity, like

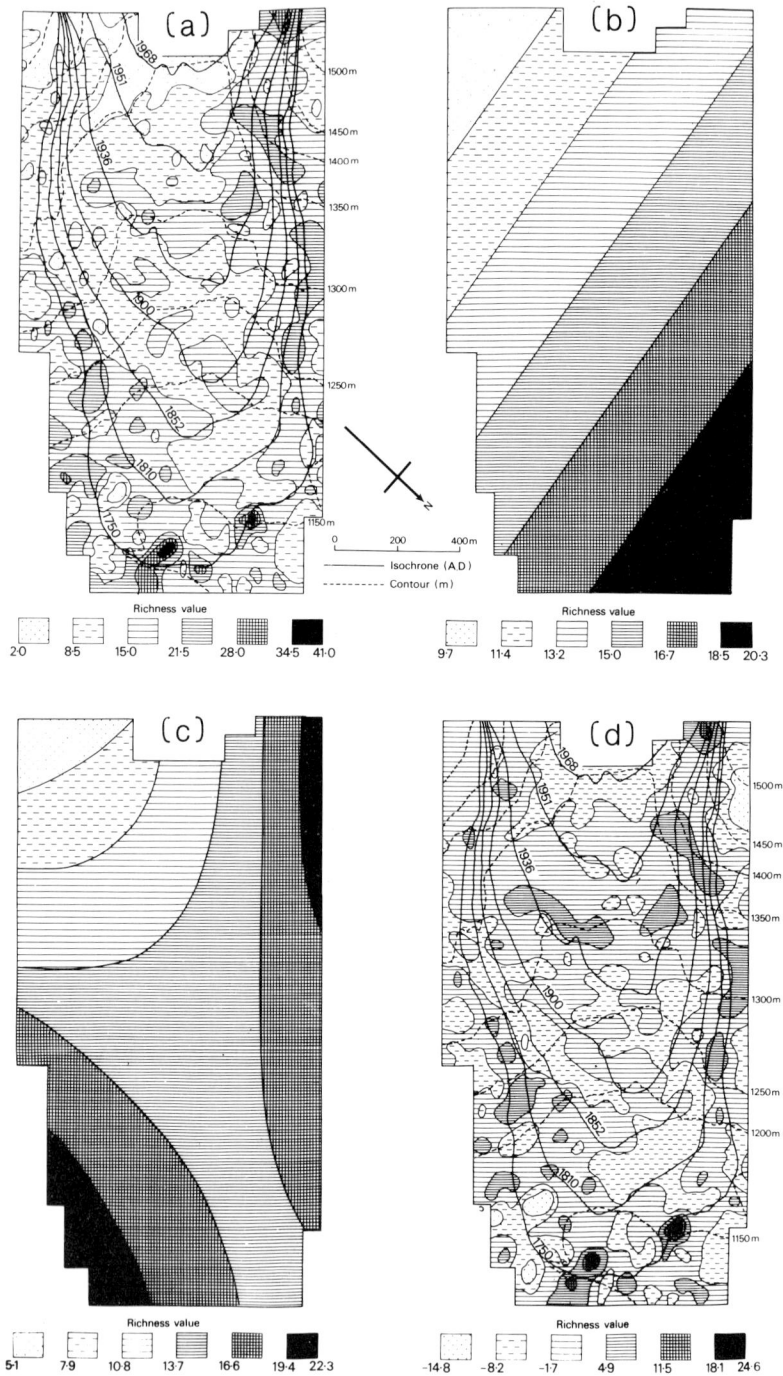

Fig. 5.22. Patterns of species diversity (richness component), Storbreen, Jotunheimen: (a) species richness data; (b) first-order trend surface (variation accounted for, 15.1%); (c) second-order (quadratic) trend surface (variation accounted for, 27.2%); (d) residuals from the second-order trend. Six equal-interval classes are mapped in each case (from Matthews, 1978b).

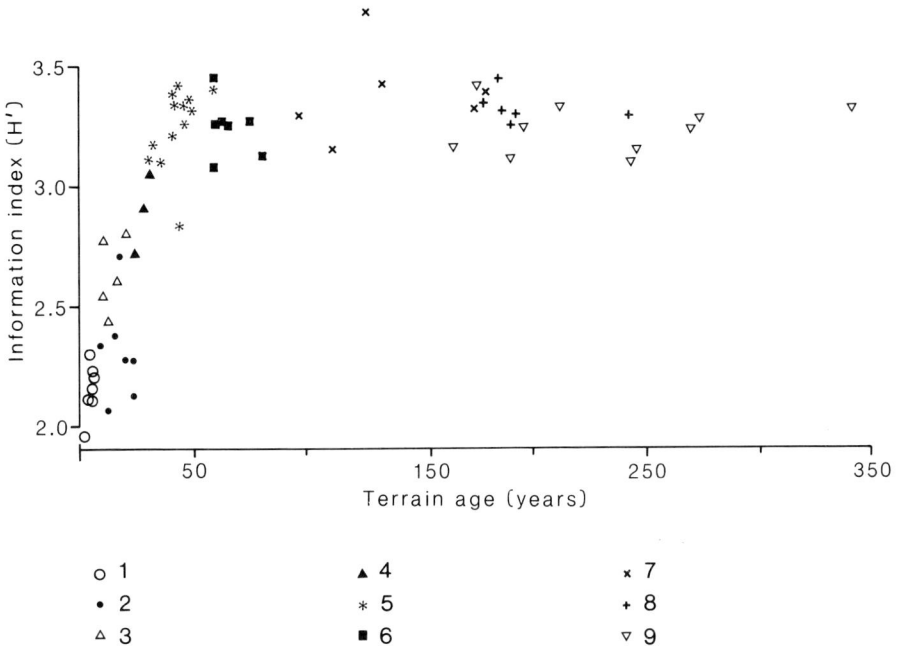

Fig. 5.23. Species diversity (Shannon-Weiner information index, H') of various plant communities (noda) with increasing terrain age, Klutlan Glacier, Yukon Territory, Canada. Noda (symbols): 1, *Crepis nana*; 2, *Dryas drummondii*; 3, *Hedysarum mackenzii*; 4, *Hedysarum-Salix*; 5, *Salix-Shepherdia*; 6, *Picea-Salix*; 7, *Picea-Arctostaphylos*; 8, *Picea-Ledum*; 9, *Picea-Rhytidium* (from Birks, 1980a).

richness, increases rapidly at first followed by little or no further increase until the muskeg stage. Equitability varies widely with a notable minimum at the stage dominated by *Dryas*-mats (which also affect the diversity indices). With the possible exceptions of the tall-shrub stratum, the stages at which strata attain peak cover or basal area do not coincide with peak diversity values. Within a particular stratum, there is a tendency for equitability and diversity to decline in stages of high dominance, indicating a concentration of cover or basal area in a small number of species.

Birks (1980a) calculated three diversity indices for the nine communities (noda) recognized in front of Klutlan Glacier, Yukon Territory. According to Birks, the trend exhibited by the Shannon-Weiner information index (H') is typical of the other diversity indices at Klutlan Glacier (Fig. 5.23), and also of the trend at Glacier Bay. The data indicate a sharp increase in diversity between 5 and 60 years and then little systematic change from 60 to 300 years despite marked changes in species composition and abundance. At Storbreen, Jotunheimen, Matthews (1976, 1978a) found several indices of diversity to follow closely the pattern described previously for richness. Again, on the older terrain, a mosaic of high and low values indicates more than one possible trend of diversity in the later stages of succession. Both Matthews (1978a) and Birks (1980a) report little successional trend in the equitability component of diversity.

5.2.6   Species composition and successional stages

Most workers who have carried out investigations on the ecology of recently-deglaciated terrain recognize successional stages based on species composition. Cooper (1923b: 224) refers to 'the indefiniteness of vegetation units' and also to the fact that 'their distinctness is more apparent than real, being due to increasing dominance of a new growth form, as when the low-growing pioneers give place to tall shrubs, and these in turn to trees'. Thus the eight stages described from Glacier Bay (section 5.1.1) or the five stages from the foreland of the Grand Glacier d'Aletsch (Table 5.2) should not be regarded as representing rigid shifts in species composition. At best such stages represent wave-like replacements of groups of species with similar ecologies; at worst they are little more than arbitrary divisions in a continuum.

Several approaches can be used to analyse successional trends in terms of species composition. A direct approach is to construct graphs of either the presence or the abundance of each species in relation to terrain age. An example of the former is given in Fig. 5.24, which relates to terrain deglaciated over the last 110 years by the Grand Glacier d'Aletsch (Richard, 1975), whilst an example of the latter (Fig. 5.25) involves the moraines of the Klutlan Glacier deglaciated over the last 200 years (Birks, 1980a). Both representations demonstrate the difficulty of defining meaningful successional stages.

Nine species groups with similar distributions in relation to terrain age are identified in Fig. 5.24; that is, various species are inferred to enter and leave the succession at approximately similar times, the ninth group being sub-alpine forest species that colonize after 200 to 300 years. It should be noted, however, that there are differences within and similarities between these groups, and that there are other species that do not fit neatly into these categories. Similar representations have been used by Stork (1963) in northern Sweden and by Scott (1974b) in front of the Frederika Glacier, Alaska. Use of average cover values to represent the major taxa in the nine noda on the Klutlan glacier moraines (Fig. 5.25) indicates even more clearly the absence of discrete stages and the arbitrariness of the noda.

More elaborate phytosociological analyses (e.g. Richard, 1968; Zol-

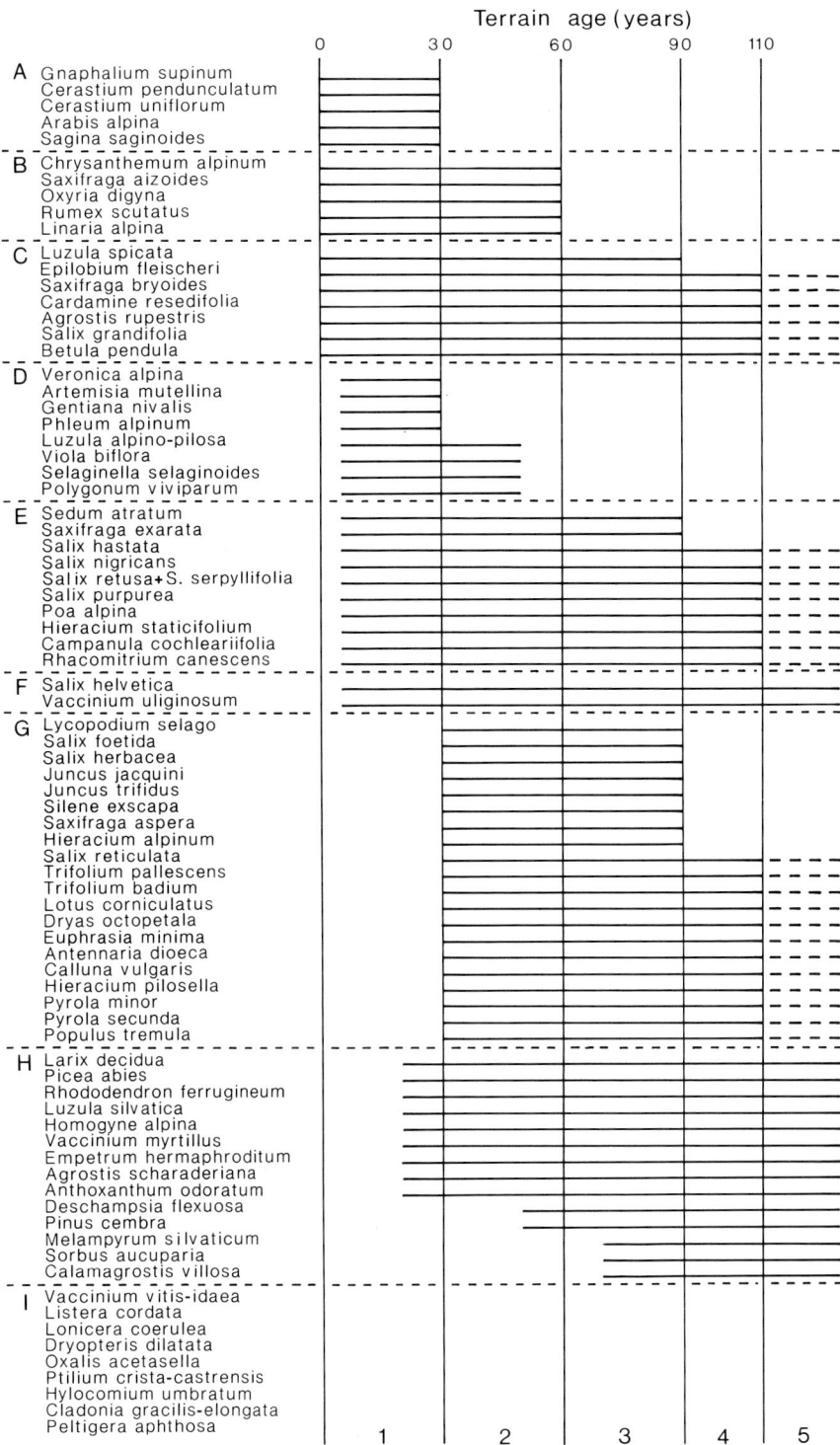

Fig. 5.24. Nine species groupings (A–I) and five successional stages (1–5), defined by the presence or absence of selected species on terrain of increasing age, Rhonegletscher, Swiss Alps (see also, Table 5.1) (from Richard, 1973).

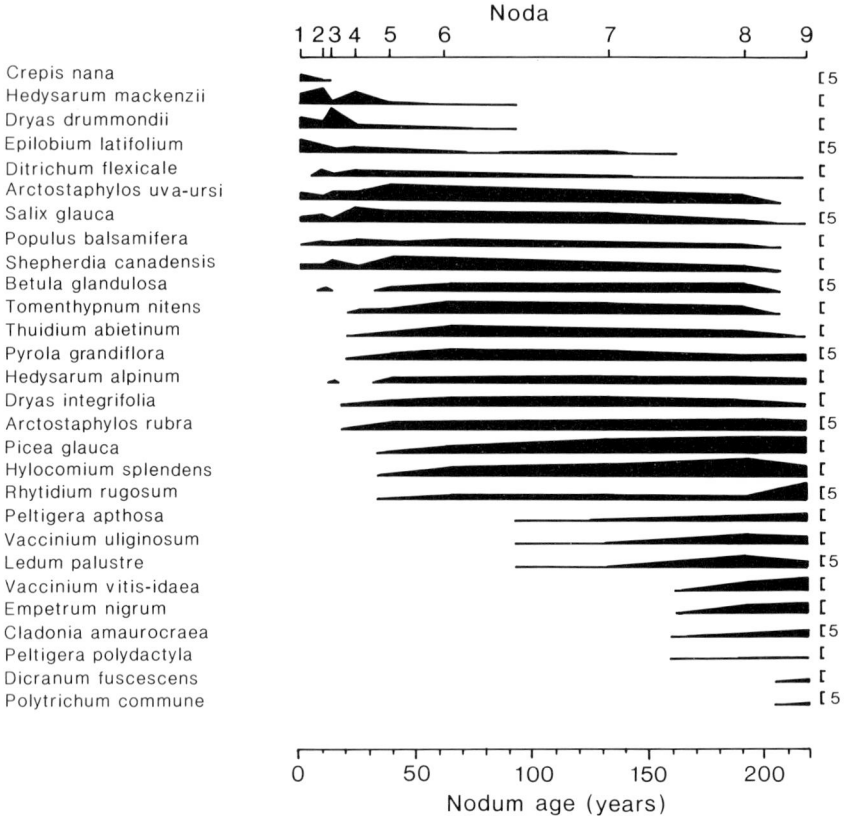

Fig. 5.25. Variation in average cover-abundance values for selected species in relation to plant community noda (numbered) and the average age of each nodum, Klutlan Glacier, Yukon Territory, Canada. Noda: 1, *Crepis nana*; 2, *Hedysarum mackenzii*; 3, *Dryas drummondii*; 4, *Hedysarum-Salix*; 5, *Salix-Shepherdia*; 6, *Picea-Salix*; 7, *Picea-Arctostaphylos*; 8, *Picea-Ledum*; 9, *Picea-Rhytidium* (from Birks, 1980a).

litsch, 1969; Elven, 1978c; Ammann, 1979; Schubiger-Bossard, 1988) or the use of numerical classification and ordination techniques (e.g. Elven, 1978b; Matthews, 1978a,b, 1979a–c; Crouch, 1991) are logical extensions of this approach. In the present context, the results of numerical approaches are particularly useful in exploring the nature of shifts in species composition. The species groups defined by Elven (1978b) at Hardanger-jökulen have already been discussed in section 5.2.2 (Fig. 5.12). On the Storbreen glacier foreland, Jotunheimen, there appears to be similar loose groups of 'pioneer', 'heath' and 'snowbed' species with relatively strong within-group similarities in their distribution patterns (Matthews, 1978a,b; see section 5.3.1).

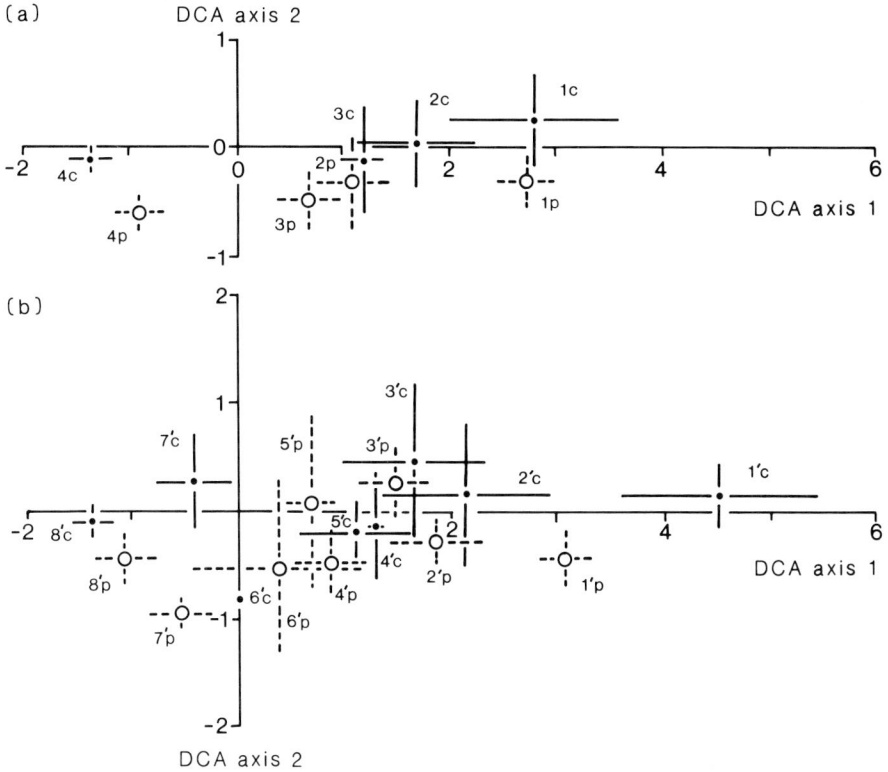

Fig. 5.26. Species groups defined by two-way indicator species analysis (TWINSPAN) and displayed in relation to ordination axes from detrended correspondence analysis (DCA), Storbreen, Jotunheimen: cryptogam and phanerogam groups at (a) the four-group and (b) the eight-group levels. Symbols represent mean species scores (group centroids) on DCA axes 1 & 2 (eigenvalues, 0.52 and 0.38, respectively); bars indicate 95% confidence intervals; filled circles represent cryptogam groups (c1–4 and c'1–8), open circles represent phanerogam groups (p1–4 and p'1–8) (from Crouch, 1991).

This type of analysis has been extended by Crouch (1991), with reference to both cryptogams and phanerogams for a subsample of 190 of Matthews' sites. Cryptogam species groups and higher plant species groups are first defined separately using two-way indicator species analysis (TWINSPAN) and then displayed together in relation to common axes derived by a detrended correspondence analysis (DCA) of the combined data set. Four- and eight-group levels of the species' classifications are shown on the same DCA axes in Fig. 5.26. Each group of species is represented by a group centroid and a 95% confidence interval. As axis 1 of the DCA is strongly correlated with terrain age ($r = -0.76$), the species groups can be interpreted in terms of a successional sequence. At both levels of division, there are higher-and lower-plant pioneer groups that are relatively distinct from all other groups within their respective classifications. Similarly, on older terrain, there are cryptogam and phanerogam groups which are relatively distinct. The overall interpretation is of three major successional stages.

In general, there are very real changes in species composition during glacier foreland successions. Except under the most severe physical environmental conditions, the species composition of the pioneer group differs substantially, if not completely, from those of later stages in the succession (see also section 5.3). Intermediate stages, which may contain a mixture of species characteristic of early and late stages, may be less well defined in terms of species composition. Although recognized stages based on species composition tend to be few in number and are not discrete, the species normally show some clustering in relation to terrain age; hence there is some justification for a community-level approach to glacier foreland succession.

The causes of such patterns are poorly understood. It seems that whilst initial species composition may depend primarily on factors determining immigration and initial establishment (Chapin, 1991), subsequent changes often involve not only competitive balance (Tilman, 1985), and differential arrival, longevity and growth rate (Chapin, 1991), but also allogenic environmental change (Matthews & Whittaker, 1987; Whittaker, 1990). The processes behind such changes are discussed more fully in chapter 6.

5.2.7   Population attributes and physiological traits

In addition to trends in gross physiognomy with increasing terrain age, differences have been recognized in the morphological attributes of species' populations and/or their physiological traits. Little physiological work has actually been carried out on glacier forelands, however, other than the recognition of possible physiological functions associated with and imputed to particular morphological attributes.

The size structures of 10 key species on the Storbreen glacier foreland have been described by Whittaker (1985). These are a reflection of age and performance effects. Data for two species, the pioneer grass *Poa alpina* and the later-colonizing shrub *Empetrum hermaphroditum*, are summarized in Fig. 5.27. *Poa* is typical of the six pioneer taxa examined (the others being *Deschampsia alpina*, *Trisetum spicatum*, *Arabis alpina*, *Saxifraga cespitosa* and *Oxyria digyna*). An early peak in mean clump diameter is followed by a rapid decline in performance, and subsequently in size and cover values. The decline in cover after about 50 years involves a decrease in size range and in mean size although, even on the oldest terrain within the foreland,

(a)

(b)

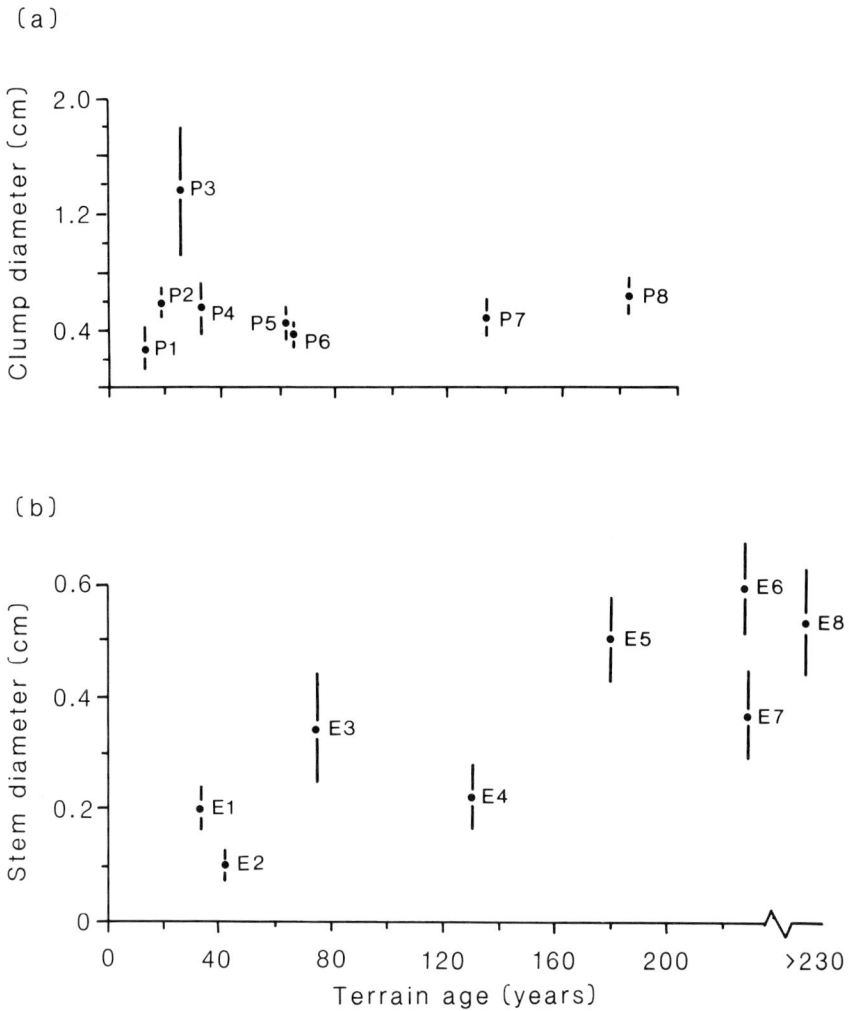

Fig. 5.27. Size structures of species' populations in relation to terrain age, Storbreen, Jotunheimen: (a) mean maximum clump diameter of *Poa alpina* at original point of rooting; (b) mean maximum stem diameter of *Empetrum hermaphroditum* at point of origin. Bars indicate 95% confidence intervals (n = 50, minimum) (from Whittaker, 1985; Matthews & Whittaker, 1987).

new individuals continue to establish as spaces arise. Four shrub taxa – *Empetrum hermaphroditum*, *Phyllodoce caerulea*, *Betula nana* and *Salix* spp. (including *S. glauca* and *S. lanata*) – exhibit more variable patterns, although all except *Betula nana* show a tendency to an increase in mean stem diameter with increasing terrain age, as shown for *Empetrum* in Fig. 5.27b. The principal discontinuity in the population behaviour of *Empe-*

trum is a considerable increase in the proportion of individuals establishing vegetatively beyond the glacier foreland boundary (4–14% on the foreland at sites E5, E6 and E7, rising to at least 44% at site E8). Despite considerable individualistic behaviour of species, similar patterns of population size-structure indicate similarity of responses to available resources and possibly similar roles in succession.

Based on a broad survey of primary succession, but including detailed reference to Glacier Bay and to flood plains in Alaska, Chapin (1991) has suggested a number of ways in which early colonizers differ from late successional species (cf. Bormann & Likens, 1979; Bazzaz, 1979; Finegan, 1984; Huston & Smith, 1987). He suggests that early colonizers in primary succession have light, wind-dispersed seeds with minimal dormancy requirements; seeds consequently germinate shortly after arrival and do not become incorporated into a buried seed pool (but see section 6.1.2). Although they have a small absolute growth rate because of small seed sizes, early colonizers have relative growth rates higher than those of late successional species and are relatively short lived. They are able to grow relatively rapidly on infertile, low-nitrogen soils because of a high potential for photosynthesis and nutrient uptake. Species able to fix nitrogen symbiotically are not amongst the first colonizers, possibly because of large seed sizes, but they quickly become important constituents of communities because of rapid growth on soils poor in nitrogen. Such differences between various successional stages may provide clues to the causal mechanisms of succession and will be discussed further in chapter 6.

5.3   Spatial variation and successional pathways

The account of broad successional trends given in section 5.2 emphasizes the evidence of vegetation patterns related to terrain age. This is at once the strength and weakness of the traditional chronosequence approach to vegetation patterns on recently-deglaciated terrain. The aim of this section is to examine in detail the additional information about successional change that can be inferred by taking into account spatial variation both within and between forelands. Later, in section 5.4, there is futher discussion of the environmental factors shown to be or suspected of being important.

5.3.1   Within-foreland patterns: mapping

Spatial variation in the vegetation on recently-deglaciated terrain has been widely recognized but rarely investigated in any detail. This and subsequent sections are therefore based heavily on studies carried out at

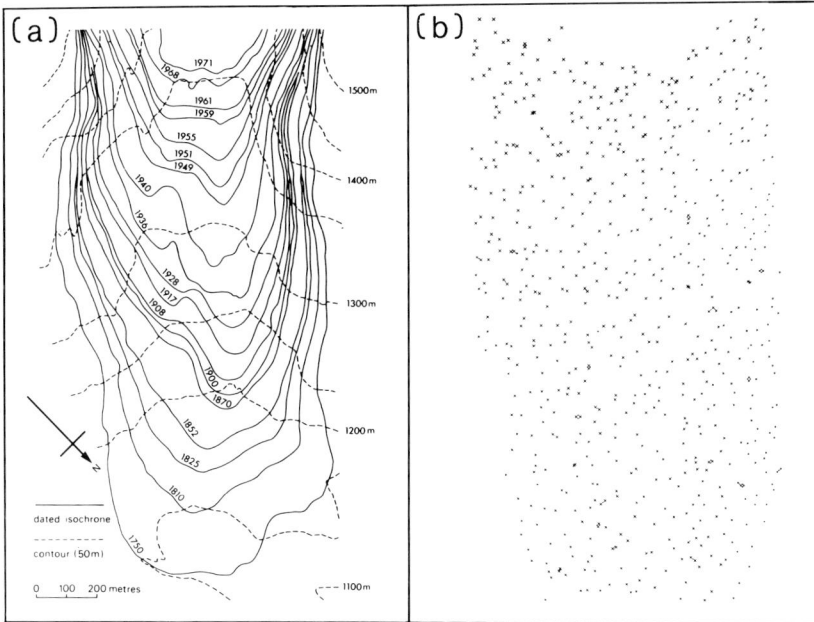

Fig. 5.28. Vegetation sampling sites at Storbreen, Jotunheimen: (a) the areal chronology; (b) the systematically-stratified random sampling design (each cross represents sixteen 1 m² quadrats) (from Matthews, 1978b).

Storbreen, Jotunheimen, where the development of a spatial approach to glacier foreland vegetation has been of explicit concern (Matthews, 1975b, 1976b, 1978b, 1979a,b; Matthews & Whittaker, 1987; Crouch, 1991). Data are particularly lacking at the species level. Apart from the work at Storbreen, published maps of individual species distributions are confined to sketch maps showing the proximal (inner) limit of selected species in front of glaciers in Iceland and on Mount Kenya. Persson's (1964) map indicates that in front of Skaftafellsjökull, the limit of the lichen *Lecidea dicksonii* corresponds with a moraine ridge dating from about A.D. 1934, whereas *Placopsis gelida* is found on younger ground with a limit corresponding to the ice margin about 1950. In front of Lewis and Tyndall Glaciers, Coe (1964, 1967) indicates the limits of eight vascular plant species and a limit for mosses and lichens as a single group. Such limits are of restricted usefulness because they are difficult to define where colonization is inherently sparse.

Detailed, quantitative maps have been drawn showing the distribution of individual species on the Storbreen glacier foreland. Each map depicts the local (shoot) frequency of a species at 638 sites located according to a systematically-stratified random design (Fig. 5.28). At each site of 16 m²,

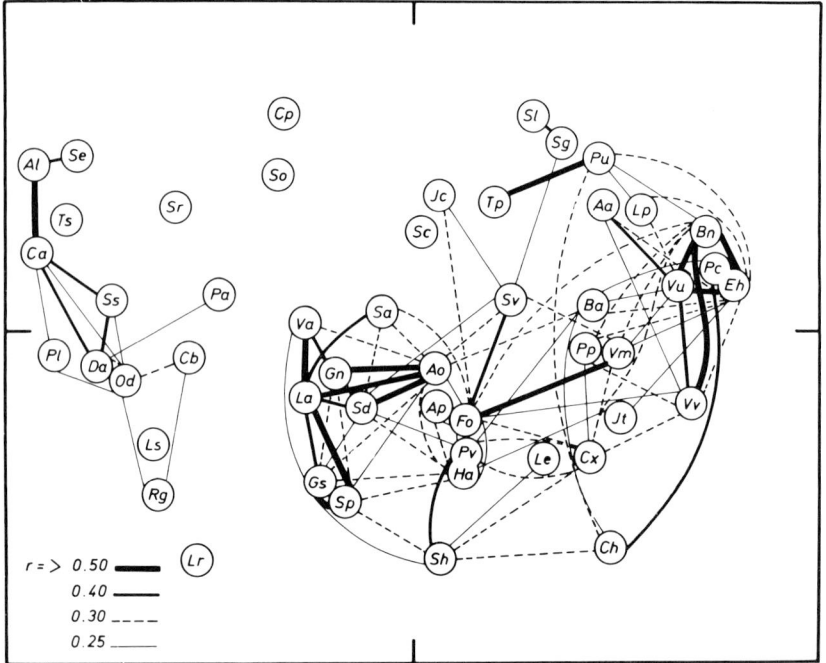

Fig. 5.29. Species plexus produced by non-metric multidimensional scaling (NMDS), Storbreen, Jotunheimen. Species symbols: *Aa, Arctostaphylos alpina; Al, Arabis alpina; Ao, Anthoxanthum odoratum; Ap, Antennaria alpina; Ba, Bartsia alpina; Bn, Betula nana; Ca, Cerastium* spp.; *Cb, Cardamine bellidifolia; Ch, Cassiope hypnoides; Cp, Cardaminopsis petraea; Cx, Carex* spp.; *Da, Deschampsia alpina; Eh, Empetrum hermaphroditum; Fo, Festuca ovina; Gn, Gnaphalium norvegicum; Gs, G. supinum; Ha, Hieracium* spp.; *Jc, Juniperus communis; Jt, Juncus trifidus; La, Leontodon autumnalis; Le, Lycopodium selago; Lp, Loiseleuria procumbens; Lr, Luzula arcuata; Ls, L. spicata; Od, Oxyria digyna; Pa, Phleum alpinum; Pc, Phyllodoce caerulea; Pl, Poa alpina; Pp, Pedicularis lapponica; Pu, Pinguicula vulgaris; Pv, Polygonum viviparum; Rg, Ranunculus glacialis; Sa, Saussurea alpina; Sc, Silene acaulis; Sd, Sedum rosea; Se, Saxifraga groenlandica; Sg, Salix glauca; Sh, S. herbacea; Sl, S. lanata; So, Saxifraga oppositifolia; Sp, Sibbaldia procumbens; Sr, Saxifraga rivularis; Ss, S. stellaris; Sv, Solidago virgaurea; Tp, Tofieldia pusilla; Ts, Trisetum spicatum; Va, Veronica alpina; Vm, Vaccinium myrtillus; Vu, V. uliginosum; Vv, V. vitis-idaea* (from Matthews, 1978a,b).

local frequency is determined on a scale of zero to 400 (the number of 20×20 cm subdivisions of the site in which the species occurs). Species representative of the three species groups defined by multidimensional scaling (Fig. 5.29) are respectively shown in Figs. 5.30a–c. The species plexus (Fig.5.29) summarizes the similarities and differences between the distribution patterns of the 50 most commonly occurring species. The strongest correlation coefficients (links between species in the plexus) suggest the 'core' species of each group. Three species – *Betula nana, Empetrum hermaphroditum* and *Vaccinium uliginosum* – form the 'core' of the heath species group, members of which tend to be most frequent not only on relatively old terrain but also at relatively low altitudes, and on southerly aspects. A suite of other species belong to this group but their distribution patterns differ in various ways from those of the 'core' species. In particular, many species are differentiated by their relative frequencies with respect to the glacier foreland boundary (Matthews, 1978a).

Species comprising the 'core' of the snowbed species group – *Gnaphalium supinum, G. norvegicum, Leontodon autumnalis, Veronica alpina, Anthoxanthum odoratum, Sedum rosea* and *Sibbaldia procumbens* – are also relatively frequent on older terrain but in these cases peak frequencies are found at higher altitudes, and at north-facing sites. 'Core' pioneer species – *Deschampsia alpina, Cerastium* spp. and *Saxifraga stellaris* – are confined to relatively young terrain. Differences between these species, other species from these groups, and various 'outlying', 'transitional' and 'rare' species, which do not fit conveniently into any group, are considerable and are discussed fully by Matthews (1976b, 1978a).

Two important conclusions follow from a consideration of the distribution patterns of individual species at Storbreen. First, whilst the patterns can be interpreted in terms of waves of colonization, the differences between the species indicate considerable individualistic behaviour. Only a small number of species combinations exhibit correlation coefficients of $\geqslant 0.5$ (Fig. 5.29). This reinforces the notion of indistinct successional stages. Second, any waves of colonization are environmentally conditioned. That is, the spatial distribution patterns invariably indicate at least an element of environmental dependence. This applies at the meso-scale, with reference to altitude and aspect, and also at the micro-scale, where microtopography is a significant influence (Matthews, 1978a). Such physical environmental factors condition both the nature and rate of succession at particular sites.

There is a much longer tradition of mapping community types on glacier forelands than there is of mapping individual species. Cooper (1923a) produced a map of Glacier Bay showing three broad classes of vegetation – 'alder-willow thicket', 'young climax forest' and 'old climax forest' – whilst Decker (1966) provides a more elaborate map of eight successional stages (described in section 5.1.1) for part of the same area. Oliver & Adams (1979) have made a similar generalized map of six 'major plant habitats' and a

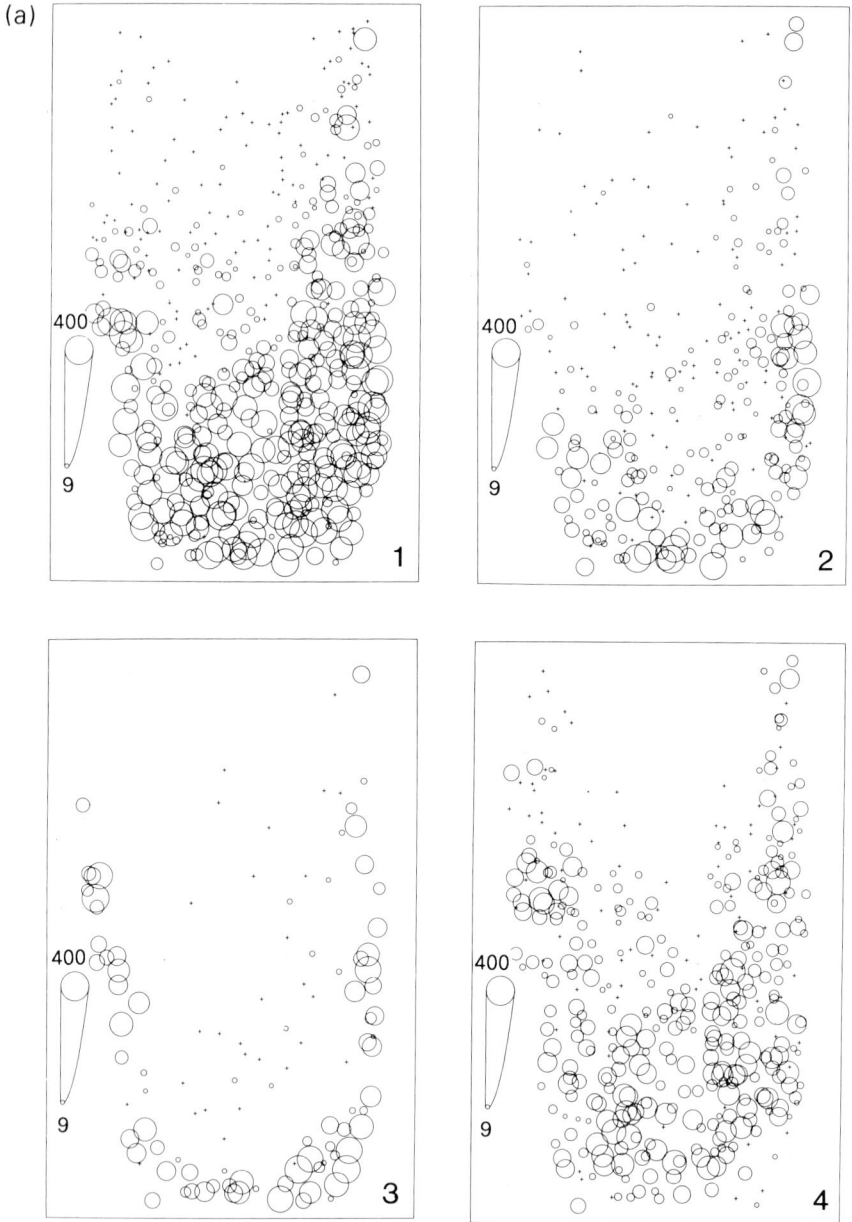

Fig. 5.30. Maps of selected species, Storbreen, Jotunheimen: (a) 'heath species' –
(1) *Empetrum hermaphroditum*, (2) *Vaccinium uliginosum*, (3) *V. vitis-idaea*, (4)
Phyllodoce caerulea; (b) 'snowbed species' – (1) *Gnaphalium supinum*, (2) *Salix
herbacea*, (3) *Hieracium* spp., (4) *Polygonum viviparum*; (c) 'pioneer species' – (1)
Deschampsia alpina, (2) *Cerastium* spp., (3) *Arabis alpina*, (4) *Saxifraga groenlandica*. Circles are proportional to local shoot frequency (Scale: 0–400; + = <9)
(from Matthews, 1978b).

(b)

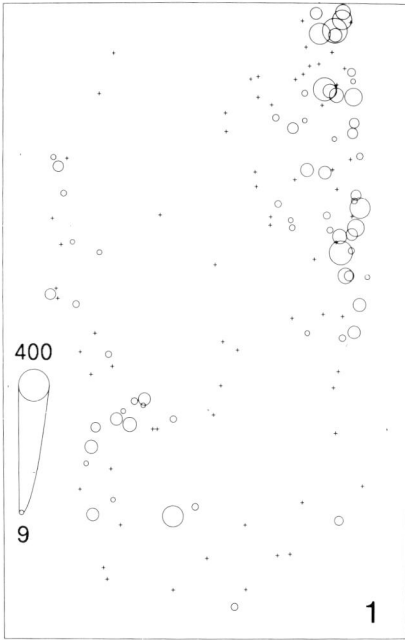
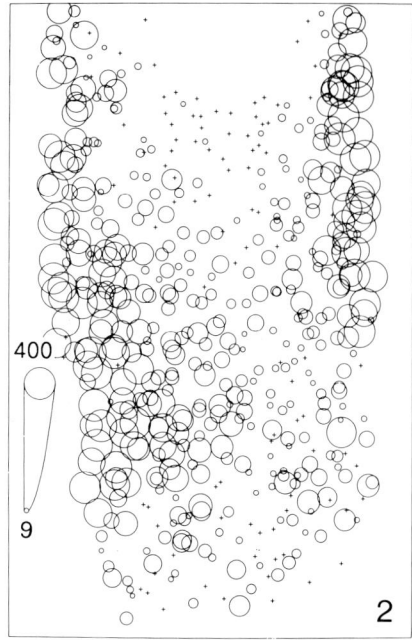
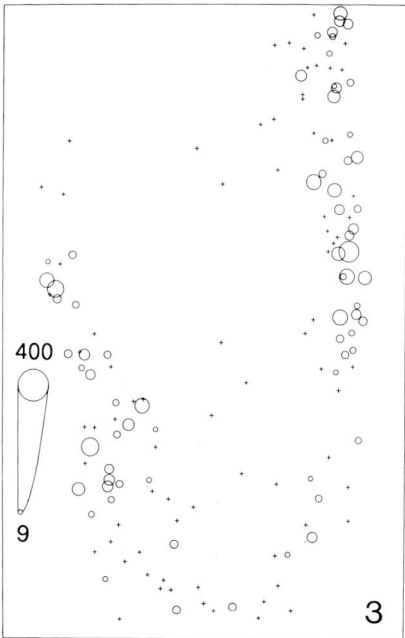
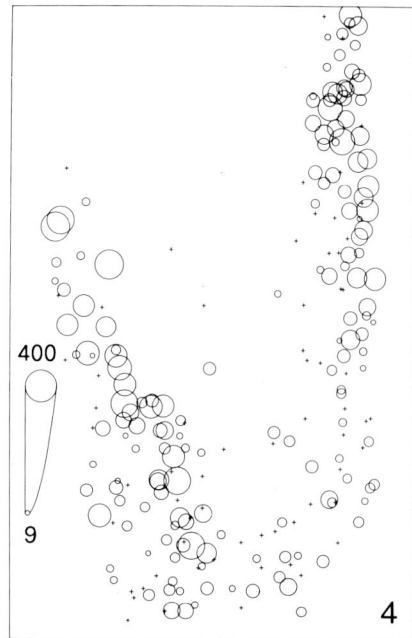

(c)

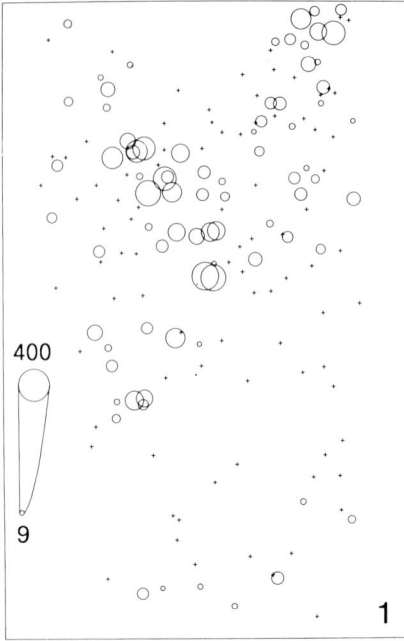

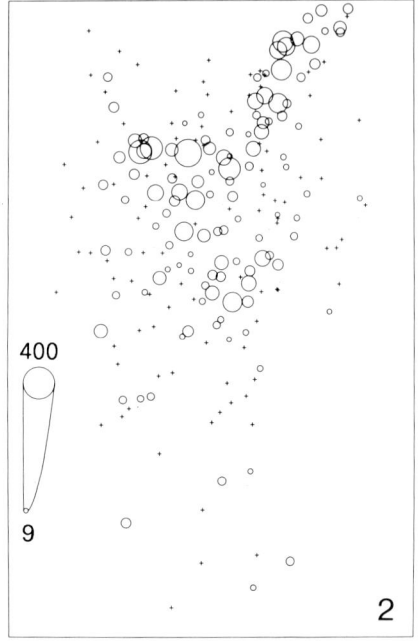

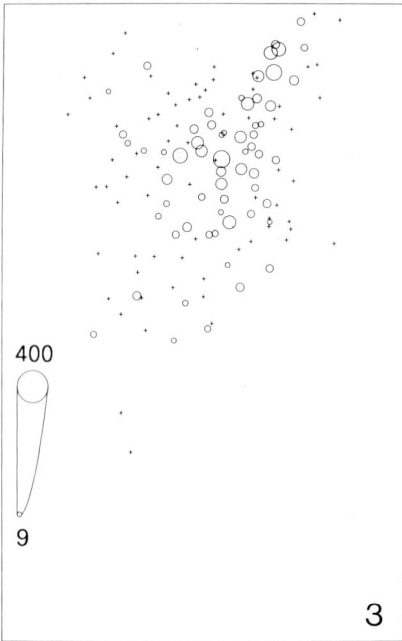

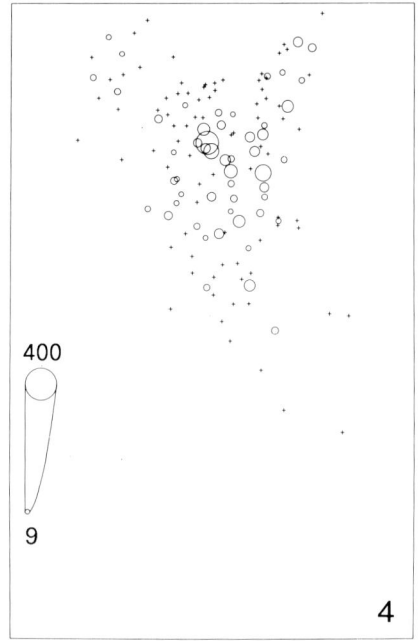

more detailed map depicting 25 community types in front of Nooksack Glacier, Washington, the units of the latter being defined by a form of cluster analysis. The earliest detailed map from the Alps appears to be that of Friedel (1938) covering the foreland of the Hintereisferners. His map depicts three vegetation zones corresponding to successional stages together with certain non-zonal vegetation types. Several workers have since produced more elaborate maps covering at least major parts of glacier forelands in the Alps (Friedel, 1956; Richard, 1968; Jochimsen, 1970, 1972; Ammann, 1979; Schubiger-Bossard, 1988) and northern Norway (Elven, 1978c). These maps involve for the most part community units defined by following formal phytosociological procedures.

One of the simpler of the phytosociological maps is shown in Fig. 5.31, redrawn from the original, which is in colour at a scale of 1 : 5000 (Jochimsen, 1970). Nine plant communities and unvegetated areas are shown, which can be related to terrain age and major physical environmental features. At the same scale, and also in colour, Friedel (1956) has published a vegetation map covering a much larger area and depicting 48 different plant communities on and near the forelands of the Pasterze Gletscher and neighbouring glaciers in the Großglockner massif (Hohe Tauern) Austria. Schubiger-Bossard's (1988) map (also 1 : 5000) of the foreland of the Rhonegletscher depicts 40 plant communities (associations, subassociations and variants). These represent 15 of the 120 associations recognized for phyto-ecological mapping in Switzerland. The largest scale map is that of Ammann (1979) showing part of the foreland of the Oberaargletscher at a scale of about 1:1370. This map is extremely detailed. It represents, for example, narrow zones of distinctive vegetation associated with moraine ridges and stream banks, and individual thickets of seven different species of *Salix*, in addition to the more common community types.

Despite the subjectivity of the methods involved in defining the communities and delimiting their boundaries, these phytosociological maps are extremely valuable. They substantiate the view that vegetation patterns on the glacier foreland are not explicable solely in terms of terrain age. They provide, moreover, strong circumstantial evidence of the environmental factors likely to be important controls on succession. They also provide the basis for reconstructing various alternative successional pathways within the vegetation landscape (section 5.3.3).

5.3.2 Quantitative community analysis at Storbreen, Jotunheimen

In order to obtain a more objective description of within-foreland community distribution patterns, various numerical classification and

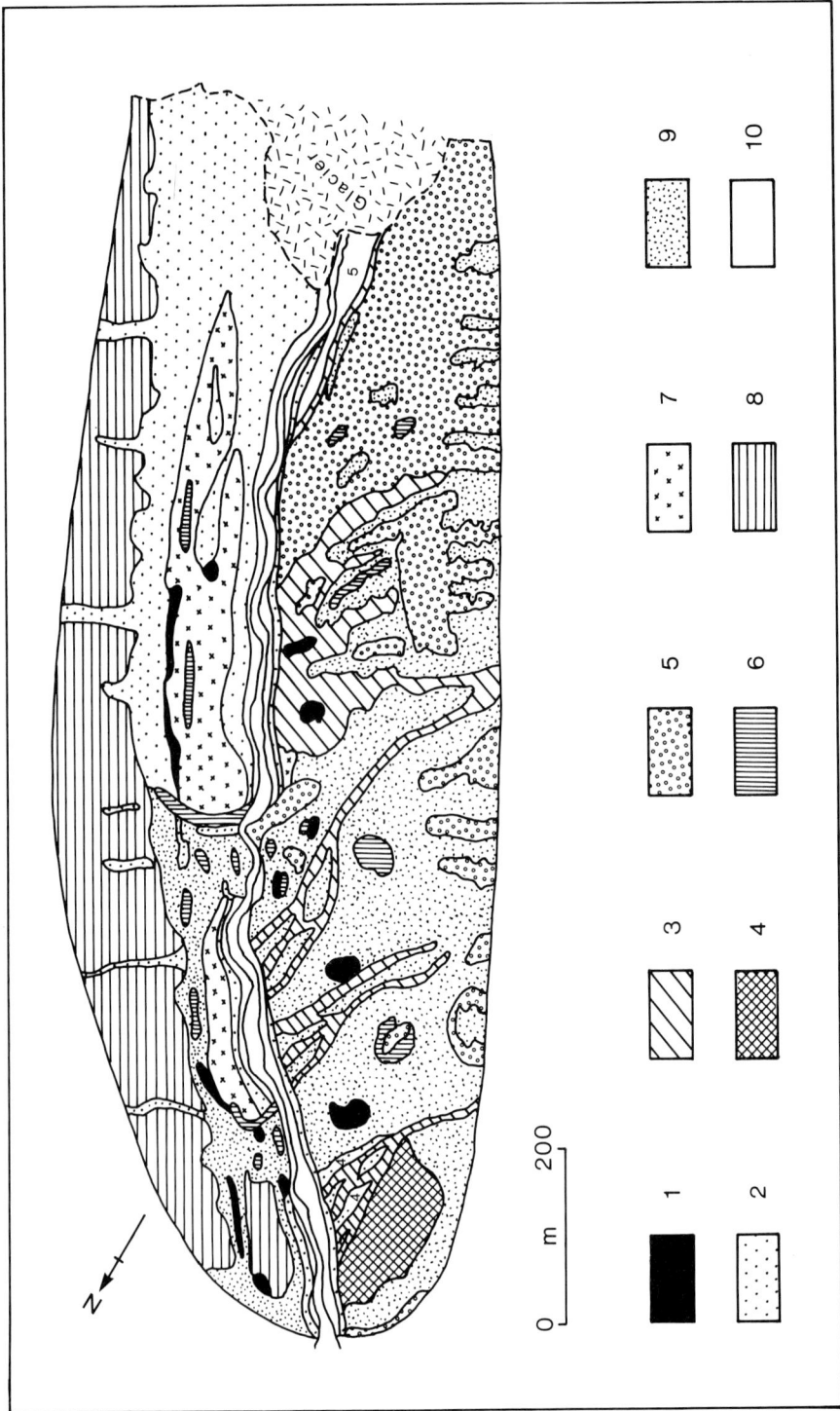

Fig. 5.31. Phytosociological map of plant communities, Rotmoosferners, Austrian Alps: (1) *Salix herbacea*; (2) *Saxifraga aizoides*; (3) *Pohlia gracilis*; (4) *Festuca violacea*; (5) *Cerastium uniflorum*; (6) *Rhacomitrium canescens*; (7) *Trifolium pallescens*; (8) *Elyna myosuroides*; (9) *Poa alpina*; (10) unvegetated (from a coloured map at a scale of 1 : 50 000 in Jochimsen, 1970).

ordination procedures have been applied to data from the Storbreen glacier foreland as a preliminary to mapping community types and community gradients (Matthews, 1979a,b; Crouch, 1991). Three classification techniques are applied by Matthews (1979a): first a simple procedure based on the most frequently occurring (dominant) species; second, divisive information analysis (DIVINF), a monothetic, divisive, hierarchical technique based on the presence-absence of species (Lance & Williams, 1968); third, minimum variance agglomeration (AGVAR), a polythetic, agglomerative, hierarchical technique employing interval scale data (Ward, 1963; Wishart, 1969). In addition, the AGVAR types were subjected to an iterative reallocation procedure (RELOC), designed to check for and correct any misclassification resulting from a long series of analytical steps (Crawford & Wishart, 1967).

The resulting AGVAR/RELOC types are depicted at the 4-group level of the classification in Fig. 5.32. A pioneer *Poa alpina – Cerastium* spp. type is replaced on older terrain at relatively high altitudes and especially on north-facing slopes by a *Salix herbacea–Polygonum viviparum* snowbed type. At lower altitudes and on south-facing slopes, the results indicate a late-successional *Phyllodoce caerulea–Salix* spp. type (on terrain deglaciated for about 100 years) and a *Betula nana–Vaccinium* spp. type, confined largely to terrain beyond the glacier foreland boundary (deglaciated for much longer than 220 years). Meso-scale patterns within the vegetation are clearly dependent on altitude and aspect as well as terrain age, from which high-altitude (mid-alpine) two-stage and low-altitude (low-alpine) three-stage successions can be inferred. Similar differences in the number of inferred successional stages are apparent at the more detailed 8-group level of classification.

Maps of site scores from ordination techniques, which search for community gradients, provide an alternative approach to the mapping of community types. Indeed, the concept of a community gradient is highly appropriate in the context of glacier foreland vegetation patterns, where it might be expected that gradual zonal transitions would accompany successional change. Gradients defined by principal components analysis (PCA) have been mapped in front of Storbreen (Matthews, 1979b). Fig. 5.33 includes maps of scores on the first two principal components derived from a variance-covariance matrix. The pattern exhibited by the first component (Fig. 5.33a), which accounts for 26.6% of the variability in the data, shows highest scores on relatively old terrain at low altitudes and reflects the low-altitude succession described above in terms of community types. Likewise, the high-altitude snowbed succession underlies the pattern of the second

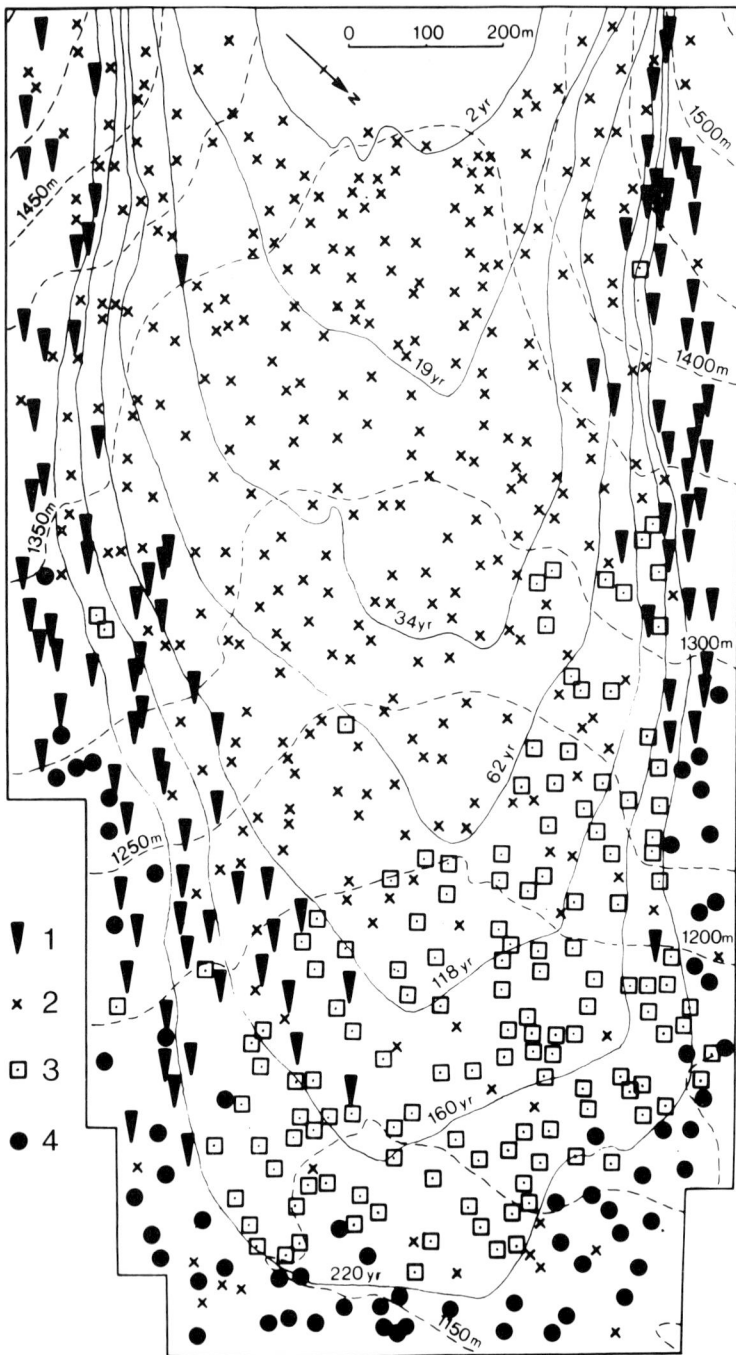

Fig. 5.32. Map of community types (AGVAR/RELOC) at the four-group level, Storbreen, Jotunheimen: (1) *Salix herbacea-Polygonum viviparum*; (2) *Poa alpinum-Cerastium* spp.; (3) *Phyllodoce caerulea-Salix* spp.; (4) *Betula nana-Vaccinium* spp. (from Matthews, 1979c).

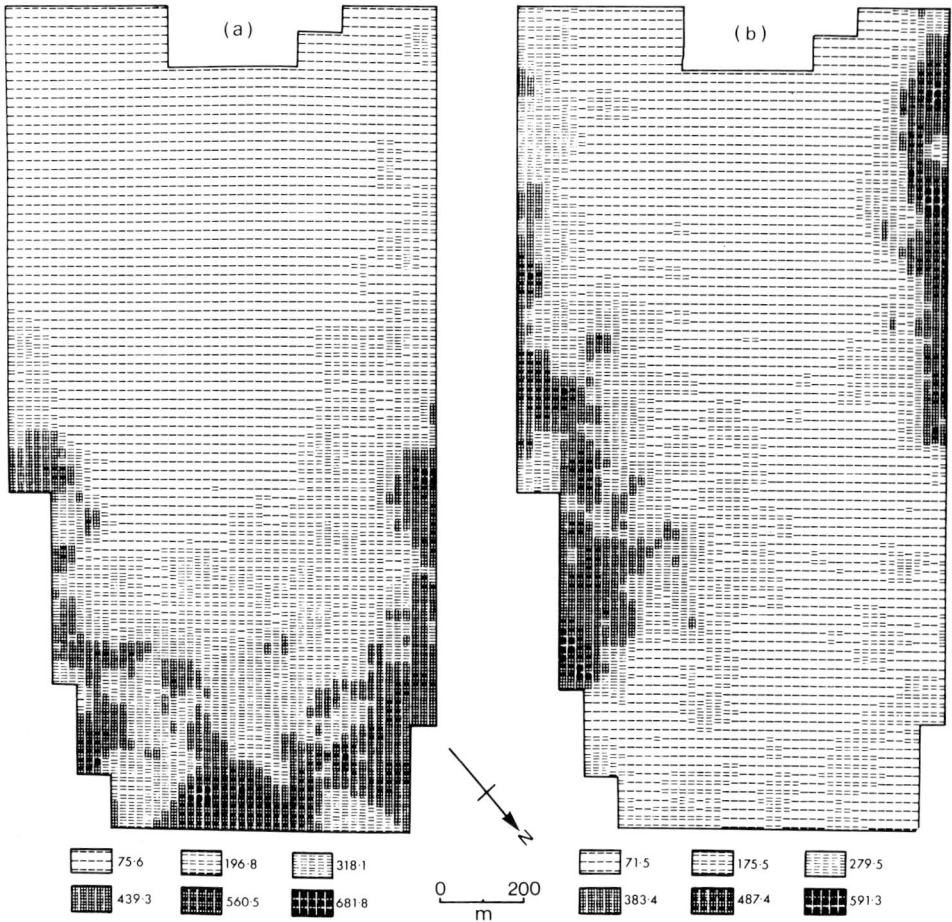

Fig. 5.33. Map of principal component (PC) scores, Storbreen, Jotunheimen: (a) first principal component; (b) second principal component. Both maps employ an equal-interval scale (from Matthews, 1979b).

component (variability accounted for, 19.4%) (Fig. 5.33b). The third component (variability accounted for, 11.0%) reflects the large differences in the vegetation landscape either side of the glacier foreland boundary. Whilst confirming the absence of clearly-defined successional stages, the community gradients are interpretable in more-or-less similar terms to the community types of Fig. 5.32.

Classification and ordination approaches are combined in Fig. 5.34, in which centroids of the community types (defined by classification) are ordinated by non-metric multidimensional scaling (NMDS) to reveal inter-type relationships in a comprehensive and objective manner (Matthews,

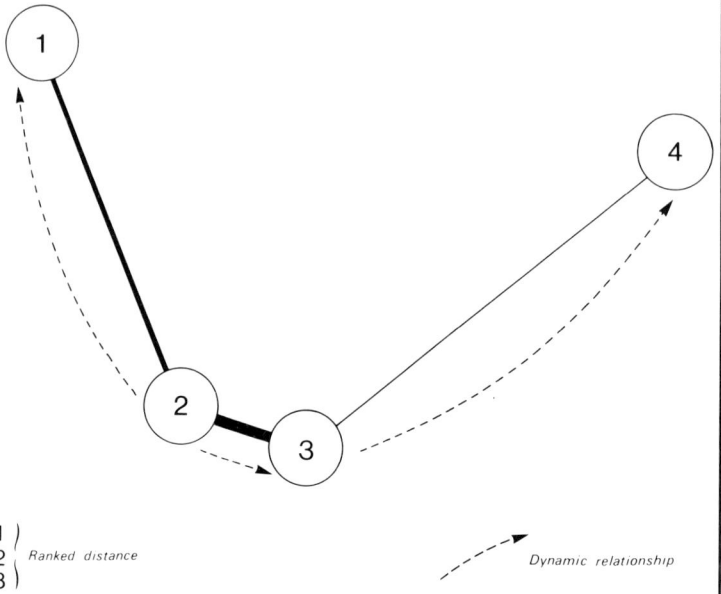

(a)

Ranked distance

Dynamic relationship

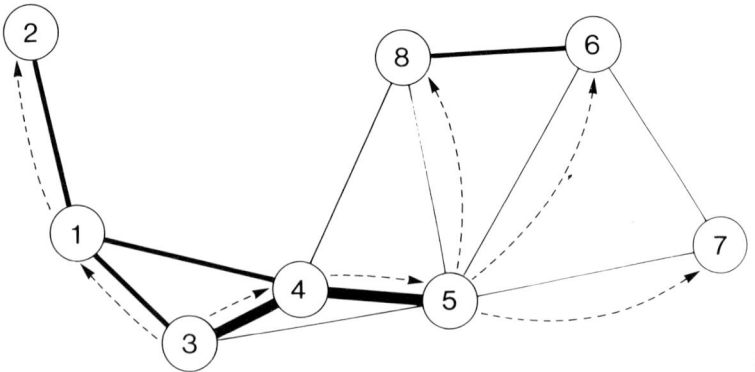

(b)

1979b). Such inter-type relationships are lost in a simple classification. In combination with the maps of the community types (Fig. 5.32) and community gradients (Fig. 5.33), from which the inferred successional and environmental relationships have been derived, the ordination diagrams (Fig. 5.34) confirm the existence of two major environmentally-conditioned successional trends. At higher altitudes, a relatively simple sequence can be conveniently described in terms of a two-stage (Fig. 5.34a) or three-stage (Fig. 5.34b) succession. At lower altitudes, a three stage (Fig. 5.34a) or four-stage (Fig. 5.34b) succession exists with up to three different pathways or trajectories. The alternative pathways also appear to be strongly conditioned by environment. In this case there is strong microtopographic control, the *Arctostaphylos alpinus–Loiseleuria procumbens, Betula nana–Juniperus communis*, and *Vaccinium myrtillus–Salix glauca* types (types 6, 7 and 8, respectively) representing a gradient within low-alpine heath vegetation from chionophobous ridge-top communities to more chionophilous communities on lower slopes and in shallow depressions (Matthews, 1979b).

Between-type differences within the vegetation landscape at Storbreen provide clear evidence of a simpler succession, with fewer stages and smaller vegetational differences between early and late successional stages, in the more severe mid-alpine environment. They also indicate successional divergence. That is, a relatively monotonous landscape in the early stages becomes a relatively heterogeneous one, with a relatively clear altitudinal zonation of vegetation and relatively distinct vegetation zones related to microtopography on older terrain. Both underlying patterns of between-type variability are reproduced using other ordination techniques, including principal components analysis (PCA) and multiple discriminant analysis (MDA) (Matthews, 1979b,c,d).

Application of MDA provides evidence to indicate that the communities of later successional stages and less severe environments are also characterized by greater within-type variability (Matthews, 1979c). Within-type variability at the 4-group and 8-group levels is summarized in Table 5.4

Fig. 5.34. Non-metric multidimensional scaling (NMDS) ordination of community types (AGVAR/RELOC group centroids), Storbreen, Jotunheimen. (a) 4-group level: (1) *Salix herbacea-Polygonum viviparum*; (2) *Poa alpina-Cerastium* spp.; (3) *Phyllodoce caerulea-Salix* spp.; (4) *Betula nana-Vaccinium* spp. (as in Fig. 5.32). (b) 8-group level: (1) *Cassiope hypnoides-Phyllodoce caerulea*; (2) *Salix herbacea-Sibbaldia procumbens*; (3) *Poa alpina-Cerastium* spp.; (4) *Salix lanata-Cassiope hypnoides*; (5) *Pinguicula vulgaris-Tofieldia pusilla*; (6) *Betula nana-Juniperus communis*; (7) *Arctostaphylos alpina-Loiseleuria procumbens*; (8) *Vaccinium myrtillus-Salix glauca* (from Matthews, 1979b).

Table 5.4. *Within-type variability of community types on the Storbreen glacier foreland, Jotunheimen (after Matthews, 1979c).*

Group no.*	No. of sites in group	Significant members* No.	%	Significant non-members No.	%
4-group level:					
1	108	49	45.4	36	33.3
2	323	237	73.4	55	17.0
3	140	94	67.1	25	17.9
4	65	22	33.9	21	32.3
8-group level:					
1	92	53	57.6	23	25.0
2	31	8	25.8	12	38.7
3	277	209	75.5	45	16.3
4	98	71	72.5	19	19.4
5	71	34	47.9	22	31.0
6	27	14	51.9	5	18.5
7	12	6	50.0	3	25.0
8	28	0	0.0	19	67.9

*For group names see Figs. 5.34a & b; see text for definition of significant members and non-members.

and, at the 4-group level, in Fig. 5.35. Here, individual sites within each AGVAR/RELOC type are positioned in relation to discriminant axes, which maximize between-type differentiation (Healy, 1965; Webster & Burrough, 1974). Fig. 5.35 is a good two-dimensional representation, because the first two discriminant axes account for 80.7% of the possible differentiation between groups. Confidence circles centred on the group centroids, are expected to enclose 95% of the sites within the respective groups. In Table 5.4, 'significant members' of each group include those sites where the probability of belonging to the group is $p > 0.95$; 'significant non-members' are those sites where the probability of belonging to the group is $p < 0.05$. Tight groups with little within-site variability, have a large number of significant members and a relatively small number of significant non-members.

It is clear that the *Poa alpina–Cerastium* spp. pioneer community type (group 2), with over 73% significant members is the tightest cluster and exhibits least within-type variability. The *Phyllodoce caerulea–Salix* spp. type (group 3), with significant members about 6% less numerous and about the same proportion of significant non-members (17–18%), exhibits

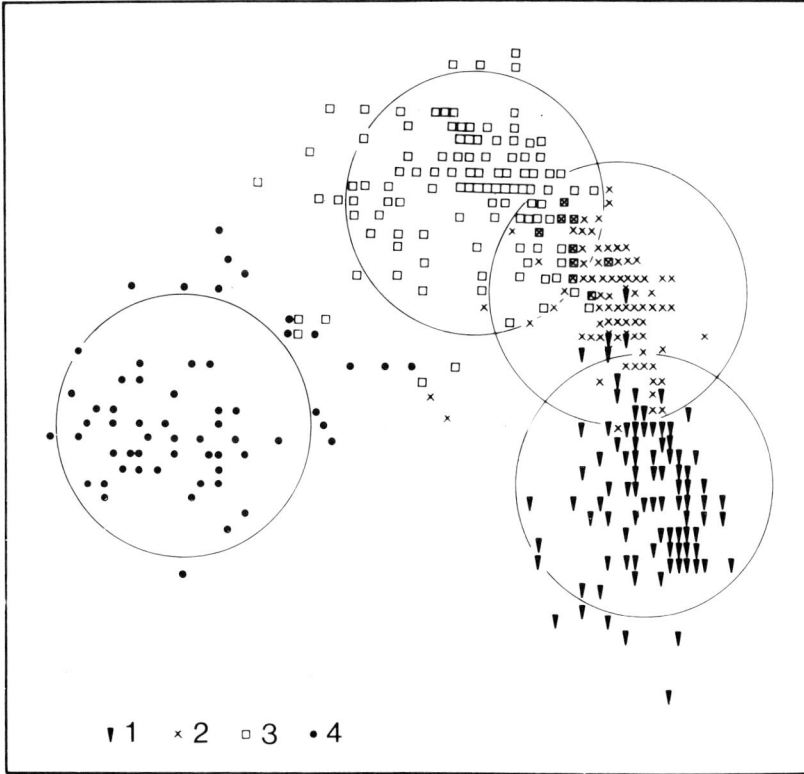

Fig. 5.35. Multiple discriminant analysis (MDA) ordination of community types (AGVAR/RELOC groups as in Fig. 5.32): (1) *Salix herbacea-Polygonum viviparum*; (2) *Poa alpina-Cerastium* spp.; (3) *Phyllodoce caerulea-Salix* spp.; (4) *Betula nana-Vaccinium* spp. 95% confidence circles are shown (from Matthews, 1979c).

slightly more within-type variability. There are substantially fewer significant members in the remaining two groups, which are characteristic of relatively old terrain. At low altitudes, the *Betula nana–Vaccinium* spp. type (group 4) has lowest within-type variability with about 34% significant members and almost the same proportion of significant non-members. Loose clusters, with relatively high within-type variability, therefore appear to be characteristic of the less severe, lower-altitude environment and, especially, the later successional stages. Similar patterns of within-type variability in respect of successional stage and environmental severity are apparent at the 8-group level (Table 5.4).

The overall pattern of successional pathways at Storbreen includes, therefore, divergence in the sense of both between-type and within-type variability. A relatively distinct vegetational zonation develops in response to altitude, aspect and microtopography. The discreteness of types is dependent on relationships involving within-type and between-type variability. It appears that the discreteness of types, together with both aspects of variability, tends to increase during succession and from high to low altitudes. However, the unity of the early successional stages owes much to low within-type variability, whereas the unity of the later stages tends to be more dependent on relatively great between-type variability (Matthews, 1979c).

5.3.3   Inferred successional pathways elsewhere

Comprehensive, objective analyses of multiple successional pathways of the type described above from the Storbreen glacier foreland have not been attempted elsewhere. Nevertheless, many investigators have considered such pathways explicitly, basing their conclusions largely on perceived spatial variations within chronosequences. The remainder of this section is concerned with the nature and interrelationships of the pathways that have been inferred from the most intensively investigated chronosequences.

Based on a combination of numerical and phytosociological classifications, strongly divergent successional pathways have been inferred by Elven (1974) on the glacier forelands of Hardangerjökulen and Omnsbreen in the alpine zone of southern Norway. The pioneer vegetation here covers about 30% of the landscape and is 'completely independent of differences in snow cover and soil structure' (Elven, 1975: 383). On older terrain, moisture, wind and snow cover are important controlling environmental factors. On older terrain in front of Blåisen, an outlet glacier of Hardangerjökulen, species groups and stand groups defined by association analysis can be interpreted largely in terms of the complex edaphic, microclimatic and hydrologic effects of snow (Elven, 1978b).

On till and glacio-fluvial sequences in front of Flatisen and Østerdalsisen, outlet glaciers of the Svartisen icecaps, northern Norway, Elven (1978c) has proposed similar patterns. At Flatisen, the youngest till surfaces at about 300 m above sea level are characterized by an open herbaceous (*Oxyria digyna*-grass) pioneer community which diverges on older terrain eventually to form a variety of sub-alpine communities (Fig. 5.36). After about 40 years, an *Oxyria–Stereocaulon* community is found on coarse, dry substrates, whilst an *Oxyria–Rhacomitrium* community occupies areas with finer texture and more moisture. Further differentia-

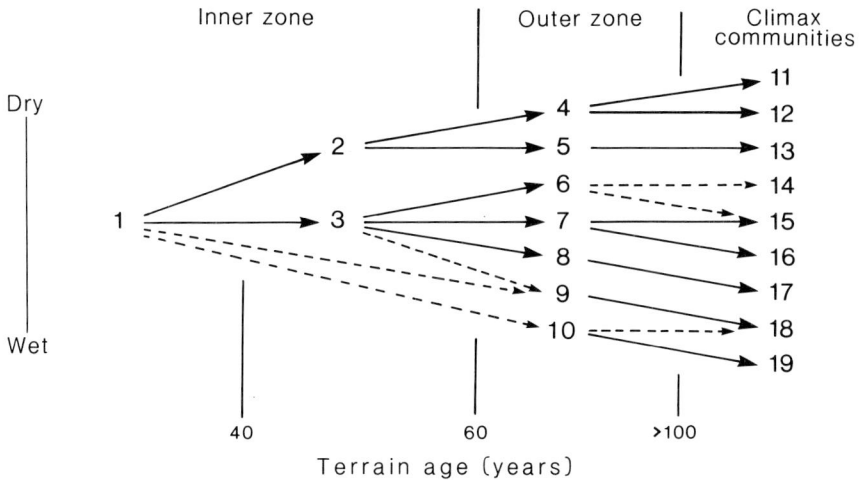

Fig. 5.36. Successional pathways, Flatisen (Svartisen), northern Norway, in relation to moisture and time. Uncertain pathways are indicated by broken lines. Communities: 1, *Oxyria digyna*-grass pioneer ground; 2, *Oxyria-Stereocaulon* pioneer ground; 3, *Oxyria-Rhacomitrium* pioneer ground; 4, *Vaccinium*-heath initials; 5, Fern snowbed initials; 6, *Dryas*-heath initials; 7, Tall-herb meadow initials; 8, *Salix herbacea*-snowbed initials; 9, Damp meadow initials; 10, Rich fen initials; 11, *Vaccinium* heath; 12, *Vaccinium-Betula* woodland; 13, Fern snowbeds; 14, *Dryas* heath; 15, Tall-herb meadow; 16, Tall-herb *Betula* woodland; 17, *Salix herbacea* snowbeds; 18, Intermediate fen; 19, Rich fen (from Elven, 1978c).

tion occurs when, after about 60 years, the initial stages of heath, tall-herb meadow, snowbed, mire and birch (*Betula pubescens*) woodland communities can be identified. Drainage and snow conditions appear to be the most important controls, mature ('climax') communities of these types requiring > 100 years to develop.

There are many similarities between the pathways at Flatisen and those at lower altitudes in front of Østerdalsisen (Elven, 1978c). However, fewer mature communities are recognized and all of these are types of birch woodland or mire. Certain heath communities in which, for example, *Empetrum hermaphroditum* and *Vaccinium uliginosum* or *V. myrtillus* are important species, are recognizable in the intermediate stages but, unlike at Flatisen, these do not survive to the later stages.

In the sub-alpine zone of southern Norway, generally convergent successional pathways are identified by Fægri (1933) in front of four Jostedalsbreen outlet glaciers. His scheme for the glacier foreland of Nigardsbreen is shown in Fig. 5.37. A relatively large number of predomi-

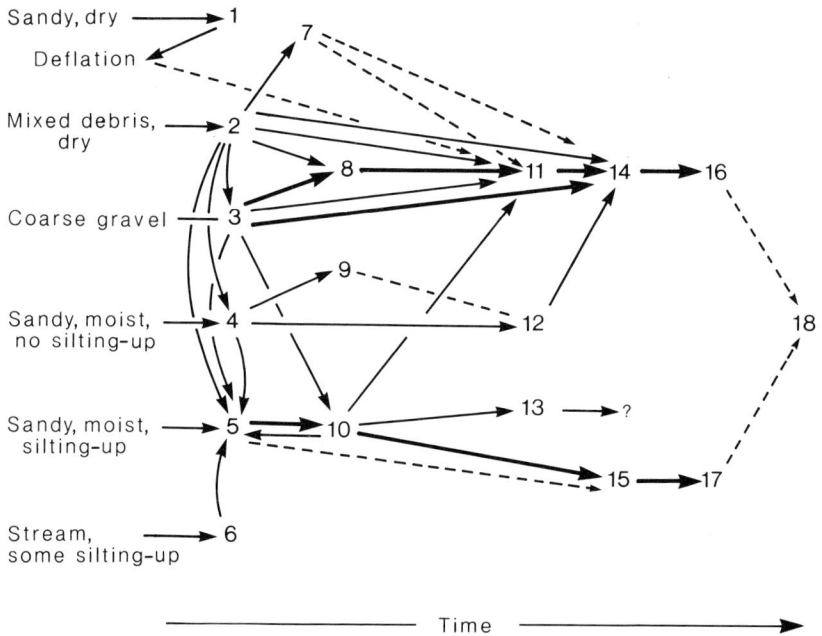

Fig. 5.37. Successional pathways, Nigardsbreen (Jostedalsbreen), southern Norway, in relation to substrate characteristics and time. Larger arrows represent the main series; broken lines indicate uncertain pathways. Communities: 1, *Polytrichum piliferum*; 2, Black liverwort association I; 3, *Rhacomitrium* association I; 4, Black liverwort association II; 5, *Pohlia-Philonotis*; 6, *Calliergon sarmentosum*; 7, *Stereocaulon rivulorum*; 8, Stony heath; 9, *Scirpus-Trichophorum*-liverwort association; 10, *Rhacomitrium-Polytrichum* association II; 11, Dwarf-shrub heath; 12, *Trichophorum* variant; 13, *Nardus*-dwarf shrub heath; 14, Dwarf shrub-*Betula* thicket; 15, *Salix* thicket; 16, Grass-*Vaccinium myrtillus-Betula* woodland; 17, Herb-rich *Betula* woodland; 18, Climax (from Fægri, 1933).

nantly cryptogamic pioneer communities is distinguished on the youngest terrain, which largely correspond with differences in substrate texture, moisture and the extent to which silting-up occurs from glacio-fluvial meltwater streams. Many of these pioneer communities do not differ greatly from each other and possible interchangeability is indicated. According to Fægri (1933: 224) the initial associates at all four forelands are similar and are centred around the '*Rhacomitrium* association I'. Considerable differentiation appears to occur in the intermediate successional stages. Little indication of convergent pathways is indicated in Fig. 5.37 until the emergence of two major communities on relatively old terrain, both dominated by *Betula pubescens*.

On the outermost moraine ridge in front of Bödalsbreen, a northern outlet of Jostedalsbreen, Vetaas (1986) also recognizes two birch-dominated communities. Crest and proximal slopes are characterized by a *Betula–Empetrum* woodland whereas, on distal slopes, a *Betula–Vaccinium* woodland occurs. On progressively older moraine ridges, birch scrub is described as expanding up moraine slopes (Vetaas, 1986: 140), covering the whole moraine with woodland within about 230 years. Previous stages and rates of succession vary according to microtopography (Fig. 5.38). The precise pathway depends on the interaction of glacier wind and slope position effects (see also section 4.3.2).

The evidence from Norwegian glacier forelands is consistent with the concept of successional pathways that are strongly divergent in the alpine zone, but constrained or at least partially convergent in the sub-alpine zone. Only in the later, tree-dominated stages of the latter does convergence become the dominant tendency. Even here, major differences in the shrub layer persist on the oldest moraines and seemingly into the more mature communities beyond the glacier foreland boundary.

Additional evidence for convergence in association with tall shrubs and trees is provided by Birks (1980a) from Yukon Territory, Canada. Although there are vegetational differences associated with drained lakes, fens, tephra hills and drainage channels, and complications due to the melting of underlying ice, the vegetation of the Klutlan moraines largely represents a linear successional sequence (Fig. 5.39). The spatial zonation results from a unidirectional vegetational succession progressing from the youngest to the oldest moraines (Birks, 1980a: 79).

The initial pioneer *Crepis nana* stage, develops on flat or gently sloping terrain to *Salix–Shepherdia* scrub via *Dryas drummondii* mats. On steeper ground ($\geqslant 5°$) the successional pathway is similar except that the pioneer nodum is replaced by the *Hedysarum mackenzii* nodum, which develops into the *Hedysarum–Salix* nodum and thence into *Salix–Shepherdia* scrub. On both flat and sloping ($\leqslant 10°$) ground, *Shepherdia canadensis*, *Salix glauca*, *S. arbusculoides*, *S. alaxensis* and *Populus balsamifera* form the 1–2 m tall shrub-dominated vegetation where vegetation has been developing for at least 30 to 60 years. With increasing terrain age, *Picea glauca* becomes more prominent, first overtopping the dense scrub vegetation on areas that have been vegetated for at least 60 to 80 years. With further increases in terrain age, the successional sequence proceeds with the spruce trees becoming increasingly large and dense. Thus, as appears to be the case at Glacier Bay, where the relevant species are *Salix* spp., *Alnus crispa* and *Picea sitchensis*, relatively rapid convergence of successional pathways occurs at stages dominated by dense tall shrubs and coniferous trees. At

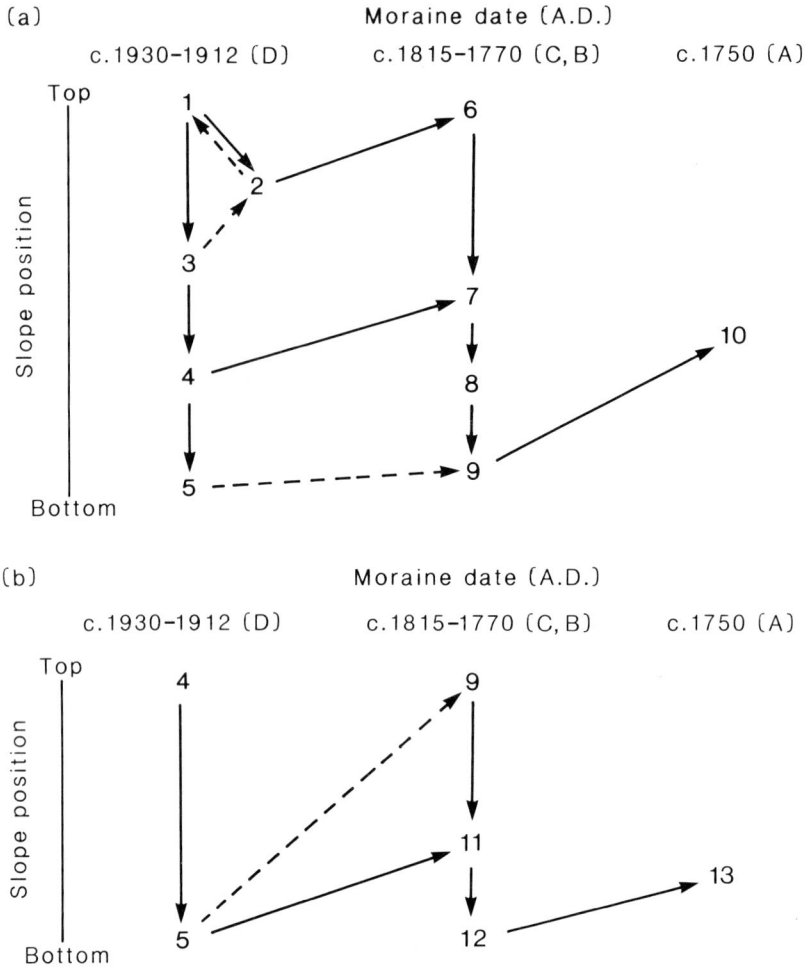

Fig. 5.38. Successional pathways on moraines (A–D), Bödalsbreen (Jostedals-breen), southern Norway, in relation to slope position and moraine age: (a) moraine crests and proximal slopes; (b) distal slopes. Broken lines indicate uncertain pathways. Communities: 1, eroded ground; 2, *Rhacomitrium lanugino-sum-Cetraria nivalis* mat; 3, *Stereocaulon-R. canescens*; 4, *Calluna-Empetrum* dwarf-shrub heath (sub-type with *R. canescens*); 5, *Betula-Salix* thicket; 6, *Rha-comitrium lanuginosum-Cetraria nivalis* mat; 7, *Calluna-Empetrum* dwarf-shrub heath (sub-type with *R. lanuginosum*); 8, *Betula-Empetrum* thicket (variant with *R. lanuginosum*); 9, *Betula-Empetrum* thicket; 10, sub-alpine *Betula-Empetrum* woodland; 11, *Betula-Vaccinium myrtillus* thicket; 12, *Betula-V. myrtillus* thicket (variant with trees); 13, Sub-alpine *Betula-Vaccinium* woodland (from Vetaas, 1986).

Bare ground

↓

Crepis nana ○
2–6

↓

Hedysarum mackenzii △
10–20

↓

Hedysarum-Salix ▲
24–30

Dryas drummondii •
9–23

↓

Salix-Shepherdia ✳
32–58

↓

Picea-Salix ■
58–80

↓

Picea-Arctostaphylos ×
96–178

↓

Picea-Ledum +
177–240

↓

Picea-Rhytidium ▽
>163–>339

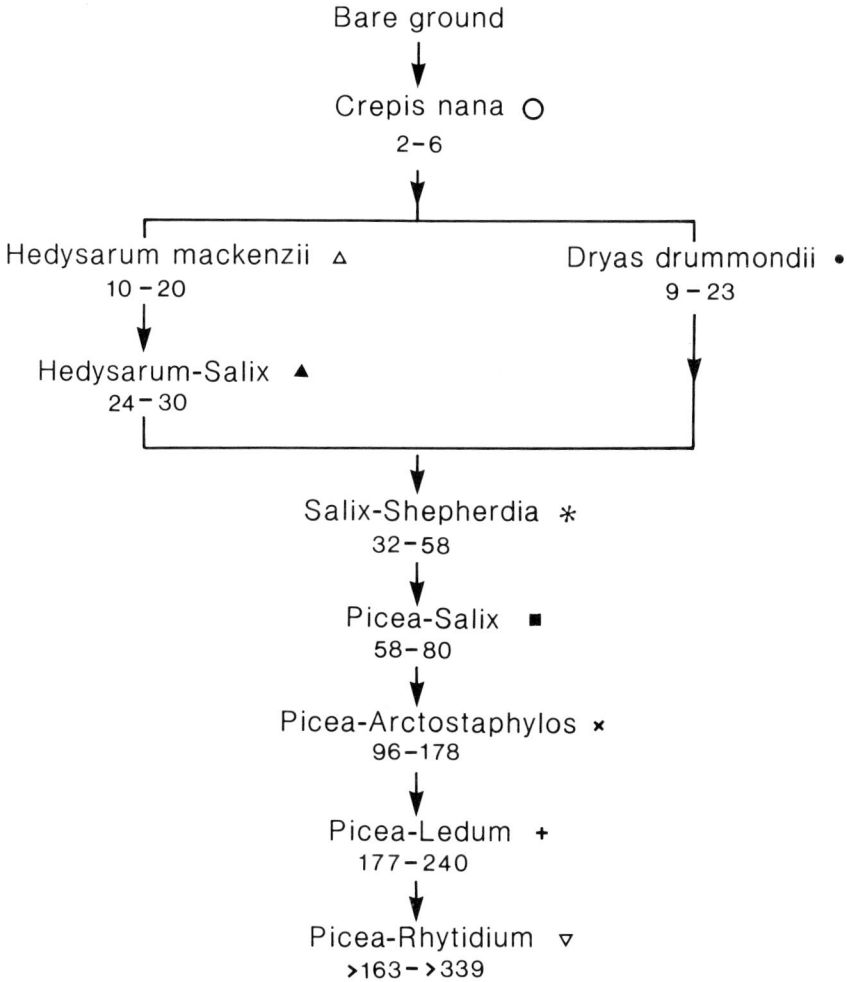

Fig. 5.39. Successional pathways, Klutlan Glacier, Alaska. Symbols denote communities (noda) as in Figs. 5.23 & 5.25. Noda age is given in years (from Birks, 1980a).

Glacier Bay, Cooper (1923b) points out that even bedrock surfaces are eventually covered by spruce trees growing on an organic mat.

A very different pattern of parallel successional pathways, with little indication of either convergence or divergence, has been inferred from the foreland of the Rhonegletscher (Fig. 5.40). Although not all pathways are found on all terrain ages, parallel development appears to be characteristic for at least 350 years (Schubiger-Bossard, 1988). Each pathway is deter-

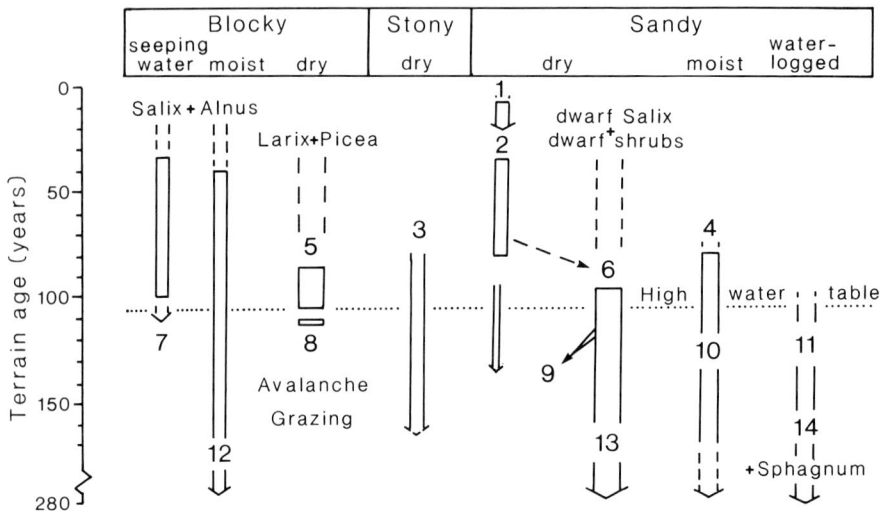

Fig. 5.40. Successional pathways, Rhonegletscher, Swiss Alps, in relation to substrate characteristics and terrain age. Additional factors are indicated in the proximal part of the foreland. Communities: 1, Oxyrietum; 2, Epilobietum fleischeri, subassociation rhacomitrietosum, *Poa nemoralis* variety; 3, as (2), *Cladonia* variety; 4, as (2), *Poa alpina* variety; 5, Dwarf-shrub heath, *Calluna-Juniperus-Rhododendron* ferrugineum, *Salix helvetica* variety; 6, as (5), *Hieracium pilosella* variety; 7, *S. daphnoides-S. pentandrae*; 8, Young *Larix* woodland; 9, *Nardus stricta*; 10, *P. alpina-S. retusa*; 11, Caricetum fuscae, *S. hastata* variety; 12, Alnetum viridis; 13, Rhododendro-Vaccinietum initials; 14, as (11), *Philonotis seriata* variety (from Schubiger-Bossard, 1988).

mined by local habitat conditions, particularly substrate texture and moisture. Some sites in the area are developing towards an open forest of *Larix decidua* with *Picea abies* and a Rhododendro–Vaccinietum understorey. Growth of trees is, however, considerably impaired by avalanches, grazing cattle, and high water tables on the distal part of the foreland. Nevertheless, *Larix* trees up to 10 m high are found on terrain deglaciated for 85–95 years. Other pathways are developing towards willow scrub (dominated by *Salix daphnoides* and *S. pentandra*), *Alnus* woodland, lichen-dominated communities, *Sphagnum* mosses, dwarf-shrub heaths and meadows.

Rapid, strong divergence, followed by very slow convergence is the general tendency suggested by a complex pattern of successional pathways in front of Pasterze Gletscher in the Alps (Fig. 5.41). Four major pathways are depicted (A–D), each consisting of nine stages which correspond with nine zones (1–9) on the glacier foreland (Zollitsch, 1969). Pathways A to D represent successions characteristic of increasingly moist, sheltered and undisturbed sites; zones 1 to 9 represent terrain ages of, respectively, < 5, 5–10, 12–30, 35–45, 80, 85–100, 350, 3400 and 6200 years. The changing pattern of cover and species richness associated with the four pathways is shown in Figs. 5.10, 5.14, and 5.20.

The pattern of successional pathways inferred by Zollitsch (1969) is particularly interesting because of the importance attached to changing environmental conditions, which form an integral part of his interpretation (Fig. 5.41). Newly-deglaciated terrain supports the initial stage of what becomes, after 5 to 10 years, a fully developed herbaceous pioneer community (Saxifragetum biflorae). This early development occurs under uniformly very moist and highly disturbed conditions. Three pathways are first distinguishable between 12 and 30 years, whilst four are recognizable after 35 to 40 years.

Pathway A is followed at disturbed sites that dry out in exposed situations, developing into a *Salix serpyllifolia–Dryas octopetala* community within 85 to 100 years and then more slowly towards a lichen-rich heath (Loiseleurio–Cetrarietum). Pathway D occurs in very wet to moist sites that are sheltered and only disturbed by running water. A moss-rich herbaceous community (Cratoneuro–Arabidetum) is characteristic in the early stages until at least 35–40 years. After 85 to 100 years this develops into a willow thicket dominated by *Salix waldsteiniana* and eventually heathland dominated by *Rhododendron ferrugineum* (Rhododendro–Vaccinietum).

Pathways B and C are intermediate in terms of environmental conditions: pathway B involves sites that are exposed, moderately moist and disturbed at the surface in the early stages, becoming more stable and drier; sites in pathway C are sheltered, stable and moderately moist. These pathways are less clearly defined in that after initial divergence they exhibit convergence at an intermediate stage and eventually diverge. They differentiate after 35–40 years from an initial stage of a grass-rich Trisetum spicati association into two subassociations, which then converge (before 85–100 years) into a *Kobresia–Festuca* grass community. This in turn develops more slowly into a *Kobresia–Festuca–Helictotrichon* grassland after about 350 years.

It would seem, therefore, that early, very rapid divergence is inextricably linked to changing site conditions, whereas rapid convergence of pathways B and C before about 85–100 years is closely associated with the development of a grass sward. Later slow divergence of these two pathways towards the same mature heath types as those characteristic of the final stages of pathways A and D appear to reflect the re-assertion of environ-

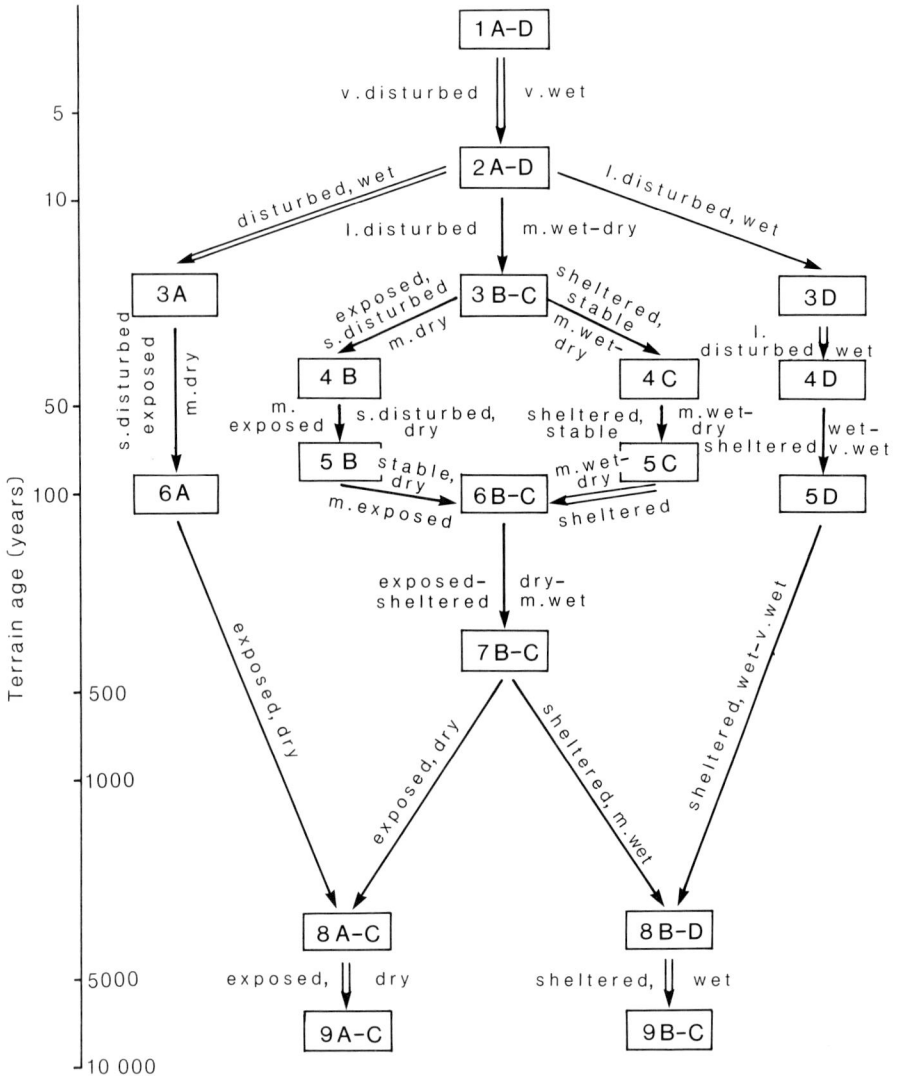

Fig. 5.41. Successional pathways, Pasterze Gletscher, Austrian Alps, in relation to habitat type (A–D), habitat change and terrain age. Abbreviations: *v*, very; *m*, moderately; *l*, little; *s*. disturbed, disturbed only at the surface. Open arrows indicate the same community. Communities: 1A–D, 2A–D, 3A, Saxifragetum biflorae; 3BC, Trisetum spicati (initial stage); 3D, 4D, Cratoneuro-Arabidetum; 4B, Trisetum spicati, subassociation typicum; 4C, Trisetum spicati subassociation kobresietosum myosuroidis; 5B, *Kobresia-Festuca* grassland (open); 5C, 6BC, *Kobresia-Festuca* grassland (closed); 6A *Salix serpyllifolia-Dryas octopetala*; 6D, *Salix* thicket; 7BC, *Kobresia-Festuca-Helictotrichon* grassland; 8A–C, 9A–C, Loiseleurio-Cetrarietum; 8B–D, 9B–D, Rhododendro-Vaccinietum (from Zollitsch, 1969).

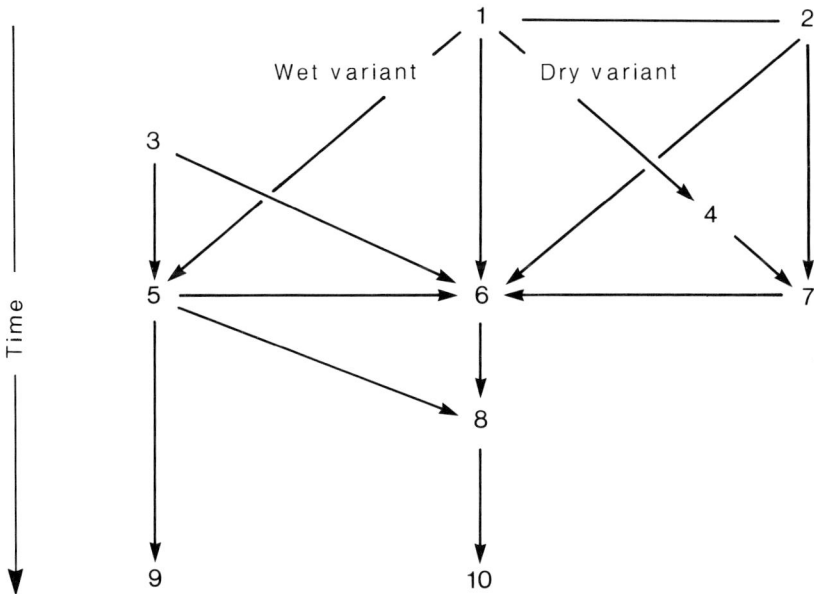

Fig. 5.42. Successional pathways, Rotmoosferners and Gaisbergferners, Austrian Alps. Communities: 1, *Saxifraga aizoides*; 2, *Cerastium uniflorum*; 3, *Pohlia gracilis*; 4, *Trifolium pallescens*; 5, *Salix herbacea*; 6, *Poa alpina*; 7, *Rhacomitrium canescens*; 8, *Elyna myosuroides*; 9, Nardetum; 10, Elynetum (from Jochimsen, 1970).

mental control. Overall, between about 350 years and 3400 years, there is nevertheless a slow net convergence from three pathways to two relatively distinct heath communities. It is possible, therefore, for any particular site to follow alternative pathways depending on changing combinations of both biological and physical environmental factors. Only at those sites characterized by the most extreme physical conditions is there a relatively simple successional sequence of communities without the possibility of alternative pathways.

Successional pathways of comparable complexity are proposed by Jochimsen (1963, 1970) from the glacier forelands of Rotmoosferners and Gaisbergferners (Fig. 5.42). Based on maps (e.g. Fig. 5.31) and on observation of permanent quadrats, she stresses in particular the importance of water relations in controlling rates of succession and precise pathways, greater independence from physical environmental factors being apparent when vegetation cover approaches 100%. Initially, there is again comparatively little variation within a *Saxifraga aizoides* community, found near streams and on moist slopes, and a *Cerastium uniflorum* community of quite similar species composition on drier sites and coarse till. A *Pohlia gracilis* community develops slightly later as a specialized pioneer community on fine sandy substrates inundated by running water.

The *Poa alpina* community develops on a wide variety of sites, from any of the pioneer communities or from other secondary communities if site conditions change. For example, under extremely dry conditions, the *Cerastium uniflorum* community develops into a *Rhacomitrium canescens* community which, if conditions remain dry, may undergo little further development. The latter can also develop from the *Saxifraga aizoides* community via a *Trifolium pallescens* community if the site dries out. Under improving moisture conditions however, the *R. canescens* community converges on the *P. alpina* community. A wide variety of convergent or divergent pathways are therefore possible in the early stages, driven in particular by changes in the moisture regime. The general trend in front of Rotmoosferners is for initial divergence, followed by convergence towards the *Poa alpina* community and *Elyna myosuroides* meadow.

In concluding this section, there appears to be a widespread tendency for divergent successional pathways associated with early stages and relatively severe physical environments. Additional support for this view comes from New Zealand, where Archer (1973) and Wardle (1980) indicate divergence followed by later convergence; from Mount Kenya, where Spence (1989) and Harmsen, Spence & Mahaney (1990) suggest divergence in the early stages; and from Svalbard, where Kuc (1964) recognizes divergence from the pioneer stage towards three different tundra communities. Divergence seems to be favoured by the existence of relatively strong physical environmental controls, convergence by relatively strong biotic controls.

5.3.4   Between-foreland patterns: a comparative approach

Similarities and differences in vegetation patterns and successional sequences from neighbouring forelands were noted in early studies of the ecology of recently-deglaciated terrain. Coaz (1887) made a brief comparison of species composition in the early stages of succession in front of the Rhone Gletscher with those of four other Alpine forelands. Davy de Virville (1929) compared and contrasted the floras associated with the Glacier du Mont-Aigu and Glacier du Péguère in the Pyrenees. Cooper (1923b) pointed out the great difference in rates of succession between the two main branches of Glacier Bay. Fægri (1933) not only discussed in considerable detail successional pathways on the forelands of four outlet glaciers of the Jostedalsbreen ice cap but also compared them with sites in the Alps and

Alaska. However, comparisons between forelands, even in the same region, have tended to be a secondary consideration in most investigations.

Factors of both the physical and biological environment lead to differences in successional pathways between forelands, many of which will be considered in detail in sections 5.4 and 6.1. Here, attention focuses on major variations in the pattern and rate of succession that have been detected at a regional scale.

In Scandinavia, the position of the foreland with respect to altitudinal zone appears to be the most important factor. Sub-alpine successions (Fægri, 1933; Elven, 1978c; Vetaas, 1986) proceed relatively rapidly towards mature woodland whilst successions in the alpine zone proceed towards heath and snowbed communities (Stork, 1963; Elven, 1975, 1978b; Matthews, 1978a, 1979b). Within the sub-alpine zone, oceanicity appears to be of paramount importance, the succession at Åbrekkebreen (Brenndalsbreen) to *Alnus incana* woodland being the most rapid and that to *Betula pubescens* woodland at Nigardsbreen being slowest (Fægri, 1933). At Åbrekkebreen, where the vegetative period is about one month longer, the pioneer communities, which contain fewer alpine species, have been replaced within 20 years and *Alnus* is completely dominant within 70 years. At Nigardsbreen (Figs. 5.11 & 5.37), pioneer communities are still found on terrain deglaciated for 40 years and mature *Betula* woodland requires 180 years to develop. In northern Norway, largely similar successions occur from pioneer communities, via *Rhacomitrium-* and *Stereocaulon-*rich cryptogamic types, to heath initials and *Betula* woodland. According to Elven (1978b) the biggest difference between the succession on the moraines at Flatisen (Fig. 5.36) and Nigardsbreen (Fig. 5.37) is the slower development of the former towards *Betula* woodland.

Within the Scandinavian alpine zone, successions are slower and shorter in the sense of fewer stages and smaller vegetational differences between pioneer and mature stages. Pioneer communities have the same physiognomy and some of the same species as pioneer stages in the sub-alpine zone, but most communities have not reached a mature state after 220 years (Elven, 1975). Successions generally simplify from low- to mid- to high-alpine belts, and mature communities at higher altitudes often resemble earlier successional stages at lower altitudes (Matthews, 1979a,b).

In the low-alpine belt, dwarf-shrub heaths, lichen heaths, meadows and mires are amongst the mature communities represented. In the mid-alpine zone, the growing season is too short for shrubs, except for *Salix herbacea* which is important in late-snowbed communities. Herbaceous perennials and cryptogams are more important in mature communities of the mid-alpine belt but there are fewer communities than in the low-alpine belt. In the high-alpine belt of northern Sweden, in front of Tarfalaglaciär, Stork (1963: 10) describes the vegetation as true moss and lichen desert with only small patches of hardy phanerogams in sheltered sites. Similar extreme situations exist in eastern Jotunheimen, southern Norway, where the predominantly herbaceous and cryptogamic vegetation is sparse on moraines deglaciated for at least 230 years, and there is little or no difference in species composition between surfaces of differing age (Matthews, unpublished).

A large number of studies from the Alps reveal broadly similar regional variations in the pattern and rate of succession (e.g. Friedel, 1938, 1956; Lüdi, 1945, 1958; Jochimsen, 1963, 1970; Richard, 1968, 1973; Zollitsch, 1969; Ammann, 1975, 1979; Burtscher, 1982; Burga, 1987; Ellenberg, 1988; Schubiger-Bossard, 1988). Successional communities in the Alps are comparatively rich in species. Although this richness decreases with increasing altitude, it is also highly dependent on substrate composition, calcareous material leading to particularly rich floras. Thus, in front of Hüfigletscher, Lüdi (1958) reported 107 species of flowering plants and pteridophytes (and 10 mosses) on limestone-rich terrain deglaciated for only 12–17 years. Only 86 species were reported by Zollitsch (1969) from the silicious foreland of the Ödenwinkelkees, whereas 170 were found in front of the Pasterze Gletscher where limestone is present. Perhaps the most floristically-rich foreland is that of the Rhonegletscher where 382 higher plant species have been recorded, which represents > 13% of the Swiss flora (Schubiger-Bossard, 1988). Although species composition on Alpine forelands differs considerably from those in Scandinavia, broadly equivalent stages are recognizable on the basis of physiognomy. In addition, particularly in early successional stages, taxonomically close species can often be recognized as occupying similar positions in the successions. For example, species of *Pohlia, Rhacomitrium, Stereocaulon, Poa, Trisetum, Saxifraga, Cerastium, Oxyria* and *Epilobium* are important in the early stages in both regions.

Under favourable conditions, rates of succession in the various altitudinal zones in the Alps tend to be greater than their equivalents in Scandinavia, although local environmental conditions often cause greater differences than those attributable to latitude. A damp, Atlantic climate and locations in deep valleys are particularly favourable for early colonization by woody plants (Lüdi, 1958). In front of the Oberer Grindelwaldgletscher, 3–4 m high thickets of *Alnus incana* with *Salix daphnoides* are present after 14–20 years, whilst dense *Alnus* woodland (in wet sites) and *Picea abies* forest (in dry sites) develop within 100 years of deglaciation (Lüdi, 1958).

An open woodland of *Betula pendula* and *Larix decidua* is found on terrain of equivalent age in front of the Grand Gletscher D'Aletsch, which is expected to develop into a *Pinus cembra–Larix decidua* forest with a *Rhododendron ferrugineum* and *Vaccinium myrtillus* understorey (Richard, 1968, 1973; Table 5.2). Young *Larix decidua* and *Picea abies* are found on terrain deglaciated for < 20 years in front of the Aletsch Gletscher and after 34 years at the Rhonegletscher, developing into a young forest at the latter foreland in 85 to 95 years (Schubiger-Bossard, 1988). However, the open nature of woodland on many sub-alpine forelands is primarily attributable to relatively heavy grazing, which favours the development of meadows and heath communities.

Increasing continentality from west to east, with increasing temperature contrasts between summer and winter, lower precipitation totals and greater susceptibility to summer drought, is also a significant influence in the Alps. This causes slower development towards dwarf-shrub heathland and long-lasting grassland, as exemplified by the forelands of Morter-atschgletscher and Roseggletscher, where the mean annual precipitation is about 815 mm (450 mm less than at Grindelwald). After 100 to 150 years there are no signs of woodland development, although young trees are present (Lüdi, 1958). In the sub- and low-alpine belt there are significant variations in the dominant shrubs; the Atlantic species, such as *Calluna vulgaris*, *Empetrum hermaphroditum* and *Rhododendron ferrugineum* disappearing or becoming less important towards the east.

The most rapid successions to forest in western North America are equivalent to or exceed the highest rates recorded in western Norway and the Alps. At Glacier Bay, closed *Alnus crispa* thicket develops within about 30 years and young *Picea sitchensis* forest can develop within 75 to 90 years (section 5.1.1). Spruce forest seems to establish even faster near the neighbouring Mendenhall and Herbert Glaciers, where the seed source is closer (Lawrence, 1958), and comparable rates of succession are also implied by the rapid establishment and growth of a variety of tree species on glacier forelands in maritime locations of Washington, British Columbia and Alaska (Viereck, 1968; Sigafoos & Hendricks, 1972; Oliver & Adams, 1979; Heikkinen, 1984a; Oliver, Adams & Zasoski, 1985). Farther inland, (Heusser, 1956; Tisdale, Fosberg & Poulton, 1966; Fraser, 1970; Sondheim & Standish, 1983; Blundon, 1989) the development of forest generally requires longer. For example, sub-alpine forest of the *Lupinus–Abies–Tsuga* association on glacier forelands in Garabaldi Park, British Columbia, seems to require about 150 years (Fraser, 1970). This may be primarily due to longer establishment times and/or slower growth rates.

On the relatively continental and higher altitude moraines of the Klutlan Glacier, Yukon Territory, Birks (1980a) points out that *Picea glauca* forest develops within about 100 years of initial plant colonization. However, an additional 75 years is required following deglaciation for stabilization of the ice-cored moraines, and it is a further 40 years before spruce colonization occurs. Other differences between the inferred successions at Klutlan Glacier (Figs, 5.25 & 5.39) and Glacier Bay include the prominence of mosses such as *Rhacomitrium canescens* in the early stages at Glacier Bay, the lack of an important role for *Alnus crispa* at Klutlan, and the possibility of succession proceeding to muskeg at Glacier Bay. The most striking floristic similarity is the presence of a *Dryas drummondii* stage at both forelands. This is not a general feature of successions in North America, however, only two of the 12 forelands examined by Heusser (1956) in the Canadian Rockies being characterized by extensive *Dryas*-mats. According to H. J. B. Birks (pers. comm., 1990), although *Dryas drummondii* is abundant on river terraces and sandar throughout the Canadian Rockies, it is curiously patchy on forelands and avoids those that are not flat.

Many of the important species on the Klutlan moraines are also important in the boreal-alpine ecozone in front of Frederika Glacier, in the neighbouring Wrangell Mountains of Alaska (Scott, 1974b). However, the rate of succession is slower and there is no evidence of the attainment of mature spruce forest, even though *Picea glauca* occurs as scattered individuals on the oldest moraines and stablized glacio-fluvial deposits. An *Arctagrostis latifolia–Poa arctica* nodum occurs on surfaces deglaciated for less than about 20 years. *Dryas octopetala, Shepherdia canadensis, Oxytropis campestris* and *Hedysarum alpinum* are all important in the intermediate stages, but not *Alnus*. On surfaces older than about 80 years a *Salix alaxensis–Shepherdia canadensis–Cladonia pyxidata* association is typical, the physiognomy of which varies from dense willow thicket to sparse willows with a well-developed understorey. *Salix alaxensis* thicket is expected to persist, without the attainment of boreal forest.

Although succession rates vary greatly from site to site, successional sequences in south-east Alaska and adjacent Yukon Territory appear to have much in common in terms of species composition from sea level to the low-Alpine belt (Scott, 1974b). A low-alpine succession is briefly described by Rampton (1970) from ice-cored moraines in front of Natazhat Glacier, which lie at an altitude of about 200 m above the adjacent Klutlan moraines. Early stages with *Dryas, Salix, Hedysarum* and *Shepherdia canadensis* develop into shrub tundra dominated by *Betula glandulosa*. There is also some similarity, more especially in the early stages, with successional sequences in the interior, where the duration of each stage is longer than in coastal Alaska or at the maritime-continental boundary. On glacio-fluvial outwash gravels in front of Muldrow Glacier, Alaska, at an altitude of about 100 m above the regional tree line, Viereck (1966) recognizes a pioneer phase with isolated mats including *Dryas drummondii*, *D. integrifolia* and *Oxytropis campestris*. Within 100 years, this develops into a meadow stage dominated by *Elymus innovatus*, which is followed by *Betula glandulosa* scrub on terrain ages of 150–300 years. Eventually, this probably develops into a low shrub-sedge, tussock-moss tundra, consisting of shrub birch and ericaceous shrubs interspersed with *Eriophorum vaginatum* tussocks growing through *Sphagnum warnstorfianum* and other mosses.

Little information has been published from North American glacier forelands relating either to specialized communities such as sub-alpine meadows (Sandgren & Noble, 1978; Lawrence, Noble & Tilman, *in prep.*), or to the mid- and high-alpine belts (Butters, 1914; Given & Soper, 1975). The more detailed work of Spence in the Teton Range, Wyoming (Spence, 1981, 1985; Spence & Shaw, 1983), is largely floristic and unfortunately does not include data on successional sequences. It would appear that relatively simple successions are characteristic of these high altitude continental sites, where the distribution of a relatively small number of species is closely related to site conditions. At 12 sites in the Teton Range, involving 10 Neoglacial moraines and two related features, Spence (1985) recorded 109 species of vascular plants of which only seven were found at all sites and 25% were restricted to a single site. Like the vascular plants (section 5.2.5), terricolous cryptogams do not exhibit a significant relationship with terrain age. The distribution of saxicolous lichens is most strongly related to substrate age and type, whilst the terricolous cryptogams are strongly related to aspect (Spence, 1981).

Beyond Scandinavia, the Alps and North America, data are not sufficient to allow detailed statements about regional variations in the pattern or rate of succession. Indications of major regional variations are, however, available from the Tien Shan, New Zealand, and the Antarctic and sub-Antarctic. Solomina (1989) reports great variations in successional sequences from valley to valley in the Tien Shan (see Fig. 5.19), citing as important controls, variations in altitude, topography, exposure, glacier climates, foreland size, lithology, ability to reproduce vegetatively or by seed, and anthropogenic effects. In New Zealand, sub-alpine and alpine successions are clearly different and slower than those to forest at lower altitudes (Wardle, 1977, 1980), and within the alpine zone, successions are slower and simpler at higher altitudes (Archer, 1983; Sommerville, Mark & Wilson, 1982). Although successional sequences are generally simple also in polar regions, data from the Antarctic are important in showing that single-stage successions are only found in the most severe areas; they are not, for example, generally characteristic of the sub-Antarctic (Lindsay, 1971; Smith, 1972, 1982, 1984; Collins, 1976; Wynn-Williams, 1985; Scott, 1990).

This brief survey of regional variations is consistent with the generalization that relatively complex seres are characteristic of relatively favourable physical environments. The more favourable the environment, the greater the floristic and structural differences between the early stages and the mature stages. In relatively favourable environments there are a larger number of stages, which in turn tend to be of shorter duration. Although many other factors may be important locally, regional-scale patterns are related in particular to climatic gradients; the most unfavourable environments from this point of view being extreme polar, continental or high-altitude climates, where successions may consist of a single stage.

5.4   Environmental controls on successional sequences

In this section a closer examination is made of the nature of the control exerted on vegetational chronosequences by physical environmental factors. In *actual* chronosequences, environmental factors are effective both in the form of initial site conditions and as influx variables (see section 5.1.1). In the former role, environmental factors may be considered as passive influences on the pattern and rate of succession; in the latter guise, they are actively involved in the processes of successional change (cf. section 4.1.1). A second theme is the importance of environmental factor complexes; that is the interaction of many environmental variables in the landscape, which exists as a dynamic multivariate system. The reader is referred also to the role of environmental factors in soil development (section 4.3).

5.4.1   Initial site conditions

Various physical environmental factors, to different degrees and in different combinations, provide successional sequences with different starting conditions. Many of the within- and between-foreland patterns described in section 5.3 can be attributed to such differences. Substrate characteristics (including texture and lithology), topography, (including altitude, aspect and microtopography) and climate (including temperature, moisture, snow distribution and exposure) are the most frequently cited factors in this context.

Substrate characteristics clearly have a major effect on the ability of pioneer plants to colonize recently-deglaciated terrain. Cooper (1923b) recognizes three types of substrate – bare rock, till surfaces and glacio-fluvial deposits – at Glacier Bay. Earlier he compared the succession on moraines and sandar in front of Robson Glacier, British Columbia (Cooper, 1916). Although the stages appear similar, the rate of succession is much slower on the sandar. Butters (1914) gives an interesting example of the effect of lithology in a comparison of two lateral moraines of the Sir Sandford Glacier in the Selkirk Mountains of British Columbia. The main difference between the two moraines is in chemical composition, the northern lateral moraine comprising largely granitic rocks and mica schist, the southern moraine containing limestone and dolomite. Of the 110 species of higher plant found (68 on the northern moraine and 76 on the southern), only 34 (30.9%) occurred on both moraines and, of those, 13 differed greatly in abundance between the two moraines. Thus, only 21 species (19.1%) occurred with even approximately similar frequency on the two substrates.

On glacier forelands in the Tien Shan, it seems that succession is slower on gravel than on sand, and most rapid on loam (Solomina, 1989). In front of Skaftafellsjökull, Iceland, differentiation of vegetation on terrain of the same age depends mainly on texture (Persson, 1964). Vegetation is best developed on substrates rich in small stones, finer textures being subject to wind erosion. Grubb (1986, 1987), based partly on observations in the sub-alpine belt of southern Norway, suggests that trees and shrubs (with mosses) are the main pioneers in rock crevices and in bouldery areas, whereas forbs are more important on till, and grasses (with mosses) characterize silty substrates; a pattern possibly related to moisture and/or nutrient supply.

Textural differences are often responsible for the differentiation of pioneer communities (Jochimsen, 1963; Sigafoos & Hendricks, 1969; Worley, 1973; Elven, 1974, 1978c; Lawrence, 1979; Wardle, 1980; Schubiger-Bossard, 1988). Whilst the effects of these and other substrate differences are most obvious in early successional stages, they may remain of importance in later stages. Tisdale, Fosberg & Poulton (1966) note that coarser-textured till ('crevasse dumps'), which covers about 20% of the area investigated in front of Robson Glacier, exhibits a relatively slow vegetation succession that is detectable even on terrain deglaciated for 180 years. Birks (1980a) recognizes different variants of *Picea glauca* forest that are related to substrate texture in the later stages of succession on the ice-cored moraines fronting Klutlan Glacier. Thus, even where there are strongly convergent seres, initial substrate differences may remain effective influences. The mixture of grain sizes and the abundant fines in tills generally provides a more favourable substrate than well-sorted glacio-fluvial sediments, especially where the latter are coarse gravels.

Initial conditions in terms of topographic factors and related climatic differences are clearly of paramount importance in accounting for the major differences between successional sequences at a regional scale. Controls exerted by altitude, latitude and continentality have all been thoroughly discussed in section 5.3.4. At the within-foreland scale, aspect and microtopography are likely to have a longer-lasting influence than substrate differences, particularly where these topographic effects are large in relation to any biological processes leading to convergent successional pathways.

Aspect is of particular significance in deep valleys (Friedel, 1938; Jochimsen, 1970; Ellis, 1975; Matthews, 1979a,b; Turmanina & Volodina, 1979; Spence, 1981; Schubiger-Bossard, 1988; see, for example, Figs. 5.30–5.33). On the glacier foreland of the Hintereisferners, Austria different successions proceed at different rates on sunny and shady valley sides towards different mature soils and plant communities (Friedel, 1938). The successional sequences begin with a common *Poa laxa–Cerastium uni-florum* community. On the south-facing aspect, this is replaced by an *Agrostis rupestris–Polytrichum juniperinum* community at a distance of about 100 m from the glacier, which changes into a *Trifolium pallescens–Polytrichum juniperinum* community at a distance of about 350 m. On the north-facing aspect, an *Agrostis rupestris–Rhacomitrium canescens* community appears at a distance of about 350 m and is replaced by a *Silene acaulis–Polytrichum piliferum* community at a distance of about 600 m. On older terrain, *Festuca varia* and *F. halleri* communities respectively characterize the sunny and shady aspects.

Microtopography is an effective control on succession on most glacier forelands. Microtopography influences successional sequences through microclimatic gradients and effects on the drainage and stability of slopes. Many of the successional pathways identified in section 5.3.3 are related closely to microtopography, particularly through exposure, disturbance, snow distribution and moisture regimes. Only where there is a strong, overriding, unifying factor does microtopography appear to be relatively unimportant. This applies to strong biological effects, such as in the later stages of succession at Glacier Bay, when the influences of microtopo-graphy that are apparent in the early stages disappear under a tree stratum (Lawrence, 1979). The same principle also applies to strong physical effects, such as the heavy precipitation and frequent fog characteristic of the alpine

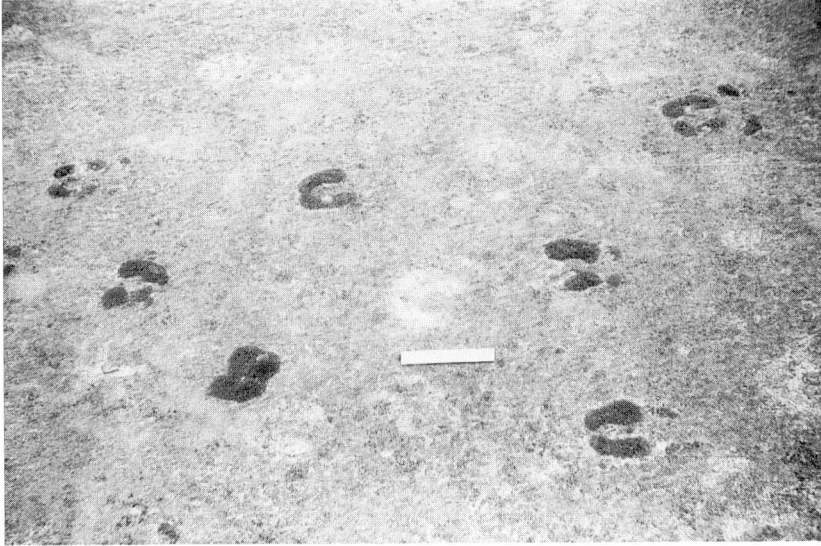

Fig. 5.43. *Aongstroemia longipes* in reindeer hoofprints, Austerdalsisen, northern Norway. Scale = 15 cm (from Theakstone & Knighton, 1979).

meadow site at Mount Wright overlooking Glacier Bay (Sandgren & Noble, 1978).

Under the influence of such unifying factors, only strong microtopographic gradients are likely to be effective. This is evident on the Klutlan moraines where Birks (1980a) identifies one major divergent phase of succession where slopes are ⩾ 5° (Fig. 5.39). He also recognizes a distinctive *Shepherdia canadensis–Phacelia mollis* nodum on the oldest moraine where slopes are too steep (⩾ 10°) and unstable, with soils too dry and shallow, for closed *Picea glauca* forest. These slopes correspond to areas of reworked, coarse, White River tephra and are characterized by a varied and open vegetation (20–60% cover) with a mixture of heliophyte species that are common on younger moraines (e.g. *Crepis nana*) or otherwise absent from the moraine sequence altogether.

Without strong unifying factors, slight environmental differences may be reflected in the vegetation. A particularly good example of this is provided by Theakstone & Knighton (1979) who describe the colonization of boot- and reindeer-hoof prints by *Aongstroemia longipes* in front of Austerdalsisen, northern Norway. The moss occurs in shallow depressions, 1–2 cm deep, in fine-grained surface sediments of a sparsely-vegetated abandoned distributory bar (Fig. 5.43), and grows to a height of 2–4 cm above the surrounding surface. Some of the prints are known to have been formed only 1–3 years before colonization. Growth of the moss in this micro-habitat is probably related to the concentration of propagules in the depressions during heavy summer rain, and to slightly improved moisture conditions for subsequent growth following water ponding. At a larger scale, a complex pattern of growth variations has been revealed in a dendroecological investigation of *Betula pubescens* on the foreland of Bödalsbreen, southern Norway (Staschel, 1989). Although the particular environmental causes remain largely unidentified, Staschel's results indicate large growth differences between sites, particularly in 'pointer years' (i.e. years characterized by abnormally narrow or broad tree rings).

Climatic gradients operate at a range of scales (see section 3.3). Individual plants may be relatively frequent to the lee-side of stones and boulders on exposed sandar or till plains. Coe (1964) states that up to 90 m from the snout of Tyndall Glacier on Mount Kenya, where *Senicio keniophytum* is the sole visible colonizer, most of the plants occur on the sheltered lee-side of boulders, protected from the glacier wind. At a larger scale, differences in vegetation cover and species composition have been found between the proximal and distal slopes of moraine ridges. Lutz (1930) has described the distribution of trees colonizing moraine ridges in front of the Sheridan Glacier, Alaska, where *Picea sitchensis*, *Tsuga mertensiana*, *Alnus crispa* and *Populus tacamahaca* (in order of decreasing abundance) avoid ridge crests and proximal slopes but favour distal slopes and depressions where there is some protection from the glacier wind. Trees growing on the exposed slopes often assume prostrate or near-prostrate forms and some specimens of *Picea* only 10 cm high are over 40 years old.

Different opinions have been expressed regarding the effectiveness of the glacier climate in general and of glacier winds in particular. According to Oke (1987: 179), the harshness of the glacier wind can cause vegetation to be absent for about 100 m from the glacier snout and stunted or deformed for a considerably greater distance. Similarly, in the western part of the glacier foreland of Skeidarárjökull, Iceland, Wójcik (1973) observed that the vegetation is much more extensive and grows closer to the glacier margin than in its eastern section. He suggests that this can only be explained in terms of climate and, in particular, shelter from the glacier wind. Tollner (1934) and Friedel (1936) both attribute importance to glacier winds in the Alps. Friedel describes the destructive effects of winds in front of the Pasterze Gletscher, such as eroded turf, injuries to shrubs, and crippled trees, and claims that the lower altitudinal limit of Elynetum communities is depressed by as much as 500 m in its vicinity. In front of Columbia Glacier,

British Columbia, trees grow in places protected from the glacier wind but are wind trained or sheared where they stand above the winter snow cover (Heusser, 1956). Furthermore, examples of the suppression of tree-growth following the approach of an advancing glacier have already been cited (see section 2.2.2).

At Skaftafellsjökull (Iceland), on the other hand, the occurrence close to the glacier of a number of 'southern' species (namely *Sedum annuum, S. acre* and *Galium verum*), indicates that factors other than climate may be limiting (Persson, 1964; Lindröth, 1965). Interestingly, Lindröth points out that for nocturnal soil and surface animals with their limits of activity close to $+4$ °C, conditions at night near the glacier under clear skies may be relatively favourable. In addition, the survival of vegetation on the surface of glaciers and in front of advancing glaciers supports the notion that the glacier climate is not a limiting factor for many species.

5.4.2   Environmental factors as influx variables

The concept of an influx variable is based on Jenny (1961, 1980) and has been defined and discussed in considerable detail in relation to soil development (see sections 4.1.1, 4.3 & 4.4). In the context of vegetation, environmental factors are influx variables when they actively affect the direction or rate of progress along successional pathways. Influx variables therefore include the environmental causes of allogenic successional change (Tansley, 1935).

These are particularly in evidence in the paraglacial zone where physical landscape processes are highly dependent on the previous existence of glacier ice (see section 3.4). Initially, substrates are likely to be saturated with water and tend to be unstable due to mass movement, cryogenic processes and/or glacio-fluvial activity. Some degree of stabilization of the substrate is a prerequisite for the initiation of plant succession. This can be demonstrated by the following examples. Succession may be delayed by disturbance for 20 years in wet sites in front of the Pasterze Glacier in the Alps (Zollitsch, 1969). Greater delay is likely on steep moraine slopes or where there is extensive buried ice. In the Tien Shan, colonization tends to occur on terminal moraines before lateral moraines because of the unstable steep slopes of the latter (Solomina, 1989). According to Spence & Shaw (1983) moraine crests are stable but moraine slopes are unstable in the Teton Range, Wyoming. This leads to the unusual occurrence of moraine crests being the more heavily vegetated (about 50% vegetation cover). A major lag in colonization is reportedly caused by the melting of buried ice on the glacier foreland of Hyrnebreen, Svalbard (Kuc, 1964). On the ice-cored moraines in front of Klutlan Glacier, Yukon Territory, it has been estimated that 75 years is necessary for moraine stabilization, although continued disturbance, including several types of slope collapse and the formation of thaw lakes, continues on older terrain (Birks, 1980a).

In the early stages of succession, local environmental changes are clearly able to alter both the rate and direction of succession. Jochimsen (1963, 1970) has stressed the importance of the changing moisture regime on glacier forelands in the Austrian Alps. This influences succession both directly and, through disturbance, indirectly (see section 5.3.3. & Fig. 5.42). The successional scheme of Zollitsch (1969) also indicates that changes in moisture, disturbance and exposure produce multiple successional pathways (Fig. 5.41). Thus, particular sites may follow divergent or convergent trajectories depending on such changes. As the glacier retreats, however, there is a strong directional environmental change as the substrate dries out, slopes stabilize, and sites become less exposed to glacier winds.

Textural changes characteristic of the early stages of soil development (see section 4.2.1) are also of potential importance in the early stages of vegetation succession. Pervection (section 3.2.3), frost-heave and frost-sorting (section 3.2.5) and deflation (section 3.2.8) all combine to produce the pavement effect noted by several authors as characteristic of till surfaces within a few decades of deglaciation. These processes alter the nature and availability of micro-sites for plant establishment. Loss of fines from surface horizons can in extreme cases produce very inhospitable conditions for plant growth and delay or halt succession for decades, if not centuries. The development of 'stony heath' in front of Nigardsbreen, southern Norway (Fig. 5.37), is attributed by Fægri (1933) principally to drought produced by the removal of fine sediment by water. Extensive areas of this vegetation type, which consists almost entirely of xerophile mosses and lichens, persist on terrain deglaciated for over 150 years where the water table is deep.

Fægri (1933) also recognizes a distinct community, dominated by *Pohlia gracilis*, that develops where moist sites are frequently inundated by glacio-fluvial meltwater containing high concentrations of silt. Similar communities have been recognized by Elven (1975, 1978c) elsewhere in Norway and by Jochimsen (1970) in the Alps. Such communities enter and leave the succession in response to the shifting course of meltwater streams. Aeolian deposition may similarly cause vegetation changes. Birks (1980a) notes that wind-blown silt trapped around taller plants supports various acrocarpous mosses (such as *Ditrichum flexicaule*, *Barbula icmadophila*, *Bryum caespiticium*, *Tortella fragilis* and *Encalypta rhapdocarpa*) on moraine K–II in front of the Klutlan Glacier. Viereck (1966) suggests that the accumulation of fine wind-blown material is unfavourable for further growth of *Astragalus nutzotinensis* and other clump-forming pioneers on gravel outwash in front of the Muldrow Glacier, Alaska.

The decline of certain pioneer species may be caused by environmentally-driven changes in chemical properties of the substrate of which the rapid decrease in pH with increasing terrain age is indicative (see sections 4.2.4. & 4.3.1). Such changes in the chemical environment may be accelerated by the addition of organic matter from plants but, as has been pointed out in section 4.3.3, the process is likely to occur anyway. Except where highly calcareous substrates are involved, it seems that leaching removes enough of the exchangeable bases from recently-deglaciated sediments to produce appreciable acidification on most glacier forelands within about 50 years. This is likely to be at least a contributory cause of the decline of calcicolous and basiphilous pioneer groups. Several workers have identified this important element of the pioneer flora, which occurs even where rocks are non-calcareous (Richard, 1968, 1973; Moiraud & Gonnet, 1977; Elven, 1978c, 1980; Ellenberg, 1988). In front of the Grand Glacier d'Aletsch, *Linaria alpina* and *Arabis alpina* are included in this category (see Figs. 5.4 & 5.24), whilst at the Glacier de Saint-Sorlin, Moiraud & Gonnet (1977) include *Thlaspi rotundifolium*, *Trisetum distichophyllum*, *Doronicum grandiflorum* and *Papaver alpinum* ssp. *rhaeticum*.

Allogenic changes may be relatively less important in later successional stages; they do not, however, cease. They continue to influence vegetation succession via the edaphic factor (see section 4.4) and also more directly through disturbance and climate. Surface runoff from rainfall and meltwater are important components of the disturbance regime in front of the Pasterze Glacier (Zollitsch, 1969). These forms of disturbance decrease with increasing distance from the glacier (Fig. 5.41). Avalanches and grazing are important in the distal part of the foreland of the Rhonegletscher (Schubiger-Bossard, 1988; Fig. 5.40). Brink (1964) lists needle ice, snow creep, solifluction and interfacial ice (superficial freezing of the interfacial zone between a snow cover and the bare or vegetated surface beneath) as being of importance on and near the foreland of Helm Glacier, British Columbia. Coe (1964, 1967) attributes regular disturbance in front of the Tyndall Glacier, Mount Kenya, to needle ice and other frost processes.

Archer, Simpson & Macmillan (1973) describe the continuing effects of disturbance by solifluction after about 230 years on a lateral moraine of Tasman Glacier, New Zealand. Solifluction terraces, and to a lesser extent solifluction lobes, exhibit a definite zonation of plants, generally without vegetation on the highly-disturbed tread surface (the relatively flat upper surface). Four vegetation zones characterize the less-disturbed riser (the sloping front). Where the tread merges with the steep step of the riser below, a narrow vegetated band includes *Anisome flexuosa*, *Cyathodes fraseri* and *Raoulia grandiflora*. *Chionochloa pallens* and *Celmisia lyallii* dominate the upper part of the vertical riser, and *Dracophyllum kirkii* the lower part. At the foot of the riser, on a gentle slope, *Coprosima pumila* dominates.

At a larger scale, Khapayev (1978) recognizes persistent types of vegetation associated with avalanche tracks below glacierized cirques in the Caucasus. In the alpine zone, where meadow communities are typical of the adjacent vegetation (on inter-avalanche ridges) the avalanche tracks are characterized by little or no vegetation. In the sub-alpine zone meadows are found within the avalanche tracks, whereas the corresponding vegetation of the inter-avalanche ridges involves krummholz, *Rhododendron* thickets and meadows. In the boreal coniferous forest zone, the avalanche tracks are characterized by deciduous crooked-stem forests. These vegetation types are viewed as components of distinct avalanche and non-avalanche geocomplexes.

On and down valley from the foreland of Nooksack Glacier, North Cascade Range, Washington, Oliver, Adams & Zasoski (1985) have described a vegetational mosaic that reflects many different kinds of disturbance (Fig. 5.44). Forest development follows a general pattern: (1) trees, shrubs and herbaceous species invade the disturbed areas simultaneously; (2) after about 60 years, tree and shrub exclusion result from inhibition by a dense forest canopy; (3) later, the canopy opens and there is understorey regrowth, possibly involving an autogenic process; (4) an old-growth uneven-aged forest may develop. However, significant differences are detectable in forest composition depending on the type of disturbance. In particular, stands developing from avalanche and rock-fall disturbance have a higher proportion of *Abies amabilis* (mean stand overstorey composition, 69.8%) than stands developing from soil mass movements and glacier retreat (23.9%) ($p < 0.005$; Mann–Whitney test). The avalanche and rock-fall disturbances also appear to be characterized by less *Tsuga heterophylla* and *Alnus sinuata* but more *Vaccinium* spp. ($p < 0.20$ in each case).

Small-scale frost disturbance has been monitored systematically across the chronosequence at Storbreen, Jotunheimen (R. J. Whittaker, unpublished). In a simple experiment, set up for one year in August 1983, matchsticks of 4.5 cm length and 0.2 cm diameter were inserted vertically

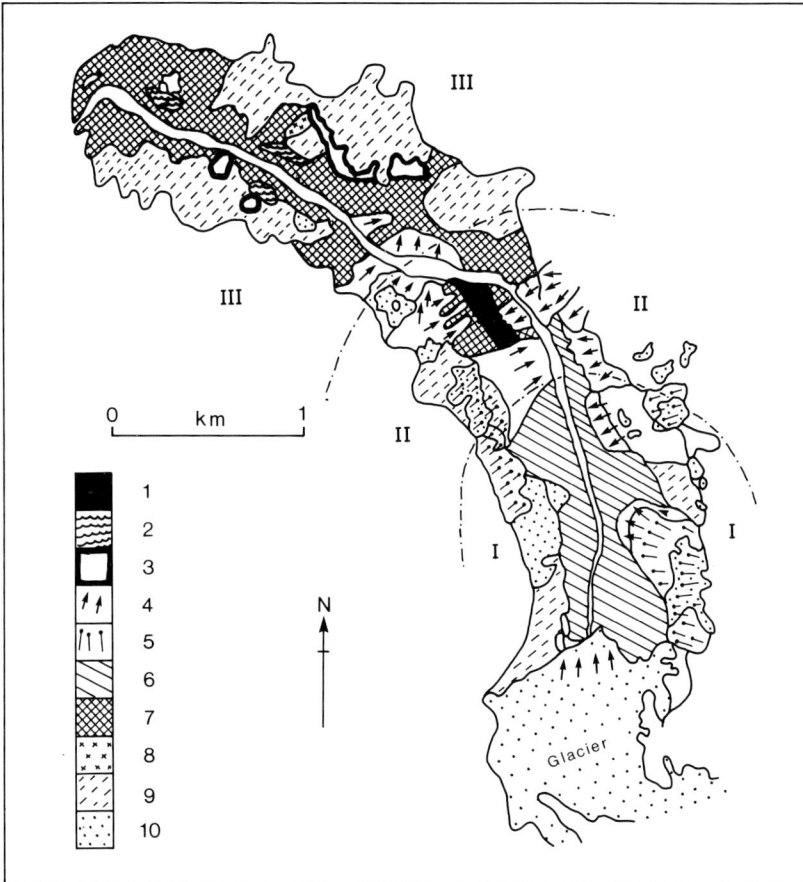

Fig. 5.44. Map of major types of disturbance near Nooksack Glacier, North Cascade Range, Washington, U.S.A.: (1) abandoned river channel; (2) flooded areas; (3) large unstable boulders; (4) avalanche areas; (5) rockslides; (6) morainal areas; (7) old-growth forests; (8) soil mass movement; (9) creeping snowfields and possible avalanches; (10) intermittent snowfields. Primary disturbance areas: I, recent glacier activity; II, primary disturbance (possibly glacial) about 600 years B.P.; III, no obvious evidence of major primary disturbance (from Oliver, Adams & Zasoski, 1985).

into the substrate to a depth of 3.5 cm at 29 sites, 27 of which were in the permanent quadrats described previously. All sites were within an altitudinal range of 1300 to 1350 m. At least nine matchsticks were located in a 10 × 10 cm grid at each site, in the flattest, fine-matrix patch available. Results indicate a marked decline in activity with increasing terrain age (Fig. 5.45), with a total of 192 (70%) of the 273 matchsticks uplifted to

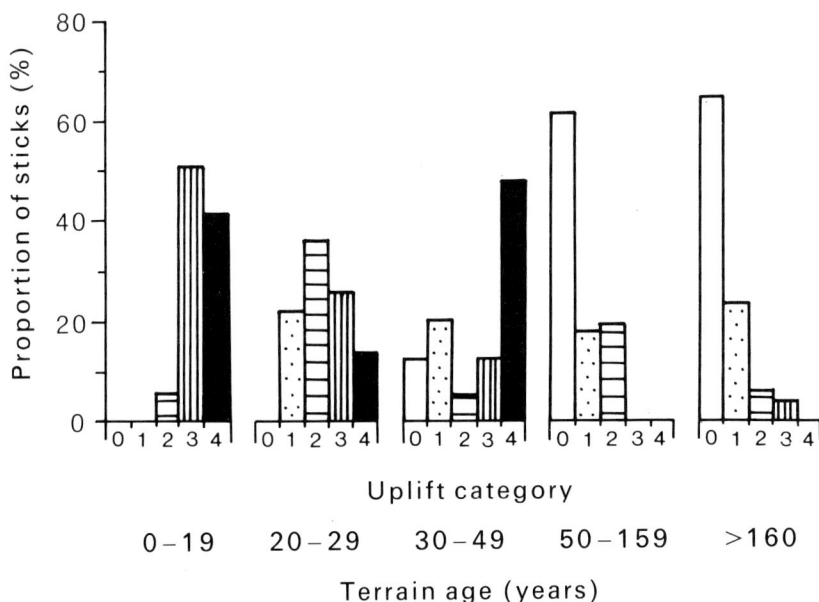

Fig. 5.45. Small-scale frost disturbance of matchsticks with increasing terrain age, Storbreen, Jotunheimen. Disturbance categories: 0, no displacement; 1, displaced $\geqslant 25\%$ of buried length; 2, $\geqslant 50\%$; 3, $\geqslant 75\%$; 4, extruded on to the surface (n $\geqslant 45$ matchsticks from at least 5 sites per terrain-age class) (from R. J. Whittaker, unpublished).

varying degrees. Kolmogorov-Smirnov two-sample tests indicate statistically significant differences between all age classes other than the two oldest ($p \leqslant 0.05$). Only at two sites (both within the 50–159 age class) was no movement detected. These results indicate the existence of forces sufficient to disrupt plant growth, most probably by the frost-pull mechanism (cf. Harris & Matthews, 1974). The pioneer grasses, such as *Poa alpina* and *Trisetum spicatum*, which possess fibrous root systems may well be favoured by such disturbance processes (R. J. Whittaker, unpublished). Although it is possible to explain the reduction in frost disturbance in terms of pervection and other physical processes, negative feedbacks between vegetation development and frost-heave processes are probably also involved.

Glacio-fluvial deposits commonly experience disturbance for a considerable time after deglaciation (section 3.2.1). This is reflected in the age of trees in abandoned meltwater channels of the Emmons Glacier, Mount Rainier, Washington, which contain trees up to 50 years younger than those on the surrounding moraines (Sigafoos & Hendricks, 1972). Not only rates of succession are affected; qualitative differences in species composition may be detectable long after stabilization (Birks, 1980a; Dale & MacIsaac, 1989). These observations strongly suggest that glacio-fluvial disturbance alters both the rate and the trajectory of succession.

According to Brandani (1983) certain trees, such as *Alnus*, *Betula* and *Populus*, are favoured by the regular disturbance regime of the glacial river valley. Relatively predictable flow patterns produce high water tables in spring to which early flowering is correlated, and successful establishment is dependent on early summer seed dispersal, lack of seed dormancy and rapid germination prior to the onset of adverse winter conditions. Large annual seed crops, small wind- and water-dispersed seeds, fast growth rates, low-density wood, short life spans, low shade tolerance, and the ability to sprout when damaged, are all seen by Brandani (1983) as additional disturbance-adapted traits in this environment.

Disturbance may therefore be necessary for the maintenance of certain species populations. Wind throw at irregular intervals may perform this function in the mature spruce forest at Glacier Bay. It has been hypothesized by Bormann & Sidle (1990) and by Ugolini, Bormann & Bowers (in prep.), that wind throw releases immobilized nutrients and improves decomposition. At the oldest site examined by Bormann & Sidle (deglaciated for about 210 years), 91% of the nitrogen in the ecosystem (excluding the C horizon) exists in the soil O and Bh horizons and stem components. From an examination of a chronosequence of earthy mounds produced by wind throw in mature spruce-hemlock forest, Spaltenstein & Ugolini (1988) report that although total phosphorus in the system remains at about the same level over a period of 500 to 1000 years, the amount available to plants decreases as the Bh horizon develops. In the absence of disturbance it would appear that the gradual operation of other environmental processes (such as long-term leaching, or rising water tables in response to climatic change) contributes, together with autogenic processes (Ugolini & Mann, 1979), to paludification and muskeg formation (see section 6.1.7). However, wind throw may itself contribute to muskeg formation by creating pits and disturbing the moss carpet, permitting *Sphagnum girgensohnii* to colonize the forest floor (Noble, Lawrence & Streveler, 1984).

Periodic disturbance of a different kind has been proposed as a causal factor in vegetation succession and soil development under the influence of meltwaters from snow and ice in the alpine zone. Based on recently-deglaciated moraines in the Ben Ohau Range, New Zealand, Archer (1973) envisages that water table fluctuations can occur due to both local oscillations of glacier snouts and variations of the perennial snow line in response to minor climatic changes. A rise in the water table from excessive melting rejuvenates the soil from suspended sediment. Subsequent lowering of the water table results in greater plant growth, an increase in organic matter production, and an increase in leaching. This form of disturbance differs from that in locations with drier north and west aspects where, in the absence of a strong meltwater influence, disturbances are in the form of cycles of aeolian erosion and deposition.

Finally, as pointed out in section 3.3.4, climatic amelioration is likely to have been a general feature of glacier forelands since the peak of the 'Little Ice Age'. Particular sites will therefore have experienced the global warming trend reinforced by the local climatic changes that follow glacier retreat. At Storbreen, it is likely that these climatic changes were sufficient to produce a vegetational response (Matthews, 1979b), as mean annual temperature changes experienced by a particular site over the last 230 years have been of the same order as the 3.0 °C temperature difference that corresponds with the altitudinal range of the foreland (*ca* 400 m). The latter difference in climate has been sufficient to produce an altitudinal vegetation zonation on terrain of the same age. In addition, the widespread retrogressive vegetational changes measured directly over a 12-year period on this foreland (see section 5.1.2; Figs. 5.6 & 5.7) may well have been influenced by this kind of environmental change.

5.4.3   Environmental factor complexes

Various combinations of environmental factors account for much of the spatial variation encountered between sites on terrain of the same age and influence the differences in vegetation found on terrain of different ages. These environmental factors operate individually and collectively to define the initial conditions for succession and, as influx variables, exert considerable control on subsequent successional pathways. Numerous examples have been cited in this chapter. However, there has been little detailed investigation of the ways in which these environmental factors interact with each other and with the developing vegetation. In this section detailed consideration is given to one particular study, the explicit objective of which was to analyse the interaction of environmental variables in the context of succession.

Whittaker (1989) has developed a strategy for the analysis of vegetation-environment relationships and has applied it to the Storbreen glacier foreland, Jotunheimen. Vegetation and environmental factors are analysed within the chronosequence at a sub-set of 108 of the 638 permanent plots

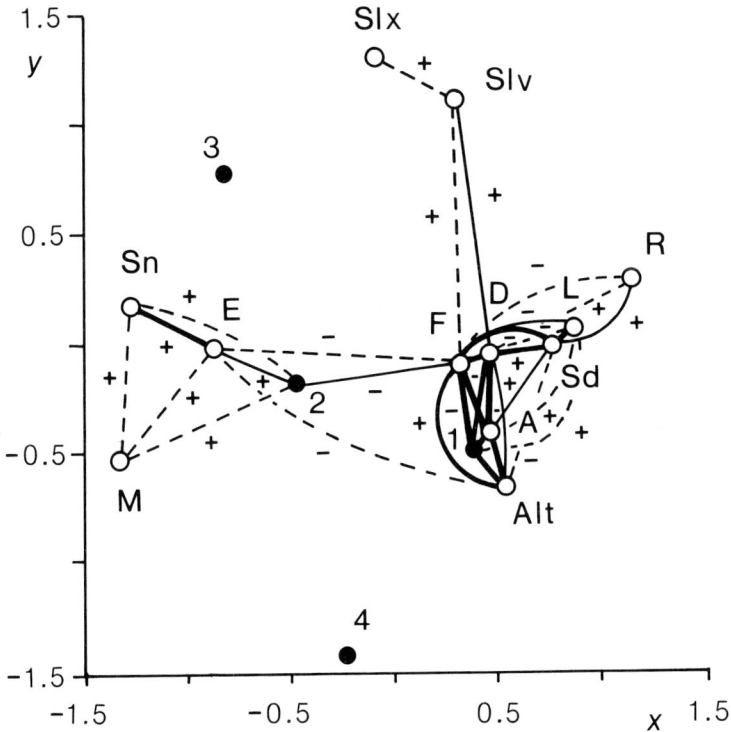

Fig. 5.46. Vegetation-environment plexus defined by non-metric multidimensional scaling (NMDS), Storbreen, Jotunheimen. Numbers indicate vegetation ordination axes (DCA axes 1–4). Symbols indicate environmental variables as defined in Table 5.5. Correlation coefficients (τ): $> \pm 0.3$ (- - -); $> \pm 0.4$ (——); $> \pm 0.5$ (▬▬) (from Whittaker, 1987).

previously described (see section 5.3.1). Twelve environmental variables, including terrain age, were measured and these are related to four vegetation gradients (ordination axes) defined by detrended correspondence analysis (DCA). Central to the strategy is the construction of a plexus diagram, which relates the vegetation axes to environmental factor complexes (Fig. 5.46).

Definitions of the environmental variables are given in Table 5.5. It should be noted that whereas some are measured on an interval scale, others are ordinal-scale estimates. Soil variables (soil, litter and root depths) refer to the apparent optimum point for soil development within each plot . Disturbance variables (frost churning and slope movement, components of the overall disturbance regime) were estimated on the basis

Table 5.5. *Definitions of 12 environmental variables used in the analysis of vegetation-environment relationships on the Storbreen glacier foreland, Jotunheimen. Symbols are those used in Figs. 3.16 and 5.46 (after Whittaker, 1987, 1989).*

Variable	Definition
Terrain age (A)	Site age at time of sampling
Altitude (Alt)	Altitude in metres above sea level
Soil depth (Sd)	Depth of organic staining or of mixed organic-inorganic material in cm, whichever is the greater
Litter depth (L)	Depth of surface organic horizon (A_o & A_{oo}) in cm
Root depth (R)	Average depth of root penetration in cm
Slope maximum (Slx)	Maximum within-plot slope angle in degrees
Frost churning (F)	Estimate of the degree of frost disturbance (cryoturbation) on a scale of 1 to 5
Slope movement (Slv)	Estimate of the degree of small-scale slope movement (e.g. solifluction) on a scale of 1 to 5
Disturbance (D)	Estimate of the overall disturbance regime (including F, Slv, grazing, trampling & stream flow) on a scale of 1 to 5
Exposure (E)	Estimate of exposure based on micro- and meso-scale topography (principally indicating exposure to wind) on a scale of 1 (exposed) to 4 (sheltered)
Moisture (M)	Estimate of drainage conditions on a scale of 1 (dry) to 5 (wet)
Snowmelt (Sn)	Estimate of timing of snowmelt from the plot on a scale of 1 (early) to 5 (late)

of geomorphological evidence and plant damage on a scale of 1 (least disturbed) to 5 (most disturbed). The moisture estimate makes no distinction between wetness derived from different sources (e.g. running water or ground water). Snowmelt was judged on the basis of the amount of snow at the time of sampling, the proximity of the site to retreating snow beds and the phenological state of the vegetation.

Non-metric multidimensional scaling is used to produce the configuration in Fig. 5.46 from a matrix of correlation coefficients between the environmental variables and vegetation axes (Table 5.6). Use of a non-parametric correlation coefficient (Kendall's tau) is compatible with the inclusion of ordinal-scale variables and possible non-linear relationships. The addition of significant correlation coefficients ($p \leqslant 0.001$) as links between variables, reveals the underlying structure. Further details of the technique are given in Whittaker (1987). The fact that a similar configuration and structure is recorded when the vegetation axes are omitted from

Table 5.6. *Correlation matrix of vegetation-environment relationships on the Storbreen glacier foreland, Jotunheimen. The coefficient is Kendall's tau; significant correlations (p≤0.001) are in bold; variables are defined in Table 5.5 (after Whittaker, 1987).*

1	2	3	4	5	6	7	8	9	10	11	12	13	14	15	16
Terrain age															
-0.635	Altitude														
0.157	-0.059	Slope maximum													
-0.132	0.023	-0.139	Moisture regime												
-0.594	**0.452**	-0.066	0.152	Disturbance											
0.177	**-0.348**	0.088	**0.298**	-0.175	Exposure										
-0.137	0.001	0.128	**0.341**	0.086	**0.499**	Snow melt									
-0.684	**0.508**	-0.122	0.058	**0.776**	**-0.349**	-0.027	Frost churning								
-0.193	0.149	**0.361**	-0.052	**0.438**	-0.061	0.176	**0.359**	Slope movement							
0.360	**-0.230**	0.206	-0.005	**-0.378**	0.142	0.072	**-0.426**	-0.080	Litter depth						
0.493	**-0.335**	**0.211**	-0.047	**-0.497**	0.159	0.036	**-0.527**	-0.122	**0.806**	Soil depth					
0.279	-0.218	0.080	-0.021	**-0.360**	0.139	0.008	**-0.301**	-0.183	**0.365**	**0.455**	Root depth				
0.762	**-0.556**	0.134	-0.208	**-0.590**	0.131	-0.174	**-0.659**	-0.202	**0.319**	**0.488**	**0.261**	DCA axis 1			
0.240	**-0.210**	0.142	**0.311**	-0.216	**0.447**	**0.347**	**-0.417**	-0.040	**0.296**	**0.278**	**0.256**	0.171	DCA axis 2		
-0.130	0.087	**-0.211**	0.119	0.144	**-0.276**	**-0.273**	**0.234**	-0.134	-0.155	0.041	-0.156	**-0.209**	-0.156	DCA axis 3	
0.291	0.171	0.006	0.075	0.150	0.024	0.148	0.199	0.051	-0.110	-0.100	-0.019	**-0.294**	-0.109	0.109	DCA axis 4

the analysis (Fig. 3.16) suggests that the technique is robust and that the factor complexes are real.

Two of the four vegetation axes are closely associated with two environmental factor complexes. The 'terrain-age factor complex' – comprising terrain age, frost churning, disturbance, altitude, and the three soil variables – is closely associated with the most important vegetation axis (DCA axis 1; eigenvalue 0.747). The 'exposure-moisture-snowmelt (microtopographic) factor complex' is associated with DCA axis 2 (eigenvalue 0.277). DCA axis 3 (eigenvalue 0.181) and DCA axis 4 (eigenvalue 0.146) are not only isolated in the plexus but also uninterpretable in terms of species composition and are therefore considered as unimportant 'noise' axes (Whittaker, 1987, 1989).

The first vegetation axis (DCA axis 1) is clearly interpretable as a vegetation gradient. Characteristic 'heath' species, including the 'core' heath species defined by Matthews (1978a,b) (see section 5.3.1) – *Betula nana, Vaccinium uliginosum* and *Empetrum hermaphroditum* – occur at the positive end of the axis, whilst the core 'pioneer' species – *Saxifraga cespitosa, Cerastium* spp. and *Deschampsia alpina* – occur at the negative end. On DCA axis 2, these species groups are not clearly segregated. Instead, species well known as characteristic of wet or snowbed habitats are congregated at the positive end of the axis (e.g. *Saxifraga rivularis, S. nivalis, S. stellaris, Viola palustris, Sedum rosea, Oxyria digyna, Pinguicula vulgaris* and *Ranunculus glacialis*. At the negative end, *Dryas octopetala, Lycopodium annotinum, L. selago, Phyllodoce caerulea, Silene acaulis, Betula pubescens* (seedlings) and *Trisetum spicatum* are all identifiable with drier or more exposed locations of shorter snow lie.

Detailed analyses of the interrelationships within the factor complexes and in relation to the vegetation axes have been carried out by Whittaker (1989) using simple and partial correlation, together with two- and three-dimensional graphical displays. Although the variables do not comprehensively define the effective environment, they provide insights into the complex interactions involved in the major geoecological structures. Within the terrain-age factor complex, terrain age has the highest correlation of any variable ($\tau = +0.76$) with DCA axis 1, which is also highly correlated with frost churning (-0.66) and overall disturbance (-0.59) and, to a lesser extent, altitude and the three soil variables. Although frost churning and disturbance are both negatively correlated with terrain age, the former is more strongly correlated (-0.68), which suggests that the other components of the disturbance regime taken together (slope movement, trampling, grazing and glacio-fluvial erosion, etc.) do not exhibit such strong time-dependent trends. Taking DCA axis 1 as the dependent variable, first-order partial correlation coefficients indicate that soil depth (+0.14), altitude (−0.15) and disturbance (−0.15) have insignificant correlations with DCA axis 1 when terrain age is controlled. On the other hand, the partial correlation between DCA axis 1 and frost churning, controlling for terrain age, remains significant (−0.30). This suggests that, unlike the other variables in the terrain-age factor complex, frost churning may exert a time-independent effect on axis 1 (as well as its strong time-dependent effect).

The role of altitude in the plexus is somewhat problematical because there is a significant correlation with DCA axis 1 despite the small altitudinal range. Indeed, a significant correlation coefficient is also found when the analysis is confined to sites from the even more restricted altitudinal range (1300 to 1350 m) that comprise the majority of the data set (Whittaker, 1985). It is likely that a 50 m difference in altitude, which corresponds to a mean annual temperature difference of only 0.27 °C (Matthews, 1987), is ineffective . It appears, therefore, that the significant correlation coefficient in this case merely reflects simple spatial correspondence between altitude and terrain age. Significant correlations between DCA axis 1 and the soil variables are, however, entirely compatible with an intimate time-dependent association between vegetation succession and soil development.

DCA axis 2 is highly correlated with exposure ($\tau = +0.45$), snowmelt (+0.35) and moisture (+0.31). It is also correlated with some of the variables in the terrain-age factor complex but, with the exception of frost churning (−0.42), the latter correlations are relatively weak (Fig. 5.46). As pointed out by Whittaker (1989), despite the individual correlations within the microtopographic factor complex being weaker than many of those in the terrain-age factor complex, the three variables in combination provide a powerful explanation of DCA axis 2. This is convincingly demonstrated in the isometric projection in Fig. 5.47a, which shows that site scores on axis 2 increase in response to both increasingly late snowmelt and increasing shelter.

The fact that frost churning is relevant to DCA axis 2 as well as axis 1 is consistent with disturbance effects from this variable being associated with exposed microtopographic sites as well as relatively young terrain. However, there appears to be little variation in axis 2 scores with exposure until the two highest points on the frost-churning scale (Fig. 5.47b). This suggests that the impact of exposure on axis 2 is principally confined to the younger, more disturbed terrain. This in turn suggests that the axis 2 structure may be relatively sensitive to the sites on relatively young terrain.

(a)

(b)

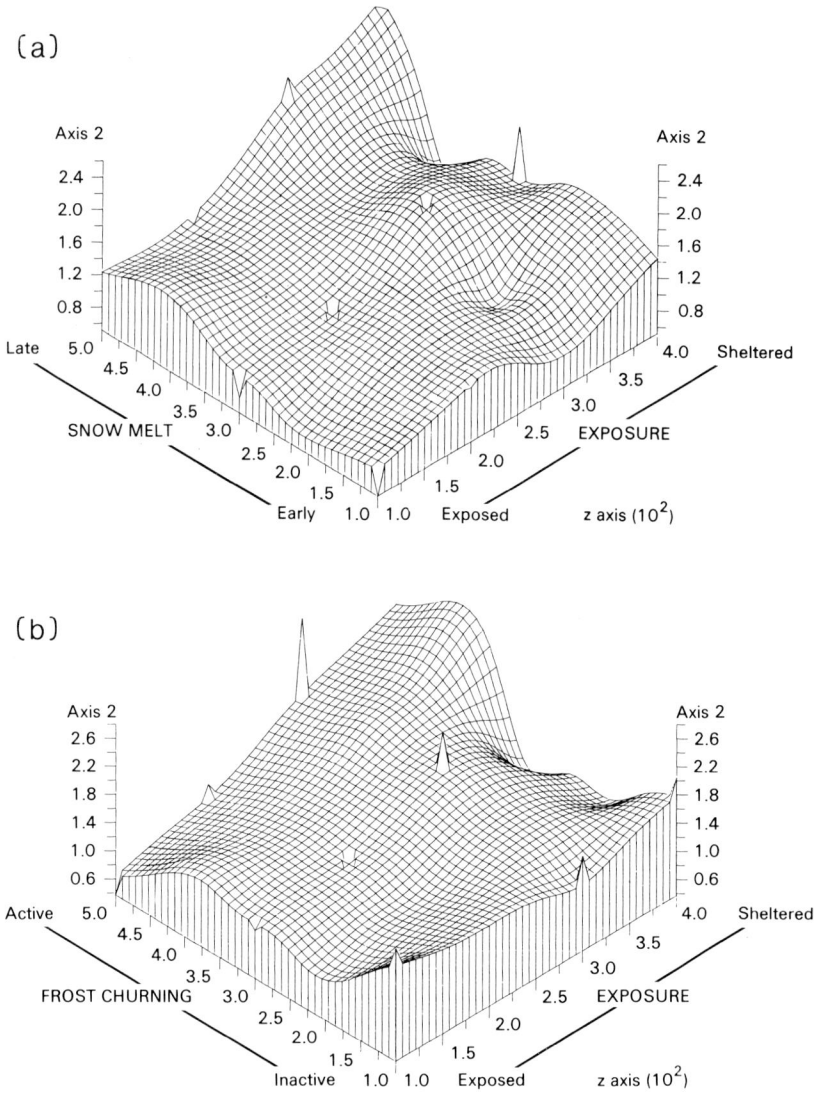

Fig. 5.47. Isometric projections of detrended correspondence analysis (DCA) vegetation axis 2 against selected pairs of environmental variables, Storbreen, Jotunheimen: (a) snowmelt and exposure; (b) frost churning and exposure (from Whittaker, 1989).

Canonical correspondence analysis (CCA), a form of direct gradient analysis, is generally supportive of the conclusions reached from the indirect gradient analyses of the plexus approach. Whittaker (1989) utilized CCA as provided in the CANOCO package of Ter Braak (1986, 1987), which extracts ordination axes that are constrained to account for a maximum amount of the vegetational variability in terms of linear combinations of environmental variables. Graphical representations of the results are presented in Figs. 5.48a & 5.48b, in each of which species and environmental variables are positioned in relation to pairs of CCA axes. Eigenvalues for the three axes are 0.664, 0.271 and 0.171, respectively. Arrow length in these diagrams represents the strength of the relationship between an environmental variable and a vegetation axis; arrow direction indicates whether the relationship is positive or negative.

The first CCA axis is closely related to terrain age ($\tau = +0.92$) and is also strongly related to disturbance (-0.78), frost churning (-0.78) and altitude (-0.73): important variables of the terrain-age factor complex. The most important vegetation axis again reflects a successional trend from young, actively disturbed sites with shallow soils (negative scores) to old, inactive sites with deep soils. The second and third CCA axes can both be interpreted as reflecting different aspects of the microtopographic factor complex. In relation to CCA axis 2, moisture is unimportant ($+0.18$) and the most important environmental variables are slope maximum ($+0.46$), exposure ($+0.39$), frost churning (-0.38), litter depth ($+0.36$) and snow-melt ($+0.31$). Moisture (-0.62) is the only important variable on CCA axis 3.

A comparison of the indirect and direct gradient analyses may be summarized thus (Whittaker, 1989). The analyses strongly confirm the importance of terrain age and associated factors, but slightly different configurations of relationships are suggested between vegetation and the microtopographic factors. It seems that the techniques produce closely comparable results where the relationships are clear and strong (i.e. correlations within the terrain-age factor complex). Where the relationships are more complex and weaker, differences arise. In the latter situation, it appears that CCA may tease out distinct effects of the interwoven variables, whereas the NMDS plexus approach appears to provide a better picture of the overall structure and the complex interactions. It should be noted, however, that a CCA analysis omitting sites beyond the glacier foreland boundary produced a result in which the three key variables of the microtopographic factor complex were the three environmental variables most strongly related to CCA axis 2 (Whittaker, 1989). It is possible,

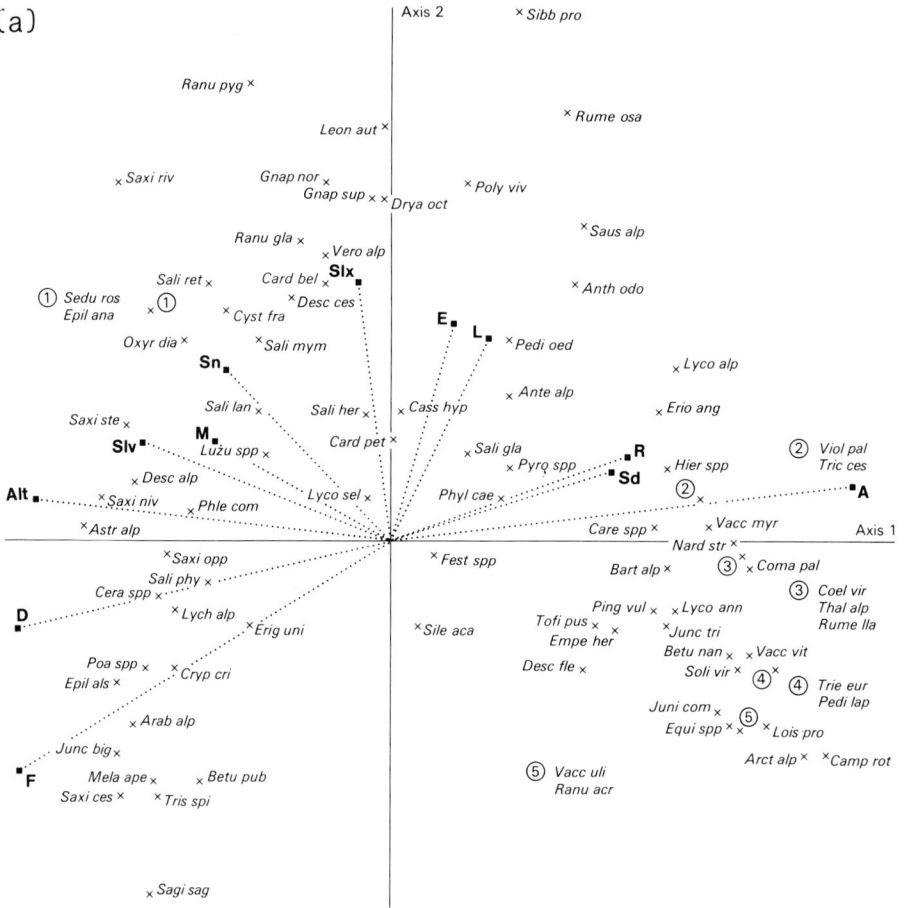

Fig. 5.48. Biplots of canonical correspondence analysis (CCA) vegetation axes displaying the position of species and environmental variables, Storbreen, Jotunheimen: (a) CCA axes 1 and 2; (b) CCA axes 1 and 3 (from Whittaker, 1989). See Table 5.5 for explanation of environmental variables. Species names:

Ante alp	*Antennaria alpina*
Anth odo	*Anthoxanthum odoratum*
Arab alp	*Arabis alpina*
Arct alp	*Arctostaphylos alpina*
Astr alp	*Astragalus alpinus*
Bart alp	*Bartsia alpina*
Betu nan	*Betula nana*
Betu pub	*Betula pubescens* (including *B. tortuosa*)
Camp rot	*Campanula rotundifolia*
Card bel	*Cardamine bellidifolia*
Card pet	*Cardaminopsis petraea*
Care spp	*Carex* species
Cass hyp	*Cassiope hypnoides*

(b)

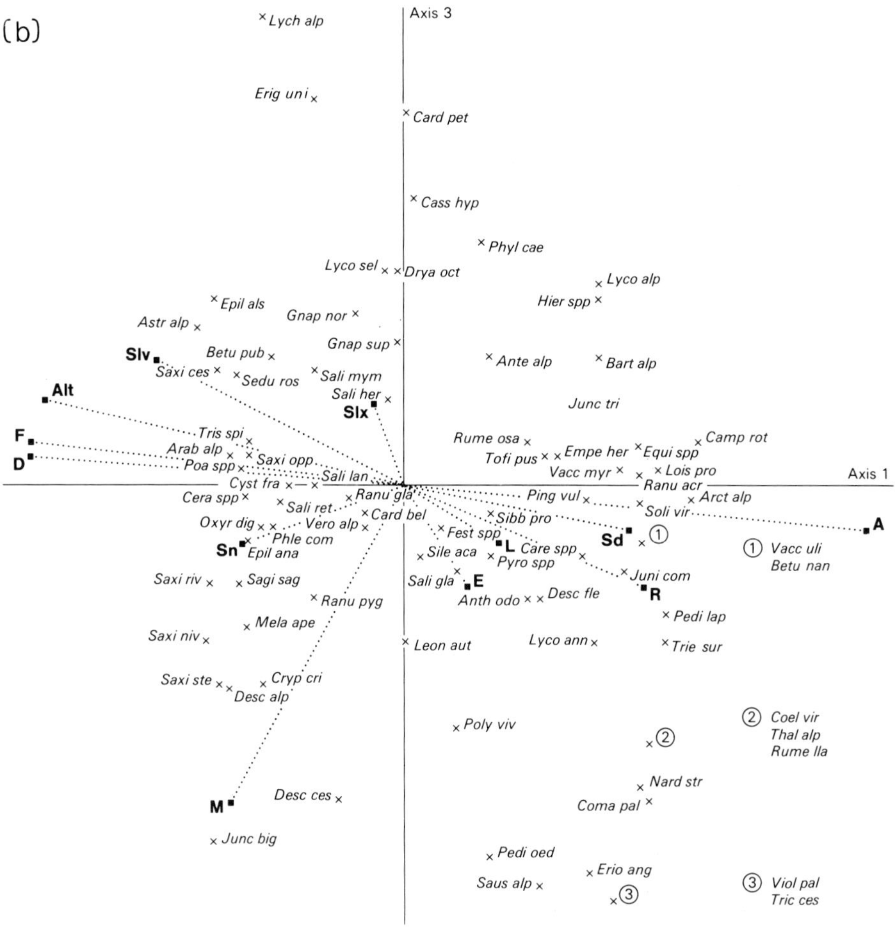

Axis 3

Axis 1

× Lych alp

Erig uni ×

× Card pet

× Cass hyp

× Phyl cae

Lyco sel × × Drya oct

× Lyco alp

Hier spp ×

× Epil als   Gnap nor ×

Astr alp ×

Slv ■   Betu pub ×   Gnap sup ×

Alt ■   Säxi ces × × Sedu ros × Sali mym

× Ante alp   × Bart alp

Sali her ×

Slx ■

Junc tri

F ■   ···Tris spi ×

D ■   ···Arab alp × × Saxi opp

Rume osa ×

× Camp rot

Tofi pus × × Empe her × Equi spp

···Poa spp × × Sali lan

Vacc myr × × × Lois pro

Cyst fra × × × Ranu gla

Ranu acr

Cera spp ×   × Sali ret × Card bel

Ping vul ×   × × Arct alp

Oxyr dig × × Vero alp ×

× Sibb pro   Soli vir ×

Sn ■ × Phle com   · Fest spp ×

Sd ■ × ①   A ■

Epil ana

× Sile aca × L Care spp ×

Pyro spp ×

① Vacc uli
Betu nan

Saxi riv ×   × Sagi sag

Sali gla × ■ E

× Juni com

× Ranu pyg

Anth odo × × Desc fle

R ■

Saxi niv ×

× Pedi lap

Saxi ste × × × Cryp cri

× Leon aut   Lyco ann ×

× Trie sur

Desc alp

× Poly viv

× ②

② Coel vir
Thal alp
Rume lla

M ■   Desc ces ×

× Nard str

Coma pal ×

× Junc big

× Pedi oed

Saus alp ×   × Erio ang

× ③

③ Viol pal
Tric ces

Cera spp	Cerastium spp. (C. alpinum and C. cerastoides)
Coel vir	Coeloglossum viride
Coma pal	Comarum palustre
Cryp cri	Cryptogamma crispa
Cyst fra	Cystopteris fragilis
Desc alp	Deschampsia alpina
Desc ces	Deschampsia cespitosa
Desc fle	Deschampsia flexuosa
Drya oct	Dryas octopetala
Empe her	Empetrum hermaphroditum
Epil als	Epilobium alsinifolium
Epil ana	Epilobium anagallidifolium
Equi spp	Equisetum spp.
Erig uni	Erigeron uniflorum
Erio ang	Eriophorum angustifolium
Fest spp	Festuca spp. (F. ovina)
Gnap nor	Gnaphalium norvegicum
Gnap sup	Gnaphalium supinum

Fig 5.48. (*cont.*)

Hier spp	*Hieracium* spp.
Junc big	*Juncus biglumis*
Junc tri	*Juncus trifidus*
Juni com	*Juniperus communis*
Leon aut	*Leontodon autumnalis*
Lois pro	*Loiseleuria procumbens*
Luzu spp	*Luzula* spp. *(L. arcuata, L. confusa, L. frigida, L. spicata)*
Lych alp	*Lychnis alpina (Viscaria alpina)*
Lyco alp	*Lycopodium alpinum*
Lyco ann	*Lycopodium annotinum*
Lyco sel	*Lycopodium selago*
Mela ape	*Melandrium apetalum*
Nard str	*Nardus stricta*
Oxyr dig	*Oxyria digyna*
Pedi lap	*Pedicularis lapponica*
Pedi oed	*Pedicularis oederi*
Phle com	*Phleum commutatum (P. alpinum)*
Phyl cae	*Phyllodoce caerulea*
Ping vul	*Pinguicula vulgaris*
Poa spp	*Poa alpina* subspecies *alpina* and *vivipara*
Poly viv	*Polygonum viviparum*
Pyro spp	*Pyrola* spp. *(P. minor, P. novegica, P. rotundifolia* and *Orthilia secunda*
Ranu acr	*Ranunculus acris*
Ranu gla	*Ranunculus glacialis*
Ranu pyg	*Ranunculus pygmaeus*
Rume lla	*Rumex acetosella*
Rume osa	*Rumex acetosa*
Sagi sag	*Sagina saginoides*
Sali gla	*Salix glauca*
Sali her	*Salix herbacea*
Sali lan	*Salix lanata*
Sali mym	*Salix myrsinites*
Sali phy	*Salix phylicifolia*
Sali ret	*Salix reticulata*
Saus alp	*Saussurea alpina*
Saxi ces	*Saxifraga cespitosa (S. groenlandica)*
Saxi niv	*Saxifraga nivalis*
Saxi opp	*Saxifraga oppositifolia*
Saxi riv	*Saxifraga rivularis*
Saxi ste	*Saxifraga stellaris*
Sedu ros	*Sedum rosea*
Sibb pro	*Sibbaldia procumbens*
Sile aca	*Silene acaulis*
Soli vir	*Solidago virgaurea*
Thal alp	*Thalictrum alpinum*
Tofi pus	*Tofieldia pusilla*
Tric ces	*Tricophorum cespitosum*
Trie eur	*Trientalis europaea*
Tris spi	*Trisetum spicatum*

Fig. 5.48. (*cont.*)

Vacc myr	*Vaccinium myrtillus*
Vacc uli	*Vaccinium uliginosum*
Vacc vit	*Vaccinium vitis-idaea*
Vero alp	*Veronica alpina*
Viol pal	*Viola palustris*

therefore, that some of the differences between the techniques may reflect data quality in relation to the more stringent statistical assumptions of CCA.

Whilst the strength of the relationship between vegetation axes and terrain age lends support to a chronosequence appproach to the study of succession, several features of the analyses confirm the necessity to take account of environmental factors. Whilst some factors appear ineffective, and hence irrelevant in this sample, others have a major influence on the vegetation. The largely time-independent factors, including the microtopographic factor complex can, in theory, be held constant. However, the variables in the terrain-age factor complex, being interdependent, cannot be ignored or controlled. Some of these, such as altitude (in this sample of restricted altitudinal range), do not appear to be effective despite high correlations with terrain age. Of the effective variables in the terrain-age factor complex the soil and disturbance variables represent potential causal influences on succession.

6

Plant succession: processes and models

Emphasis in this chapter is given to the explanation of successional patterns and to the mechanisms of change. First, the biological processes involved in the initiation and continued development of the vegetation landscape are considered. This leads on to an examination of the models that have been proposed to explain primary succession in general and glacier foreland succession in particular. These models, which have largely invoked biological processes as the driving forces of succession, are examined critically in the light of the evidence from recently-deglaciated terrain. Lastly, a geoecological model is proposed, in which biological and physical environmental processes combine to provide a more complete explanation of succession in the evolving landscape.

6.1   Biological processes of colonization and succession

In his classic work, Clements (1928) recognizes six basic successional processes: nudation, migration, ecesis, competition, reaction and stabilization. These processes still provide a valid framework for discussion (e.g. MacMahon, 1980, 1987), although Pickett, Collins & Armesto (1987) point out that stabilization is best regarded as an effect of the first five causes. It should also be emphasized here that, except in the initial stages of succession, these processes are interactive rather than successive. All but the first, the creation of bare terrain, are biological in the sense of being a function of vegetation. The biological processes will be considered in turn, with some subdivision and additions.

6.1.1   Migration

Species composition on newly-deglaciated terrain is obviously dependent, in the first instance, on an ability to reach the site. Many workers have implied that pioneer colonizers have light, wind-dispersed

Table 6.1. *Seed mass and successional status of some major species in the chronosequence at Glacier Bay, Alaska (after Chapin, 1991).*

Species	Seed mass (μg seed^{-1})	Successional status
Epilobium latifolium	72 ± 5	Early
Dryas drummondii	97 ± 18	Early
*Salix alaxensis**	140 ± 3	Early
Alnus crispa	494 ± 23	Mid
Picea sitchensis	2694 ± 26	Late

Data are means ± 1 standard error ($n =$ five groups of 50 seeds)
*from Tanana River

seeds or are otherwise favoured by possessing specific adaptations for dispersal. For example, of the 74 species present at the lateral moraine site of Malte Brun, overlooking the Tasman Glacier in New Zealand, seven have berried fruits which could be attractive to birds (Archer, Simpson & MacMillan, 1973). Most of the other species have small, fine seeds or seeds with appendages that would assist in transport by wind (see also: Cooper, 1923b; Negri, 1934, 1936; Viereck, 1966; Lawrence, 1979; Birks, 1980a). There are, however, few detailed studies and some workers have noted that dispersal mechanisms are not necessarily reflected in colonization ability (Ryvarden, 1971, 1975; Given & Soper, 1975; Elven, 1980; Spence & Shaw, 1985; Spence, 1989).

Chapin (1991) has shown that seed mass of some major early colonizing species in the Glacier Bay chronosequence is less than that of some important later colonizers (Table 6.1). He notes that there are exceptions to this pattern, including the heavy-seeded legumes, which may nevertheless be dispersed by wind over a snow surface (Cooper, 1923b). Also, the larger seeds of *Alnus* and *Picea* both bear wings, which may increase mobility especially on snow encrusted with ice (Lawrence, 1979). *Astragalus nutzotinensis*, the earliest legume to colonize gravel bars in front of Muldrow Glacier, Alaska, has a specialized adaptation, which assists seed distribution (Viereck, 1966). When its sickle-shaped pods dehisce, the two sides of the pod remain attached to the pedicel, which creates a spiralled wheel that scatters seeds as it is blown along the bar surface. However, special adaptations may not be a requisite for dispersal by strong winds. Bonde (1969) isolated 35 species of seed plant by germination from wind-blown debris collected from the surface of St Mary's Glacier in the Colorado Front Range, USA. Bulbils of *Polygonum viviparum* were by far the most numerous, followed by seeds of *Minuartia obtusiloba* and *Saxifraga rhomboidea*, none of which possesses specific adaptations to wind.

It should be borne in mind that migration involves vertical as well as horizontal transport. This is particularly noticeable on the glacier forelands of valley glaciers where propagule sources may be high on the valley sides. At Glacier Bay, Cooper (1923b) described the young forest as creeping down from the trim line, which defines the lower limit of the older forest. Lawrence (1979) states that there is also a source of seeds in the alpine tundra above the forests and, with the passage of time, there may be less distance for the pioneers to travel where upper areas are deglaciated before the valley bottoms. Lawrence, Noble & Tilman (in prep.) include data on the potential contribution of high-altitude areas to low-altitude successional assemblages below. For example, floristic studies on Mount Wright at about 700 m (Sandgren & Noble, 1978) and directly below on Muir Point near sea level (Noble & Sandgren, 1976) provide information for sites that are 2.5 km apart. The occurrence at Muir Point of plants that are also on Mount Wright ranges from 17% for bryophytes to 20% for pteridophytes, 27% for seed plants and 31% for lichens. Similar percentages are given for successional stages near Muir Point, based on species lists compiled for Mount Wright, Sebree Mountain and Muir Inlet (Decker, 1966). Shared species comprise 24% in the pioneer stage, 33% in the thicket stage, 20% in the forest stage and 14% in the beach area. Lower values for beach areas are probably accounted for by the relatively large habitat differences at sites raised from below sea level by tectonic uplift and glacio-isostatic adjustment. Seeds and whole plants may be easily transported down by water and avalanches (Cooper, 1923b; Heusser, Schuster & Gilkey, 1954; Adams & Dale, 1987), a phenomenon that is also widespread in the deep valleys of Norway and the Alps (Fægri, 1933; Friedel, 1938; Jochimsen, 1970). A specific example is recognized by Ammann (1979) in front of the Oberaar Glacier, where the initial stages of the *Caricetum frigidae* community develop along channels on the glacier foreland, propagules having been washed through gaps in the outer moraine.

A variety of methods have been used by Ryvarden (1971, 1975) to sample the seed rain in front of Kongsnutbreen and Blåisen, outlet glaciers of the Hardangerjökulen ice cap in southern Norway. His studies constitute the first and by far the most detailed systematic investigation of seed dispersal on a glacier foreland. Two types of trap were used, water-filled plastic trays and a stream trap. In addition, winter dispersal was investigated by scraping snow patch surfaces on and in front of the glacier. Over 16 000 diaspores representing 57 species were trapped over three seasons in

Table 6.2. *Number and characteristics of diaspores trapped in water-filled trays near the Hardangerjökulen ice cap, southern Norway (after Ryvarden, 1971).*

(a) Total number of diaspores and likely dispersal distances:

Year	No. traps	No. diaspores	% dispersed < 5 m*	% dispersed > 5 m*
1968	27	2477	86.8	13.2
1969	57	8937	84.2	15.8
1970	57	4683	84.4	15.6

*for explanation see text

(b) Diaspores classified according to morphological dispersal type:

Diaspore group	% anemochores	% zoochores	% no adaptation
All diaspores	74.0	—	28.0
Dispersed < 5 m	81.0	—	19.0
Dispersed > 5 m	35.5	1.0	63.5
Dispersed < 5 m, − S.h.*	11.5	—	88.5
Dispersed > 5 m, − S.h.	13.5	1.0	85.5

*S.h. = *Salix* cf. *herbacea* type. For explanation see text.

the water-filled trays (Table 6.2). The greater numbers trapped in 1969 appears to have been due to an extremely warm summer, and especially an extended growing season. This quantity of diaspores represents an annual rain of 650 diaspores m^{-2} within 1 km of the glacier in a favourable season (Ryvarden, 1971).

Using the location of the nearest plants, the diaspores are classified in Table 6.2a according to whether they could have originated within 5 m of the trap or whether they must have been transported farther. Consistent results for the three years indicate that most of the diaspores produced are probably dispersed only a short distance. In general, the content of the traps reflects the quantity of the surrounding vegetation and its floral diversity. Species-rich plots yield 11–21 species each season, whereas species–poor areas, particularly those close to the glacier, yield 2–7 species. An increase in the total number of diaspores trapped with increasing distance from the glacier (Fig. 6.1) reflects the decreasing distance from dense vegetation.

In Table 6.2b the diaspores have been classified according to whether their morphology indicates specific adaptations for dispersal. Diaspores

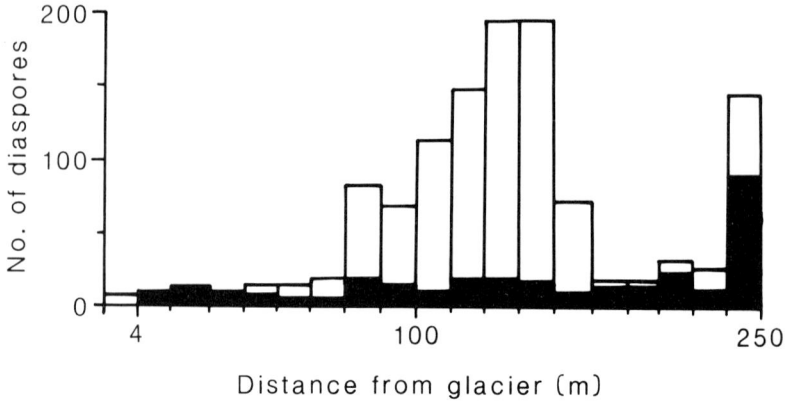

Fig. 6.1. Number of diaspores trapped with increasing distance from Blåisen (Hardangerjökulen), southern Norway. Shaded area indicates *Salix* seeds; each column represents one trap (from Ryvarden, 1971).

with long hairs or other morphological features retarding their rate of fall to < 1 second m^{-1} in a free fall of 5 m from the release point are classified as anemochores (Ryvarden, 1971). Fleshy and juicy diaspores or diaspores equipped with distinct hooks or an adhesive coating are classified as zoochores, whilst the remainder are regarded as having no apparent adaptation for dispersal.

About 74% of all the diaspores collected are anemochores, although anemochores are in the minority amongst those diaspores deemed to have originated > 5 m from the traps (Table 6.2b). However, these data are dominated by seeds of *Salix* cf. *herbacea*, which constitute 68% of all diaspores. If the seeds of this species are excluded, a much smaller proportion of the remaining diaspores are anemochores and there is little difference in the proportion of anemochore diaspores between the two groups (11.5% and 13.5%, respectively originating < 5 m and > 5 m from the traps). According to Ryvarden (1971), the *Salix* seeds are in most cases transported very short distances ($< 7\%$ are dispersed > 5 m) because they are rarely dispersed as individuals. In most cases, small balls of a few to many seeds are blown along the ground, where they easily lodge between stones.

Of the species represented in the seed rain, less than 25% are anemochores whereas $> 70\%$ produce diaspores with no apparent adaptation for dispersal. These proportions, and the small number of zoochores, are almost identical to those characteristic of the regional flora (Table 6.3), adding weight to Ryvarden's view that no great advantage is conferred by

Table 6.3. *Morphological dispersal types of diaspores trapped by different methods and sampled from snow patches near Hardangerjökulen, in comparison with those of species characteristic of the regional flora in the Finse area, southern Norway (after Ryvarden, 1971, 1975).*

Source of data	% species (morphological dispersal type)		
	anemochores	zoochores	no adaptation
Water-filled trays	23.0	4.0	73.0
Stream traps	22.5	6.5	71.0
Snowbed scraping	23.5	3.0	73.5
Finse flora	21.5	4.5	74.0

these morphological characteristics. Ten species provide over 90% of the total trapped diaspores: *Salix* cf. *herbacea, Trisetum spicatum, Oxyria digyna, Sagina saginoides, Arabis alpina, Veronica alpina, Poa jemtlandica, Gnaphalium supinum, Poa* cf. *alpina* and *Cerastium cerastoides.* Only two of these species (*Salix* and *Gnaphalium*) are anemochores according to Ryvarden's classification. It appears that these species dominate the seed rain because they are able to tolerate the edaphic and climatic conditions on the fresh soil and open terrain in front of the glacier, rather than because of the particular morphological characteristics of their seeds (Ryvarden, 1971, 1975). However, this does not rule out transport by wind of diaspores with no apparent specific adaptations to wind dispersal, particularly if seed production is great.

There seems to be considerable potential for dispersal by water. Over 1000 diaspores (35% anemochores, 1% zoochores, 64% without apparent adaptation) from 31 species were collected by two stream traps that were operational for only six days (Ryvarden, 1975). These traps were not located on the glacier foreland but were fed by snow meltwater. Fluctuating stream regimes in similar sites on glacier forelands may play a major role in picking up, transporting and depositing diaspores.

Winter dispersal, as inferred from scraping 174 m^2 of snow patch surfaces down to a depth of 5 cm, is relatively unimportant (Ryvarden, 1975). On average, 3–4 diaspores m^{-2} are deposited, ranging from about 0.6 on the surface of the glacier to 5–7 in front of the glacier (Ryvarden, 1975). These values are two orders of magnitude smaller than the summer dispersal inferred from the traps. Most of the diaspores are apparently dispersed before the snow arrives, which must cast doubt on the importance of dispersal over snow surfaces elsewhere.

Truly long distance dispersal seems insignificant at Ryvarden's sites. A small number of *Betula pubescens* seeds, which must have travelled at least 6 km, were found on the snowbeds (Ryvarden, 1975). From the water-filled trays (Ryvarden, 1971), one seed of *Polygonum aviculare* was probably carried by grazing sheep, and two seeds of *Aster sibiricus* almost certainly arrived on a student who is known to have been cultivating this species in Oslo Botanical Gardens! Elsewhere, there are other examples of dispersal by animals, such as by terns and gulls on Signy Island (Smith, 1975) or bears at Glacier Bay (Cooper, 1939; Lawrence, 1979), but in most cases these do not appear to be more than interesting, atypical cases.

The only other published attempt to trap propagules on a glacier foreland is that of Spence (1985), who located nine wooden plates coated in petroleum jelly across the end moraine of Schoolroom Glacier, Teton Range, Wyoming. Even though only one propagule was caught at moraine sites during a 24-hour period in September, of the 24 propagules caught in an adjacent meadow, 90% have some form of morphological feature that would favour efficient wind dispersal. In the Teton Range as a whole, Spence and Shaw (1983) found no significant difference between propagule types found on Neoglacial deposits (the 'Neoglacial flora') and two pools of potential colonizers: the debris accumulation flora and the entire alpine-high subalpine flora. Thus, although this 'Neoglacial flora' consists of a high proportion of species with propagules that appear to be suited morphologically to wind dispersal, these authors concur with Ryvarden (1971) in favouring the idea of chance dispersal by any species from the surrounding vegetation which can establish on the deposits. That is, ability to reach the site is not a limiting factor, provided the species grows in nearby vegetation.

It may be the case that dispersal ability is more significant where migration occurs over greater distances. Most of the glacier forelands discussed above are small, which means that newly-deglaciated terrain is not far away from well-vegetated areas and hence seed sources. Where glacier forelands are larger, as at Glacier Bay, dispersal ability may be more important. For example, according to Chapin, Fastie & Walker (in prep.), alder and spruce seed-rain measurements that have been carried out over two summers (1988–9) on 12–15-year-old terrain in Muir Inlet, revealed that only one alder seed reached the uncolonized site. Cooper (1931) suggests that the much slower rate of succession in the north-west branch of Glacier Bay (compared with Muir Inlet) is due to less effective wind dispersal. Lawrence (1979) notes, in particular, the isolation of the shores of Johns Hopkins Inlet from potential sources of propagules.

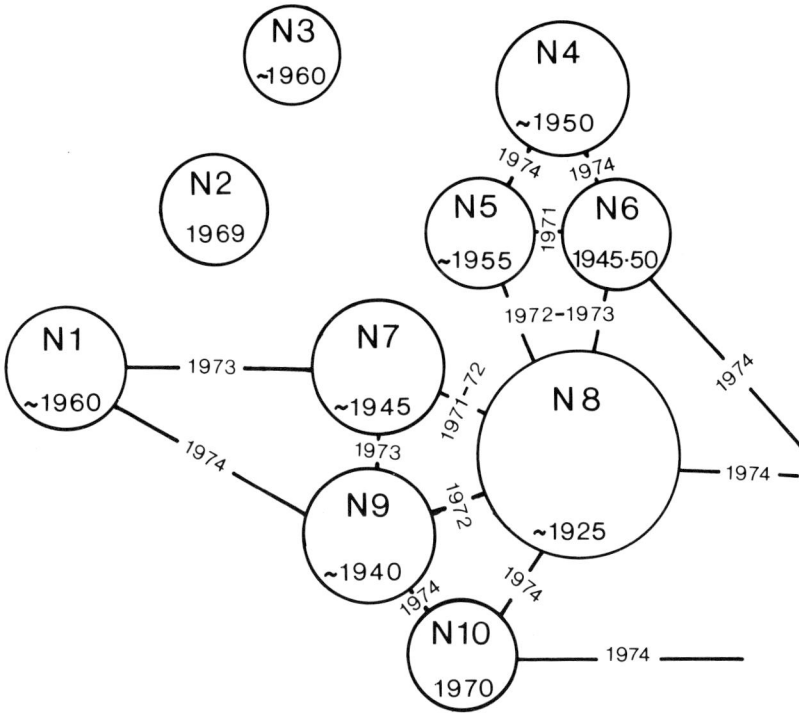

Fig. 6.2. Schematic representation of the nunatak cluster, Omnsbreen, southern Norway. Dates in circles indicate when nunataks (numbered) appeared and dates between circles indicate when they became connected (from Elven, 1980).

Further light is shed on this question by the results of studies involving nunataks isolated from potential sources of diaspores by barriers of glacier ice. The most detailed study of this type has been carried out by Elven (1980) who has investigated the colonization of 10 small nunataks uncovered during the last 50 years by the melting of Omnsbreen, a small ice cap located some 10 km to the north of Blåisen. Elven hypothesized that species sorting during nunatak invasion should depend more than the invasion of glacier forelands *sensu stricto* on effective dispersal mechanisms. The study began in 1969 when there were nine isolated nunataks, and by 1973 the 10 nunataks ranged in area from 4000 to 150000 m². Dates of appearance of the nunataks, the subsequent dates when they became connected with each other or with neighbouring terrain, and likely dispersal routes, are indicated in Fig. 6.2. Presence, abundance and fertility of phanerogam species were recorded in 1969 and 1971 and, together with mosses, hepatics and lichens in 1973. Although there is no sign of recent soil development on any

(a)   (b)

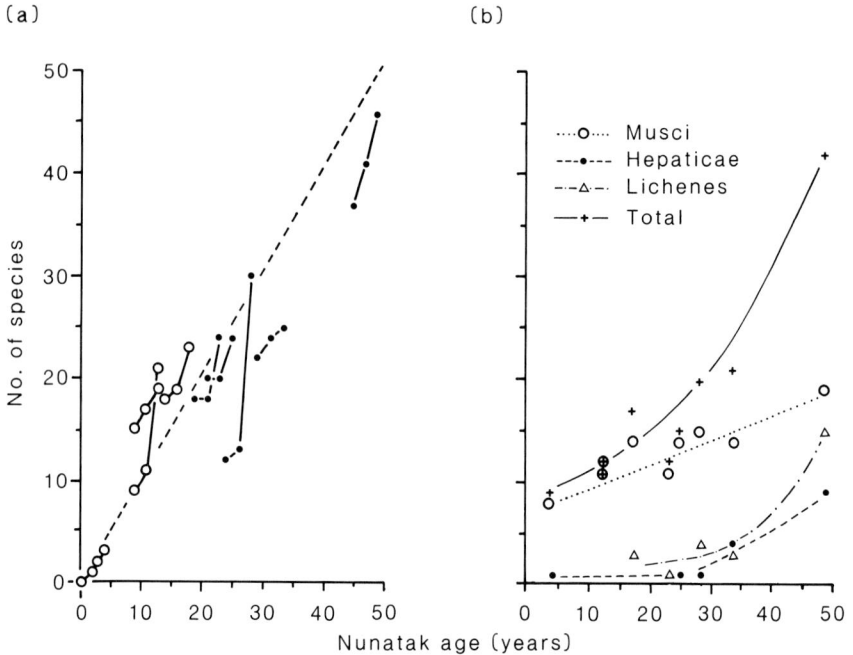

Fig. 6.3. Colonization of the Omnsbreen nunatak cluster, southern Norway: (a) cumulative number of phanerogam species for each nunatak; (b) number of cryptogam species in 1973 (from Elven, 1980).

of these nunataks, exhumed soil layers that originated before the 'Little Ice Age' (Elven, 1978a) are present in places, which presumably provide higher organic and nitrogen levels than are usual for newly-deglaciated till.

The relationship between the number of phanerogam species per nunatak and nunatak age (Fig. 6.3) reflects a mean increase in species number during the investigation period of 1.8 species per nunatak per year. This masks a change in the rate of increase from 0.7 in 1969–71 to 2.8 in 1971–3, probably caused by a combination of (1) warmer summers in 1972 and 1973, (2) an increase in the intensity of recording and (3) the opening up of new land-based dispersal routes (Fig. 6.2). Since deglaciation, the rate of increase has averaged 1.0 species per nunatak per year; a slight decline in the rate with increasing age is indicated by the rate for the five oldest nunataks of 0.8. The main trends in cryptogam invasion are illustrated in Fig. 6.3b. Mosses are the first invaders, with 8–12 species on the youngest nunataks but a slow rate of increase of about 0.3 species per nunatak per year. The hepatics and lichens are much later invaders.

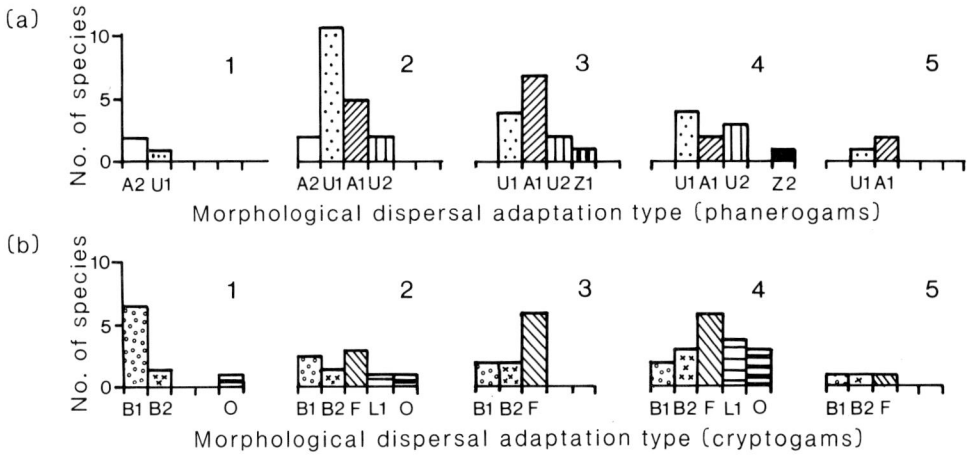

Fig. 6.4. Morphological dispersal adaptations by age group for (a) phanerogam and (b) cryptogam invaders of the Omnsbreen nunatak cluster, southern Norway. Age groups: 1, first pioneers (arrival after <5 years); 2, very early and early pioneers (5–20 years); 3, late pioneers (20–40 years); 4, very late invaders (>40 years); 5, accidentals. Dispersal adaptations: A1, anemochores; A2, tumblers; U1, unadapted but large diaspore production; U2, unadapted and small diaspore production; Z1, epizoochores; Z2, endozoochores; B1, spores (bryophytes); B2, gemmae (bryophytes); F, fragmentation (mosses and lichens); L1, soredia (lichens) (from Elven, 1980).

Only 12 (23%) of the 52 phanerogams recorded in 1969 showed no change in numbers by 1973; three more (6%) occurred sparsely in one or two years but may well have been overlooked on other occasions; 61% spread to new nunataks during the same period and increased in their abundance on the nunataks already occupied; 10% increased only in the sites they already occupied in 1969. Assuming that a species migrates from the nearest place where it is fertile, Elven has constructed the most likely dispersal routes from inter-nunatak dispersal involving *Cerastium alpinum*, *Epilobium hornemannii*, *Gnaphalium supinum*, *Oxyria digyna*, *Poa alpina* (var. *alpina* and *vivipara*), *Poa flexuosa*, *Sagina intermedia*, *S. saginoides*, *Sibbaldia procumbens*, *Saxifraga cernua*, *S. cespitosa*, and *S. rivularis*. Only in four cases, involving *Epilobium angustifolium*, *E. lactiflorum*, *Salix lapponum* and *Draba daurica*, did the diaspores certainly originate outside the nunatak cluster.

Morphological adaptations for dispersal characterizing early phanerogamic and cryptogamic invaders are shown in Fig. 6.4, in which the species have been grouped according to their rate of invasion. These 'invasion groups' are: (1) first pioneers, appearing < 5 years after deglaciation; (2) very early and early pioneers, appearing after 5–20 years; (3) late pioneers, appearing after 20–40 years; (4) very late invaders, appearing after > 40 years; and (5) accidentals. Hypothesized effective phanerogamic dispersal types, namely anemochory by means of very light, winged or hairy diaspores (A1) and zoochory (Z), are unrepresented amongst the first pioneers.

Anemochory of this type (A1) is characteristic of only 28.8% of the Omnsbreen nunatak flora but it is a more important means of dispersal in the species group invading after 20–40 years (Fig. 6.4). Pseudo-viviparous tumbling (A2) appears to be an important means of reaching the newly-deglaciated terrain. Two of the first pioneers – *Poa jemtlandica* and *Deschampsia alpina* – and two of the very early pioneers – *Festuca vivipara* and *Poa alpina* var. *vivipara* – are of this type. After this first invasion, the most important characteristic for reaching the nunataks seems to be high diaspore production (U1). Compared with species with effective means of dispersal by wind but low diaspore number (e.g. *Epilobium* spp., *Eriophorum* spp., *Gnaphalium supinum* and *Taraxacum croceum*), species with a high diaspore production per plant or shoot (e.g. *Cerastium* spp., *Sagina* spp., *Draba cacuminum*, and most *Saxifraga* spp.) are very effectively dispersed on the nunataks.

The lack of animal dispersal in this environment is easy to explain (Elven, 1980). The only common bird on the nunataks is the snow bunting (*Plectrophenax nivalis*), a species that does not normally feed on berries. Hare (*Lepus timidus*) and other mammals travel across the glacier, and reindeer (*Rangifer tarandus*) have been observed grazing on *Poa jemtlandica*, but the possibilities for animal dispersion are much less than in mature vegetation. The only endozoochorous species (Z2), *Empetrum hermaphroditum*, may well have been brought to the nunataks by birds. As the epizoochorous (Z1) *Phleum commutatum* occurs only on the most visited nunatak, it was probably transported on clothing.

Anemochorous bryophytes dispersed by spores (B1) (e.g. *Ceratodon purpureus*, *Dicranella subulata*, *Funaria hygrometrica* and *Polytrichum* spp.) do, however, predominate amongst the first pioneer cryptogams (Fig. 6.4). These, together with gemmiferous species (B2) (e.g. *Pohlia* spp., *Barbilophozia hatcheri*, *Cephaloziella rubella* and *Lophozia alpestris*) dominate the bryophyte list. Species probably dispersing by fragmentation (F) (e.g. many lichens, *Pogonatum urnigerum* and *Rhacomitrium canescens*) increase significantly in the later stages, whilst lichens with probable dispersal by soredia (L1) occur mainly on the oldest nunatak.

Excluding the cryptogams, Elven's data provide little support for a strong relationship between morphological dispersal adaptations and the timing of nunatak invasion. As proposed by Ryvarden (1971, 1975) for neighbouring glacier forelands, a large diaspore number appears much more significant than specialized adaptations in assuring rapid colonization. Investigations on more isolated nunataks in Iceland and Greenland support this conclusion (Einarsson, 1970; Frederiksen, 1971; Gjaerevoll & Ryvarden, 1977). On Jensens Nunatakker, the most isolated nunataks in Greenland (approximately 25 km from the nearest ice-free area), Gjaerevoll & Ryvarden report 24.2% anemochores and 3.2% zoochores, which is in close agreement with the Norwegian data presented above and with values for the whole Greenland flora. The data suggest that morphological adaptations for dispersal have been overemphasized on glacier forelands, where the sources of propagules are generally much closer than on nunataks and there normally appears to be a good supply of diaspores. The evidence also testifies to the efficiency of wind and water in spreading diaspores in these extreme environments. Even in the Antarctic, where recently-deglaciated soils have been cultured at the Jane Col site on Signy Island, South Orkney Islands (see section 6.1.2), there appears to be an adequate supply of propagules, mostly derived from local sources (Smith, 1987, 1991; Smith & Coupar, 1987).

6.1.2   Ecesis

Ecesis refers to species establishment and persistence as demonstrated by the ability of individual plants to complete their life cycles. Clements (1904, 1928) viewed ecesis as being primarily concerned with germination, growth and reproduction, a view that will be followed here. Evidence presented in the preceding section implies that ecesis is normally more important than migration in accounting for species composition and abundance in the pioneer stages of succession on glacier forelands. Propagules of a large number of species arrive on the newly-deglaciated terrain but not all successfully attain ecesis.

Detailed studies of (bryophyte) propagule banks in relation to recently-deglaciated terrain have been carried out only in the Antarctic. These studies have involved laboratory and field experiments on recently-deglaciated soils from Jane Col, Signy Island, South Orkney Islands (Smith, 1987, 1991; Smith & Coupar, 1987). Perhaps surprisingly, these experiments indicate a pool of viable bryophyte propagules, even at sites relatively remote from persistent vegetation.

In the first laboratory experiment, soil samples from unvegetated sorted circles and sorted stripes, and from an accumulation of dark mineral detritus washed down from the surface of an ice cap, were cultured on a thermogradient incubator (Smith 1987). Maximum growth (as % cover) was achieved at 10–15 °C; addition of NPK nutrient solution accelerated the appearance of protonemata and gametophytes but did not increase biomass or diversity. Although sub-sites with adjacent vegetation produced the greatest number of shoots, the greatest species diversity (eight of the 12 taxa recorded in all samples) was recorded from the detritus washed down from the glacier. Whilst suggesting that local vegetation supplies many of the propagules, these results also demonstrate the importance of the ice cap as a reservoir of viable propagules which eventually reach terrestrial substrata following ice melt. In a second experiment, soil samples from depths of 0.1–1.0 cm and at 4–5 cm from one of the sub-sites were cultured separately (Smith & Coupar, 1987). Although 17 bryophyte species have been found growing at the Jane Col site, only five developed in these soil cultures. However, none developed from the sub-surface samples indicating an essentially sterile (in terms of bryophytes) zone a short distance below the ground surface.

A second set of experiments was then carried out in the field over a four-year period, to stimulate *in situ* the development of the potential community lying dormant in the soil (Smith, 1991). Polystyrene cloches were placed over the soil at four of the sub-sites used in the first experiment described above, and at one cloche and one control plot per sub-site regular additions of NPK nutrient solution were applied. At sub-site 1 (deglaciated for about 10 years and receiving detritus from the ice cap) after only one year, moss cover increased from 0 to 20% beneath both cloches. By the second year cover had increased to about 40% by January and 75% by March; and by the third year cover reached 90% with numerous populations of several bryophyte species. Older but very sparsely vegetated sub-sites yielded a rather similar number of species but a lower cover, suggesting a smaller propagule pool than at sub-site 1. These results confirm the importance of the ice cap as a source of viable propagules.

Little is known about the precise nature of the propagules in these Antarctic soils. Smith (1991) considers that vegetative structures, the most common of which are detached leaves and stem apices, are by far the most important. Less than 20% of the Signy Island bryoflora produce sporophytes and only three of the common, widespread species that produce capsules in abundance (*Bartramia patens*, *Encalypta patagonica* and *Pottia austro-georgica*) have appeared in the soil cultures. In addition, cloche experiments involving the sowing of macerated and homogenized shoot

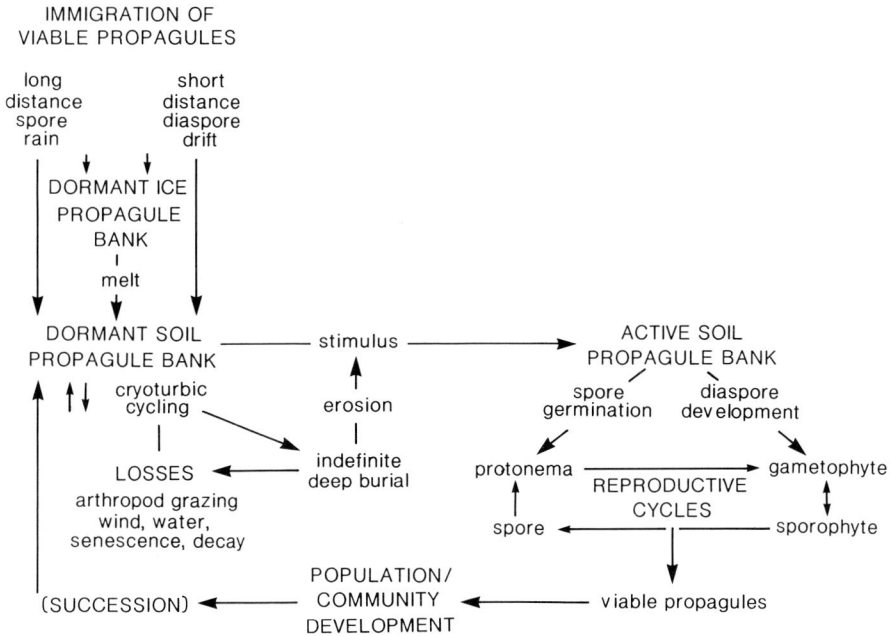

IMMIGRATION OF
VIABLE PROPAGULES

Fig. 6.5. Model of the dynamics of bryophyte propagule banks in Antarctic soils (from Smith, 1991).

apices, confirm the capability of many moss species to establish new individuals and clones from such vegetative fragments.

Based on this work, Smith (1991) has developed the idea of a dynamic propagule bank, which is summarized in Fig. 6.5. Propagules arrive in the dormant soil propagule bank either directly or via the ice propagule bank. They may be destroyed by a number of processes but may also be activated to undergo reproductive cycles and contribute to population development and succession. Smith suggests that temperature and moisture may be the most important stimuli leading to the germination of spores or the development of vegetative structures; but light quality, nutrients, substrate stability and texture may also be involved. Although there may well be differences between the situation at Jane Col and glacier forelands elsewhere, the concept may well prove more widely applicable.

Turning to spermatophytes, Chapin (1991) suggests that the early stages of succession at Glacier Bay are characterized by species without dormancy requirements. This certainly applies to those species listed in Table 6.1, late successional *Picea sitchensis* being the only one requiring a germination cue. Large numbers of small seeds with a high germination success rate and the potential to grow rapidly are envisaged as characteristic of the pioneer and early successional species. However, even with high root : shoot ratios, small plants tend to be vulnerable to adverse environmental conditions, such as desiccation. Thus the existence of favourable microsites may be particularly important for survival of seedlings in the early stages.

In a favourable microsite, a seedling can grow rapidly and quickly to produce a root system rendering it more likely to survive to maturity. This is supported by the fact that of the 49 species of phanerogam on the Omnsbreen nunataks, only six species have not been found flowering and fruiting (Elven, 1980). Four of these, *Empetrum hermaphroditum, Ranunculus glacialis, Salix glauca* and *S. lanata*, occur only as young individuals and will probably reproduce eventually. Only two – *Chamaenerion angustifolium* and *S. nigricans* – may be unable to reproduce at the altitude of the nunataks. Indeed, several species (including *Cerastium alpinum, C. cerastoides, Draba cacuminum* and *Saxifraga cespitosa*) flower and set seed twice per year on the nunataks, a phenomenon apparently unknown in mature vegetation. However, no species on the nunataks grow from diaspore to reproducing plant within one summer.

Although reproductive cycles tend to be shorter in early successional stages than in the later stages, there may be large differences between species in the time required to reach maturity. Such differences strongly influence the pattern of establishment on the Omnsbreen nunataks (Elven, 1980). For example, the pseudo-viviparous grasses, including *Poa jemtlandica*, the most frequent and physiognomically dominant species, reach maturity relatively rapidly. Growth experiments at the lower altitudes of the Blåisen glacier foreland (Elven, 1974, 1980) have established that gemmae of *Poa jemtlandica* and *Poa alpina* var. *vivipara* produce reproductive plants after one year, whilst seedlings of the non-viviparous *Poa alpina* var. *alpina* require in excess of three years. Permanent plots have shown that several seed producing plants (*Arabis alpina, Cerastium cerastoides, Draba cacuminum, Epilobium anagallidifolium, Poa flexuosa, Sagina intermedia, S. saginoides* and *Saxifraga cespitosa*) also reproduce within 1–3 years. All the above species have spread rapidly on the nunataks they have reached. Their establishment rates are much higher than species with longer reproduction cycles (including, for example, *Carex* spp., *Eriophorum* spp., *Luzula confusa*, shrubby *Salix* spp., *Saxifraga stellaris, S. tenuis, Sibbaldia procumbens, Silene acaulis, Taraxacum croceum* and *Veronica alpina*). According to Elven (1980) the same considerations appear applicable to the establishment of cryptogams; acrocarpous and gemmiferous mosses and hepatics establishing much faster than slow-growing lichens and pleurocarpous mosses.

Permanent quadrats investigated by Cooper (1923c) at Glacier Bay, by Jochimsen (1963) in the Alps and by Elven & Aarhus (1984) in Norway all indicate that more individuals germinate than survive and that there is a very high turnover in most species populations. On the Omnsbreen glacier foreland, for example, *Draba cacuminum* occurs in quadrats described in 1973 and 1979. In one of the quadrats (deglaciated for 40–50 years), the total number of individuals of this species increased from 145 to 170, largely due to an increase in the number of seedlings, and pre-reproductive plants. In 1973 these constituted 22% of the total whereas in 1979 they constituted 54%. The absolute number of reproductive plants in 1979 was 30% below the 1973 level. In 1973 only one senescent plant was found; in 1979 the number had risen to 42. According to Elven & Aarhus (1984) seed production is high in good years and probably fairly good in most years, and germination is high, so that selection is occurring in the seedling and pre-reproductive phases.

Surviving individuals tend to become nuclei for further colonization. At Glacier Bay, Cooper (1923b) has described this mode of colonization as being by 'lone individuals' forming 'isolated outposts'. If the rate of glacier retreat is rapid, a patchy vegetation cover results; with relatively slow glacier retreat, consolidation occurs rapidly and a colonization front becomes marked in the landscape. The most important persistent individuals in the early stages at Glacier Bay are *Dryas drummondii* and willows. These species are rapid colonizers, which often produce pure stands by aggregation around isolated pioneer individuals (Cooper, 1931). *Alnus crispa* generally develops from fewer individuals, producing thickets, which also expand by short-distance dispersal. This produces a mosaic of alder within a larger area of less dense willow phase (Cooper, 1931).

A number of workers have commented on the special pattern of growth of mat-forming pioneers, particularly *Dryas drummondii*. Measurements of 23 *Dryas* plants over a three-year period indicate that they grow radially at annual rates ranging from 4.6 to 12.0 cm (Lawrence, *et al.*, 1967) to form mats up to 6.1 m in diameter (Schoenike, 1984). Similar growth rates have been calculated for mats dated by growth rings on the moraines of Klutlan Glacier, Yukon Territory (Birks, 1980a), whilst Viereck (1966) estimates rates of up to 20–25 cm year^{-1} on gravel outwash of Muldrow Glacier, Alaska. As the mats grow radially they root along prostrate branches. This results in freely-flowering and fruiting shoots at the periphery of the mat, whereas the centre may be left as poorly-grown or dead (Birks, 1980a).

Whilst the prominence of some species in succession may be attributable to life cycle characteristics, other species appear unable to establish under the pertaining environmental conditions. Elven (1980) points out the conspicuous absence on the Omnsbreen nunataks of species which seem to prefer mature, humus-rich soils, both in early snowbeds (e.g. *Alchemilla alpina, Anthoxanthum odoratum, Carex bigelowii, Deschampsia flexuosa* and *Nardus stricta*) and heaths (e.g. *Arctostaphylos alpina, Juncus trifidus* and *Loiseleuria procumbens*). These species are common in the surrounding areas and their seeds undoubtedly reach the nunataks. Several of these absent species are mycorrhizal, which may also reduce their colonizing ability (cf. Persson, 1964; Fitter & Parsons, 1987). In most successions on glacier forelands, various groups of later colonizers similarly seem unable to colonize the newly-deglaciated terrain before environmental conditions are modified, either by environmental processes or the reaction of the earlier colonizers (see sections 6.1.3 & 6.1.4). Another good example is provided from below the tree line in front of the Grand Glacier d'Aletsch, Switzerland. There, sciophilous species characteristic of dense forest (e.g. *Listera cordata, Oxalis acetosella, Dryopteris dilatata* and *Adenostyles alliariae*) are still absent from the glacier foreland about 110 years after deglaciation (Fig. 5.24) (Richard, 1968, 1973).

6.1.3   Reaction

Reaction comprises those modifications of the physical habitat attributable directly to the presence of plants. This should be distinguished from the action of the habitat on plants and from any knock-on effects that reaction may have. The latter include not only the possible facilitation of other species, which Connell & Slatyer (1977) see as the basis for one major mechanistic model of succession, but also competitive inhibition, which forms the basis of another of Connell & Slatyer's models.

As soon as a plant begins to grow, it modifies its own environment. Many aspects of the reaction of plants on glacier foreland soils have already been considered in chapter 4, especially in section 4.3.3 where emphasis is given to the close association of additions of organic matter with changes in cation exchange capacity, acidity, moisture retention and nitrogen levels. Only a small number of species may play an important role in bringing about such major changes in the environment. These species, which are relatively abundant and persistent, have been designated 'builders' by Richard (1973, 1975). In front of the Grand Glacier d'Aletsch, they include the moss *Rhacomitrium canescens*, willows (including *Salix helvetica, S. retusa, S. serpyllifolia, S. hastata, S. reticulata*), clovers (*Trifolium pallescens* and *T. badium*), *Empetrum hermaphroditum, Vaccinium uliginosum* and *Rhododendron ferrugineum*.

Two other reaction processes are discussed here, namely stabilization of the land surface and modification of microclimates. Vegetation cover is very effective in stabilizing land surfaces. Such effects are particularly in evidence in the early stages of succession. At Glacier Bay, the 'black crust' phenomenon forms a tough, flexible organic mat which, where present, completely covers the mineral soil and consolidates the upper 1–2 cm (Worley, 1970). It consists of a complex mesh of intertwined shoots, rhizoids and filamentous algae. Beneath this, gametophytes and rhizomes of *Equisetum variegatum* may also be present. This structure diminishes deflation and breaks the force of rain-drop impact (cf. Kuc, 1970). It reduces particle movement in response to freezing and thawing, forms pleats or ripples when affected by minor solifluction movement and is generally resistant to breakage by surface runoff while retarding erosion.

Separate from but often adjacent to areas of 'black crust' are expanding clones of *Dryas drummondii*, the creeping growth habit and extensive root system of which protect the surface and further restrict soil movement (Lawrence *et al.*, 1967). Cooper (1931) notes that the establishment of *Alnus crispa* thickets is particularly effective in preventing the development of gulleys, and that once *Picea sitchensis* forest is present this brings practically complete stability. It seems that in general a mature vegetation cover can withstand all but large scale physical disturbances, such as the melting of extensive buried ice (Birks, 1980a), wind throw (Noble, Lawrence & Streveler, 1984), or flooding (Sidle & Milner, 1989).

It is possible that microbial colonizers play a role in the stabilization of the ground surface prior to colonization by macrobiota. Evidence favouring this idea is available from the recently-deglaciated Jane Col site in the Antarctic (Wynn-Williams, 1986, 1991). Newly-deglaciated terrain is here subject to a variety of physical disturbances from thermal, hydrostatic and freeze-thaw cycles. A combination of epifluorescence microscopy (EPM) and television image analysis (TVIA) enables the direct examination of the nature of soil crusts (Wynn-Williams, 1988). Although bacteria are abundant and fungi are present, the dominant colonizers of the mineral substrate in the centre of sorted circles are cyanobacteria and algae. Bacteria form a biofilm over the rock particles whilst the phototrophs form a heterogeneous and discontinuous crust over the soil surface. Two dominant morphological forms are involved in this crust: filamentous trichomes of cyanobacteria and aggregates of unicellular algae. Both forms are enveloped in mucilaginous sheaths and tend to form into aggregates resembling 'rafts' in a loose mosaic (Fig. 6.6a,b).

Wynn-Williams's (1991) raft analogy reflects the essentially superficial location of the microbial assemblages, which have a tenuous existence

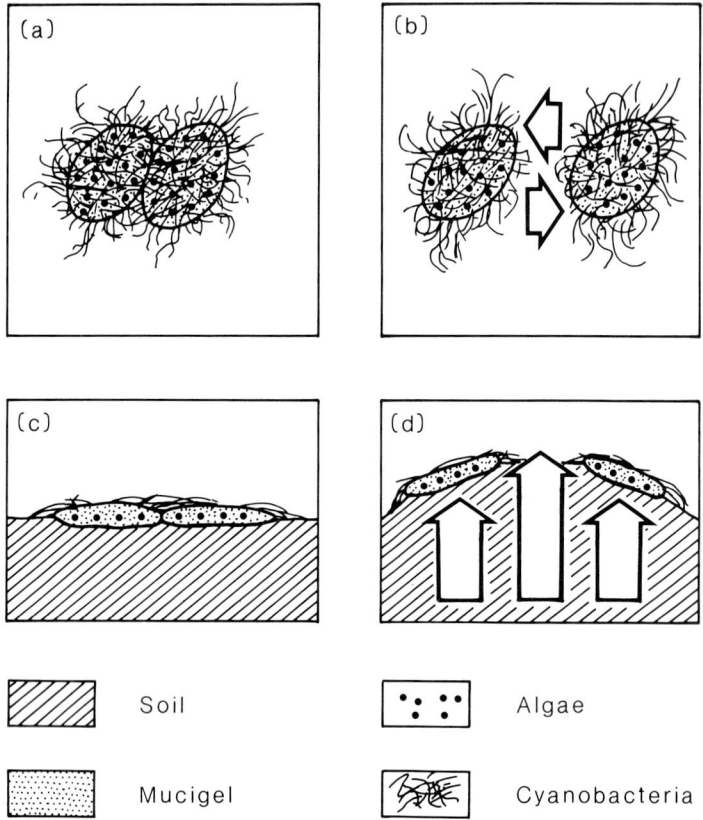

Fig. 6.6. Diagrammatic representation of aggregates of cyanobacteria and algae forming a crust on the surface of Antarctic soils: (a, c) plan and cross section; (b, d) raft-like behaviour during frost-heaving of the soil. Arrows represent vertical frost-heave and associated lateral soil movements (from Wynn-Williams, 1991).

'drifting' over the fluid substratum. Damage to the cells, caused by biologically irresistible frost-heave processes, is minimized by the raft mosaic, which can stretch, break and re-coalesce following heave and ground settling (Fig. 6.6c,d). Polysaccharide mucilage is an integral part of the life-support system. At the same time as protecting the cells from physical damage and desiccation, it acts as an adhesive layer at the ground surface.

Microclimatic effects of vegetation are legion, ranging from the reduction of wind speed near the ground to the modification of temperature, humidity and light regimes. On gravel sandur of Muldrow Glacier, Alaska, Viereck (1966) has described the effect of increasing vegetation cover on wind and snow distribution. The microclimate of the pioneer stage is severe, with no obstacles on the bare gravel bars to the intense winds that blow off the glacier. As no snow accumulates on the bars in winter, the plants are subjected to snow and sand blasting and to wind desiccation throughout the year. In later successional stages, wind speed is reduced near the ground surface by friction with the vegetation. This results in the entrainment of less sand and silt and in the deposition of snow. At the meadow stage, snow forms a complete cover over the low vegetation and many of the willow clumps have even tops that coincide with the winter snow surface. Occasional spruce trees are healthy and green up to a height of 50–75 cm, whereas above this level they show the effects of winter-kill. By the shrub stage, snow depths are increased to 1–1.5 m by the nearly continuous shrub canopy. Finally, in the 'climax' tundra, snow depth is less in conformity to the generally low height and volume of the moss-dominated vegetation. At this stage, the shrub, tree and herb species present are confined beneath a winter snow surface that lies only 20–30 cm above the moss surface.

Apart from such descriptions and occasional spot measurements of temperatures in different vegetation covers (e.g. Elven, 1974), there have been few specific investigations on glacier forelands of the microclimatic effects of vegetation. Most of the microclimatic data available, such as those described in section 3.3.3, do not permit the separation of cause and effect. However, two field experiments carried out by Lindröth (1962) in front of Skaftafellsjökull, Iceland, are worthy of mention (see also, section 3.3.3). Both experiments were performed in intermittent sunshine.

In one experiment, Lindröth compared adjacent sites within and outside of a woodland area, having first removed the field layer from both sites. Within the area of *Betula pubescens* woodland, maximum ground temperatures at 3 cm depth are lower and minimum temperatures are higher than outside the woodland (Fig. 6.7a). In the second experiment, he compared sites within the woodland, having first removed the field layer from some of them. The results indicate that sites without the field layer display higher ground temperatures than the control sites (Fig. 6.7b). In combination, these experiments demonstrate the insulating effect of the two vegetation strata. The *Betula* canopy reduces maximum temperatures by about 2 °C and elevates minimum temperatures by a similar amount at a depth of 3 cm below the ground surface. At 150 cm above the ground, the effects are less; negligible differences in minimum temperatures and a less consistent pattern in maximum temperatures. The effect of the field layer on temperatures at a depth of 3 cm is shown to be at least as important as the tree layer at these sites, maximum temperatures sometimes reaching 3 °C higher in the absence of shade from the field layer.

(a)

(b)

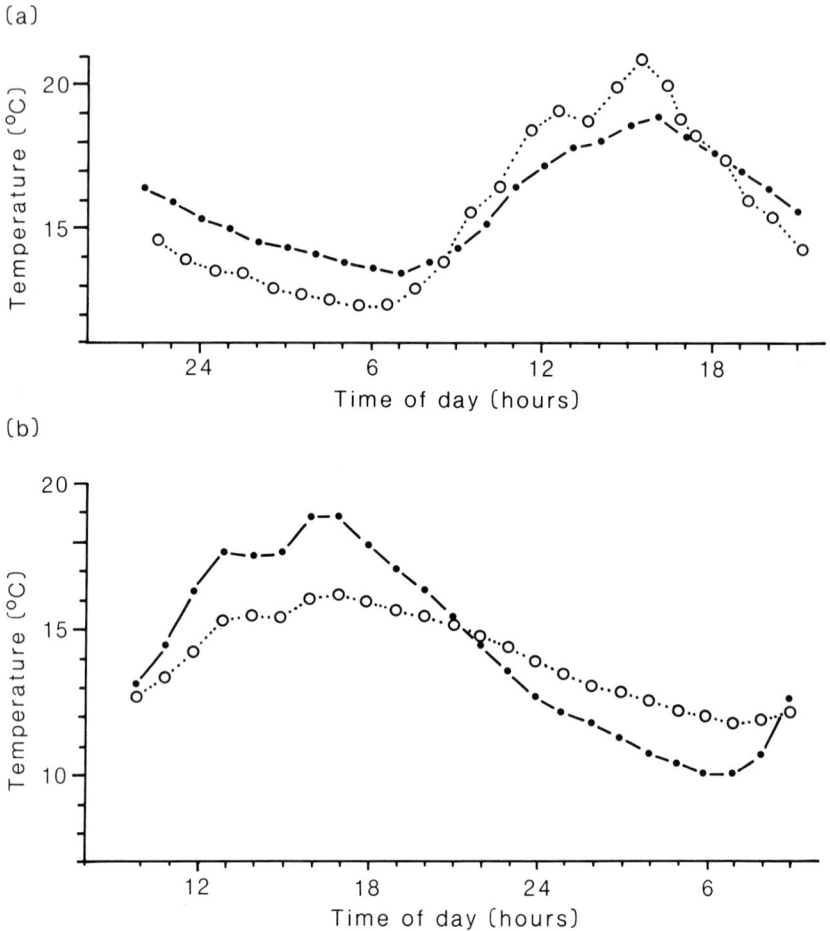

Fig. 6.7. Soil temperatures (3 cm below ground), Skaftafellsjökull, Iceland: (a) within (filled circles) and outside (open circles) birch scrub, 20–21 July, with field layer removed; (b) within birch woodland, 4–5 August, with (filled circles) and without (open circles) removal of the field layer (from Lindröth, 1962).

6.1.4   Facilitation

There is no doubt that reaction is a ubiquitous process in glacier foreland successions. What is much less certain is the extent to which reaction is the direct cause of species' replacements. Facilitation may be defined as the process by which the reaction of one species affects positively the ecesis and/or persistence of a later successional species. That is, the presence, abundance or performance of the later successional species is enhanced by the prior occurrence of an earlier colonizer.

Many workers have proposed that facilitation is an important process on glacier forelands. In no case, however, has it been demonstrated that the prior occurrence of a particular species is necessary and sufficient for the later occurrence of a different species. On gravel outwash in front of the Muldrow Glacier, Alaska, mats of *Dryas drummondii, Astragalus nutzotinensis* and *Hedysarum mackenzii* accumulate sand and silt within their centres creating favourable conditions for establishment of the acrocarpous moss *Ditrichum flexicaule* (Viereck, 1966). Either the accumulated sediment or the presence of the mosses apparently lead to the death of the mat plants, especially *A. nutzotinensis*, leaving a seed bed for other species, such as *Elymus innovatus, Astragalus tananaica, Oxytropis gracilis, Salix* spp. and *Shepherdia canadensis*. Although these species grow from the decaying stumps of the mat formers, they also appear capable of germinating and growing on bare gravel.

Birks (1980a) suggests that facilitation is involved in several species' replacements in the early stages of succession on the Klutlan moraines. On flat or gently-sloping terrain, acrocarpous mosses are again supported by wind-blown silt trapped around *Dryas drummondii* plants. Up to 6 cm of nitrogen-rich humus accumulates beneath the centre of the *Dryas*-mat, where *Salix glauca, S. alaxensis* and *Populus balsamifera* establish. These species eventually overtop the *Dryas* to form the *Salix–Shepherdia* nodum. On steeper ground ($\geqslant 5°$), the nitrophilous 'weeds', *Descurainia sophia, Funaria hygrometrica, Aongstroemia longipes* and *Marchantia alpestris* are virtually confined to the *Hedysarum mackenzii* nodum. As cover increases to 80–95% in the *Hedysarum–Salix* nodum, sufficient shade is cast to allow the first development of a carpet of pleurocarpous mosses, such as *Tomenthypnum nitens, Aulacomnium palustre, Hypnum revolutum* and *Thuidium abietinum*. In front of Robson Glacier, British Columbia, *Hedysarum* and two species of *Dryas* are described as 'centres of radiation' for other species (Sondheim & Standish, 1983: 503).

Smith (1984, 1985b, 1987, 1991) and Wynn-Williams (1986, 1991) hypothesize that the microbial colonizers of recently-deglaciated terrain in the Antarctic may be prerequisites for colonization by bryophytes and other macrobiota. It appears that the microbial phase is itself a short succession of bacterial, fungal, cyanophyte and chlorophyte populations which develop sequentially on the virgin substratum. This micro-succession is much more rapid on moist, sheltered and fine-textured soil than it is on dry, exposed and smooth rock surfaces. The pioneers or 'primary colonizers' are believed by Smith (1991) to be crucial in stabilizing otherwise more mobile surfaces (see section 6.1.3) and in laying the foundation of an organic and nutrient-enriched medium in which more complex organisms

Table 6.4. *Numbers of bacteria in subglacial silty sediment and in the rhizosphere of proglacial soils (terrain age 25 years) at the Glacier de Saint-Sorlin, French Alps (after Moiroud & Gonnet, 1977).*

Incubation temperature (°C)	No. of bacteria per gram dry weight ($\times 10^6$)		
	Subglacial sediment	Proglacial soil	Proglacial rhizosphere
2	0.4	0.1	8.0
20	1.0	0.6	10.0
28	0.05	0.1	9.0

may become established. In the Antarctic fellfield ecosystem, the 'secondary colonizers' are generally bryophytes (particularly short, acrocarpous mosses) or microlichens (especially lepraroid and placodioid forms). Smith (1991) further considers that once these secondary colonizers have established an open patchwork of transient populations, conditions are suitably advanced in terms of substrate stability, nutrients and foci for trapping propagules, for a population explosion of 'tertiary colonizers'. These include pleurocarpous mosses, liverworts and macrolichens. As yet, however, there is no evidence bearing on the extent to which the occurrence of later colonizers depends on the prior occurrence of the microbial and other early colonizers.

A possible facilitative role for bacteria is also suggested by the data of Moiroud & Gonnet (1977) who have shown by incubation experiments that large numbers of bacteria are present in the subglacial sediments of five Alpine glaciers. On terrain deglaciated for 25 years in front of the Glacier de Saint-Sorlin, bacteria are found in comparable numbers in the new soil (Table 6.4). Their numbers are commonly at least an order of magnitude greater in the rhizosphere (that is, in association with soil particles adhering closely to plant root systems). The magnitude of this rhizosphere effect varies between plant species. The ratio R/S (the number of bacteria found in association with plant roots as a proportion of the number in adjacent soil) increases from *Saxifraga moschata* (4), *Trisetum distichophyllum* (6) and *Poa alpina* (8.3) to *Leucanthemum alpinum* (11) and *Trifolium badium* (16). It must be stressed, however, that these data do not permit the differentiation of cause from effect. A change in the kinds as well as the quantity of bacteria with the appearance of plants suggests that the microbiota may be responding to the presence of the plant. Furthermore, the presence of large numbers of bacteria in the rhizosphere is likely to have negative as well as positive effects on plants (Fitter & Hay, 1987).

Other examples of facilitation are provided by the frequent association of later colonizers with acid humus and/or shade beneath a tree cover (Fægri, 1933; Ellenberg, 1988; see also section 6.1.2). In addition, nitrogen fixers are often considered to facilitate the establishment and/or growth of later colonizers. In most cases, however, the evidence is circumstantial, as in the examples above. The best known and most widely discussed example of facilitiation on glacier forelands involves the role of *Dryas drummondii* and *Alnus crispa* as nitrogen fixers at Glacier Bay. Even there, the evidence for facilitation as the driving force of successional change is equivocal (Chapin & Walker, 1990; Walker, 1991).

Several observations and experiments made at Glacier Bay are consistent with facilitation, especially by *Alnus*. First, there is the strong correlation between symbiotic nitrogen fixation by *Alnus* and the increase in soil nitrogen levels (see sections 4.2.7 and 4.3.3), which is followed by the development of spruce forest (Crocker & Major, 1955). Later, in maturing spruce stands (100–160-years-old) declining productivity is correlated with a decline in nitrogen levels in above ground biomass, which suggests a nutrient limitation to growth (Bormann & Sidle, 1990).

Second, observations by Lawrence (1958, 1979) suggest that all pioneer species except for the nitrogen-fixers show characteristics that are interpreted as symptoms of nitrogen deficiency, namely a yellowish colouration, a very slow growth rate and a prostrate growth form. It is highly pertinent that the pioneers include early colonizing individuals of *Salix* spp., *Populus balsamifera* ssp. *trichocarpa*, *Picea sitchensis* and *Tsuga heterophylla*, all of which are important in later successional stages where they grow erect. In contrast to the sickly pioneering individuals, others have a healthy blue green colouration where small natural additions of nitrogenous matter occur in the form of faeces, animal bones or exposed fossil soil layers.

Third, experimental additions of inorganic fertilizers with and without nitrogen, organic fertilizers and leaf litter were made in the summer of 1949 to six sets of 12 plots laid out on a till plain deglaciated for 12 to 20 years (Lawrence, 1951, 1953). Each plot contained a small established cottonwood (*Populus balsamifera* ssp. *trichocarpa*) and the very few alders growing in the vicinity of the plots were removed. Each set of plots consisted of two controls and the following treatments: nitrogen (N); phosphorus and potassium (PK); NPK; NPK plus traces of other essential elements (NPKT); blood meal; blood meal charcoal; fresh alder leaves; alder humus; alder nodules; 12 transplanted alder seedlings. Fertilizers containing nitrogen were applied at a rate of about 180 kg ha^{-1} of available nitrogen. Treatments stimulating shoot growth one year later were N, NPK, NPKT, blood meal, fresh alder leaves, and alder humus. Growth stimulation by alder leaves and humus was as great as that by N, but not as great as by NPK or NPKT. Thus, although a shortage of nitrogen may be the major limitation on growth, the availability of other nutrients may also be of importance.

According to Lawrence *et al.* (1967) the application of fresh alder leaves proved far more effective than any of the other treatments in continually stimulating the growth of the cottonwoods over the period 1949–55. Stimulation of cottonwood height growth by the transplanted alder seedling had just begun by the 1955 growing season. It is perhaps also significant that *Dryas* plants which were growing in the vicinity of some of the treated cottonwoods and which inadvertently received nitrogen additions, showed no signs of enhanced growth as compared to *Dryas* plants unaffected by nutrient additions (Lawrence *et al.*, 1967).

Fourth, *Lupinus* spp. and *Alnus crispa* appear to have a marked stimulating effect on other flowering plant species growing close to them. Willows, grasses and fireweed growing in association with lupins appear to bloom and grow several times faster than plants growing separately; whilst young cottonwoods surrounded by alders grow about three times as fast as plants growing separately (Lawrence & Hulbert, 1950). Similar measurements involving *Dryas drummondii* have been reported in greater detail but the stimulating effect on cottonwoods appears to be considerably less than that of alder (Schoenike, 1958; Lawrence *et al.*, 1967). In the latter case, 22 naturally-occurring saplings of cottonwood, ranging in height from 60 to 125 cm and in age from 11 to 24 years, were sampled. Mean annual height growth of 11 *Dryas*-associated plants and 11 plants growing away from *Dryas* is summarized in Fig. 6.8. The mean height growth of the *Dryas*-associated plants is greater by a factor of 1.2, although it is only during the last four years of the series that the difference becomes marked. Lawrence *et al.* (1967) cite the slowness of the decay of the semi-evergreen *Dryas* leaves as a possible reason for the apparently slow response of the cottonwoods.

Growth stimulation of *Picea sitchensis* by *Alnus* is also suggested by unpublished data collected by R. E. Schoenike in August 1952, from a site south of Goose Cove (D. B. Lawrence, pers. comm., 1990). The data relate to the height and age at breast height of 17 sprucelings growing in association with alder (generally within a 1.5 m radius, and mostly at the edges of alder thickets where there is adequate light for growth) and a similar sample of sprucelings growing away from alder (mostly on *Dryas*-mats). Average height growth of the alder-associated plants (4.115 cm yr^{-1}) was about 50% greater than the average growth of those located away from alder (2.621 cm yr^{-1}). According to a Student's t test, the

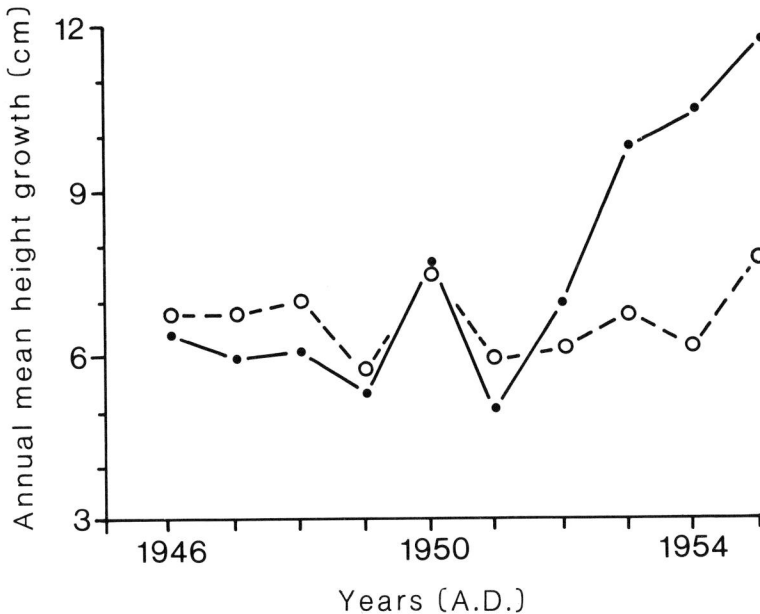

Fig. 6.8. Annual mean height growth, 1946–55, of the leading shoot of *Populus balsamifera* ssp. *trichocarpa*, Glacier Bay, Alaska. Solid line represents *Dryas*-associated individuals; broken line represents individuals growing away from *Dryas* ($n = 11$ in each case) (from Lawrence *et al.*, 1967).

differences in height between the two samples are statistically significant ($p \leqslant 0.05$), whereas their ages (15.5 yr and 17.8 yr, respectively) do not differ significantly.

In summarizing these studies at Glacier Bay, Lawrence (1984) considers that nitrogen-fixers are able to reach reproductive maturity first, they then stimulate the growth of other species which established earlier. According to D. B. Lawrence (pers. comm., 1990) this stimulation process is distinct from facilitation, which implies prior establishment of the facilitators. *Dryas* stimulation appears slight, possibly accelerating succession by 20–30 years (Schoenike, 1984). It is not an essential precursor of *Alnus*, which is able to establish in certain habitats where *Dryas* does not grow at Glacier Bay, and also in the chronosequences in front of the neighbouring Herbert and Mendenhall Glaciers, where *Dryas* is absent (Crocker & Dickson, 1957).

Lawrence *et al.* (1967) believe *Alnus* to be more effective, possibly raising the nitrogen supply to a level enabling the spruce forest to achieve dominance. Cooper (1923b) points out that the pioneering individuals of

Picea sitchensis probably do not form the mature spruce forest. They persist in a suppressed condition beneath the later arrivals which may be benefiting from the nitrogen-fixing activity of *Alnus*. There is, however, some conflict of evidence. The establishment of spruce forest does not always depend on the previous dominance of alder. F. R. Stephens (unpublished manuscript), recognizes a distinct chronosequence in which alder is absent. The chronosequence without the alder stage is characteristic of well-drained moraines, which soil tensiometer readings from near the Mendenhall Glacier confirm to be more prone to drought. *Picea* seems to invade areas of pure willow thicket as vigorously as it invades alder (Cooper, 1923b), and may even increase more rapidly in the comparatively open willow thickets (Cooper, 1931). Sprague & Lawrence (1960) nevertheless suggest that, 100 years after deglaciation, original alder-sequence areas of forest exhibit larger conifers than original willow-sequence areas. F. R. Stephens (unpublished manuscript) found both height and diameter growth of spruce to be consistently greater at sites characterized by an alder stage, and correlated this with higher levels of soil nitrogen.

Chapin & Walker (1990) are at pains to point out that the available experimental evidence from Glacier Bay relates to *Populus* rather than *Picea* (but see Schoenike's unpublished data above). On the basis of the present evidence, Walker (1991) concludes that it is premature to state that *Alnus* facilitates the growth of the later successional species. He also suggests that the direct interactions of nitrogen fixers with other species are more competitive than facilitative, a proposition that has been more thoroughly tested in relation to Alaskan riparian succession (Walker & Chapin, 1986; Walker, Zasada & Chapin, 1986). Further critical experiments are clearly required and have been initiated at Glacier Bay (F. S. Chapin III & L. R. Walker, pers. comm.).

6.1.5   Competition

Competition is used here to refer to the negative or inhibitory effects of species' interaction, which result from the common exploitation of limiting resources. It is viewed as the antithesis of facilitation. Competition, which may involve, nutrients, light, moisture, or any other resource, is likely to occur wherever plants are in close proximity to each other. In the context of glacier forelands, therefore, it probably occurs to some extent on all but the youngest terrain and in all stages of succession. As in the case of facilitation, however, proven examples of species' replacements caused by competitive inhibition are difficult if not impossible to find.

Although established wisdom suggests that competition increases during glacier foreland succession, it should not be assumed that competition is absent in the early stages. Large root : shoot biomass ratios (see section 5.2.4) suggest that roots may be in contact even where aerial parts are not. Several workers have commented on the well-developed root systems characteristic of pioneer species (e.g. Palmer & Miller, 1961; Moiroid & Gonnet, 1977). On outwash in front of Muldrow Glacier, many long willow roots occur just beneath the surface of the gravel, extending 10 to 15 m from parent plants and indicating the possibility of root competition between widely-separated clumps (Viereck, 1966).

Many species nevertheless appear to be favoured by the relatively low-levels of competition on open terrain in the early stages of succession. In front of Storglaciär, northern Sweden, Stork (1961) suggests that early colonizers thrive due to a lack of competition and high disturbance levels. Under such conditions a large number of species may be able to coexist in the same area. The early peak in species diversity that characterizes some glacier foreland successions may be explicable in similar terms (see section 5.2.5 & Fig. 5.20). Observations by Elven (1978b, 1980) indicate that species with very narrow niches elsewhere may attain local optima on glacier forelands and nunataks. This applies to *Draba cacuminum*, *Poa herjedalica*, *P. jemtlandica* and *Sagina intermedia* in the Finse area of southern Norway, and to *Carex bicolor*, *C. rufina* and *Myricaria germanica* at Svartisen in northern Norway.

At Skaftafellsjökull, Iceland, Persson (1964) considers that a combination of competition and soil development may be the cause of the disappearance of species later in succession. Zollitsch (1969) points to the close association of the decline in species diversity on the foreland of the Pasterze Glacier, Austria, with the attainment of a complete vegetation cover 80–100 years after deglaciation. Elven & Ryvarden (1974) attribute a similar decline in diversity after 30–40 years on the Bläisen glacier foreland, southern Norway, to competition. Except in habitats such as riverbanks, solifluction areas and small talus (scree) slopes, most of these first invaders appear to be 'squeezed out' later in succession. In addition, species commonly regarded as chionophilous in mature vegetation, such as *Salix herbacea*, show no preference for snowbed sites on relatively young terrain (Elven & Ryvarden, 1974; Elven, 1978b). Only on the older terrain, where competition is more intense, do these species become restricted to areas of late snowmelt. This conclusion is consistent with recent studies on the growth of *S. herbacea* in mature vegetation (Wijk, 1986a). Wijk's studies indicate that although maximum cover and biomass of this species is found in areas where snow persists longest, maxima in the size and growth of

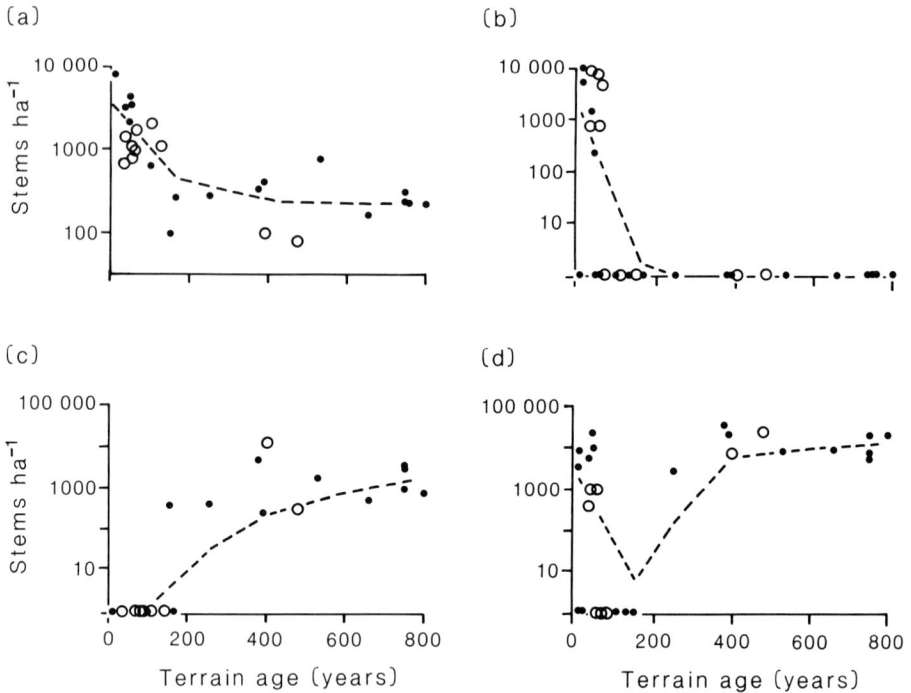

Fig. 6.9. Stem density in relation to terrain age for (a) conifer overstorey, (b) *Alnus sinuata*, (c) conifer advance regeneration, (d) *Vaccinium* spp., Nooksack Glacier, Washington, U.S.A. Open circles represent stands begun after glacier retreat; filled circles represent stands begun after other disturbances. Broken lines were fitted using a locally-weighted, scatter-plot smoothing technique (from Oliver, Adams & Zasoski, 1985).

individual shoots are found where the growing season is longer but competition is more intense.

According to Cooper (1931) and Lawrence (1958, 1979), the pioneers at Glacier Bay cannot endure shade and disappear from the succession once they are overtopped by taller species of the next stage. *Dryas*, which grows horizontally, is replaced by *Alnus* and the arborescent *Salix* spp., which grow erect. In mid-succession, Cooper (1931) envisages a mosaic of *Salix*, *Alnus* and *Picea*. Where these three species grow together they are in competition for light, with *Salix* the weakest and *Picea* the strongest. Although *Alnus* is able to establish new seedlings in areas of *Salix*, it appears unable to reproduce successfully either beneath its own canopy or beneath a canopy of *Picea* (Lawrence, 1979). In the long run, therefore, the most probable sequence of species is *Salix→Alnus→Picea*.

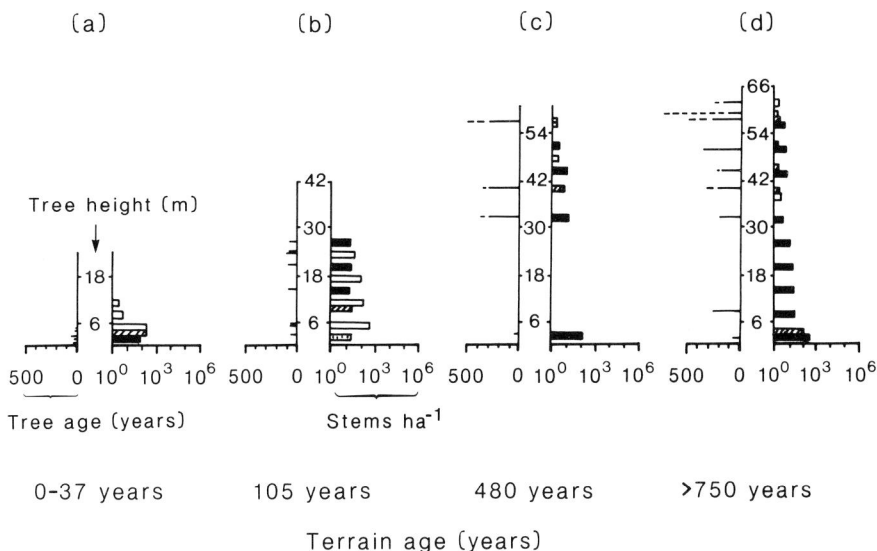

Fig. 6.10. Tree-species composition and structure (stem density, age and height), for selected stands representative of four successional stages, Nooksack Glacier, Washington, U.S.A.: (a) stand initiation stage; (b) stem exclusion stage; (c) understorey reinitiation stage; (d) old-growth stage. Terrain age is indicated. Bars (right side) represent tree species: unshaded, *Tsuga mertensiana*; solid black, *Abies amabilis*; dotted, *T. heterophylla* (bottom bar) in (b); striped, other species. Lines (left side) indicate tree age determined by increment core or cut disc (broken lines represent estimates where cores were too short to reach pith) (from Oliver, Adams & Zasoski, 1985).

In front of Nooksack Glacier, Washington, data on stand composition and structure (stem density, ages and sizes) are consistent with the existence of competitive inhibition by trees in mid-succession (Oliver, Adams & Zasoski, 1985). Fig. 6.9 shows relationships between stand age and stem density for: (a) conifer overstorey; (b) *Alnus sinuata*; (c) advance regeneration (trees distinctly younger than the overstorey and which germinate and live beneath the overstorey but grow little until released); and (d) *Vaccinium* spp. These data reflect the four successional stages defined by Oliver (1981), namely: (1) stand initiation; (2) stem exclusion; (3) understorey reinitiation; and (4) old growth (see also section 5.4.2). Tree age and size structure of stands representative of these four stages are shown in Fig. 6.10. In the stand initiation stage, which lasts for some 60 years, the number of stems of overstorey trees is high, alder is only found in very young stands and before canopy closure, and *Vaccinium* spp. are present in some young stands (Fig. 6.9). This stage has a continuum of tree sizes, and tree-age distributions reflect the entire time interval since deglaciation (Fig. 6.10a). Associated with terrain ages of 60 to 150 years (the stem exclusion stage), alder stems are no longer found, *Vaccinium* spp. stem numbers reach a minimum (Fig. 6.9) and there are relatively few young trees (Fig. 6.10b). This stage corresponds with dominance by a relatively small number of overstorey stems, which have grown considerably in height, and the elimination of others of the same and different species by intra- and interspecific competition. On older terrain (>150 years) the number of overstorey stems declines more slowly, and *Vaccinium* spp. (and other understorey species) reappear or increase at the same time as the appearance of advance regeneration (Fig. 6.9). This appears to indicate that the dominant trees are no longer able to exclude new stems completely, a conclusion supported by two distinct height classes and two peaks in the age distribution (reflecting an overstorey age class dating from deglaciation and a much younger one originating later from the understorey) (Fig. 6.10c). Barring intervening disturbances, by the old-growth stage (*ca.* >600 years) advance regeneration merges with the overstorey to produce characteristically broad height and age distributions (Fig. 6.10d).

A more complex set of processes, involving competition, facilitation and species' growth rates is proposed by Burrows (1990) to explain the relatively long succession on the foreland of Franz Josef Glacier, New Zealand (see also Stevens, 1963, 1968; Wardle 1977, 1980). The shrubs *Coriaria arborea* and *Olearia avicenniaefolia* rapidly supplant the early pioneers and modify the light, moisture and surface soil environments. Tree seedlings, which germinate abundantly beneath the shrub canopy, are at first suppressed by *Olearia* competition. Later, about 30 years after deglaciation, as shrubs age and the canopy opens, quick-growing angiosperm trees (e.g. *Melicytus ramiflorus* and *Aristotelia serrata*) take advantage of nutrients released by recycling and of reduced pH levels, to overtop the shrubs which die out. The resulting low forest of angiosperm trees is replaced by rata (*Metrosideros umbellata*) after a few hundred years. Although *Metrosideros* is relatively long-lived, it only occupies the site for one generation. Once well established, its large, dense canopy tends to inhibit the growth of most other trees except shade-tolerant species. Fallen trees are usually replaced by kamahi (*Weinmannia racemosa*), tree ferns (which also cast a dense shade) or other shade tolerant tree species. On moraines older than about 600 years, rata forest is replaced by a transitional kamahi forest. This is eventually replaced, on terrain older than about 1000 years, by a mature kamahi-mixed podocarp forest in which the slower-growing and shade-tolerant gymnosperms, miro (*Prumnopitys ferruginea*) and totara (*Podocarpus hallii*) are increasingly important. This forest probably persists for several thousand years before giving way to a mature rimu (*Dacrydium cupressinum*) – kamahi forest.

6.1.6   Allelopathy, herbivory and pathogens

Several other biological processes may be involved in glacier foreland successions but there is little information on which to judge their significance. Chemical inhibition of one species by another (allelopathy) is distinct from competition as it involves the addition of a deleterious factor to the environment rather than the reduction of a necessary resource. It is known or suspected in successions elsewhere (Muller, 1966, 1969; Rice, 1974, 1979) and the writer has observed zones of apparent chemical inhibition surrounding certain crustose lichens on rock surfaces in Jotunheimen. However, no other cases have as yet been reported from glacier forelands.

Observations on the effects of animals are more frequent, and there have been specific studies of faunal successions on glacier forelands in the Alps (Janetschek, 1949; Macfadyen, 1963), at Nooksack cirque, Washington (Weisbrod & Dragavon, 1979) and at Glacier Bay (DeLong, 1966; Trautman, 1966; Good, 1966; Rudolph, 1966; Milner, 1987; Milner & Bailey, 1989), but there is an absence of detailed research on plant/animal interactions. Animals appear to play only a minor role in diaspore dispersal and ecesis (see section 6.1.1), although individual cases are known where various groups reach at least local importance. Whilst grazing cattle, sheep and goats are of obvious importance on sub-alpine glacier forelands in Scandinavia (Fægri, 1933), Iceland (Persson, 1964) and the Alps (Ellenberg, 1988), detailed research is lacking and little is known about the effects of wild animals. It is possible that herbivores may be a major cause of plant mortality once seedlings survive the initial establishment phase (cf. Edwards & Gillman, 1987; Chapin, 1991). Reindeer, hare, ptarmigan and several species of rodent have been observed grazing on the Storbreen glacier foreland. Chamois are known to graze on the foreland of the Grand Glacier d'Aletsch (Richard, 1973) and hyrax occurs on moraines of intermediate age on Mount Kenya (Coe, 1967). On Svalbard, seabirds appear to have a significant effect on the existence of nitrophile plants and may facilitate the occupation of recently deglaciated terrain by *Cochlearia officinalis* and *Phippsia algida* (Kuc, 1964). On the glacier forelands of sub-Antarctic Heard Island, Scott (1990) considers that disturbance and nutrient enrichment by seals and penguins are major controls on vegetation patterns. The establishment of *Poa cookii* in particular seems almost invariably linked to animal activity.

Pathogens provide another possible cause of successional change on glacier forelands (cf. Connell & Slatyer, 1977; Burdon, 1987). Only two published cases are known to the writer. First, at Glacier Bay, the spruce bark beetle (*Dendroctonus rufipennis*) has caused heavy damage to *Picea sitchensis* forest in recent years (Eglitis, 1984; Bormann & Sidle, 1990) and changes in the understorey are taking place as the forest canopy dies (Lawrence, Noble & Tilman, in prep.). Second, Pignatelli & Bleuler (1988) have described effects of the larch bud moth or grey larch tortrix (*Zeiraphera diniana*) on *Larix decidua* at sites deglaciated by the Morteratsch Gletscher, Swiss Alps.

6.1.7   Stabilization

A fundamental result of the biological processes considered above is an apparent tendency for progression towards relative stability of community structure and function in the later stages of succession. Rapid directional changes in the early stages result in communities that are less likely to be replaced by different species than by similar ones. According to Clements (1928) stabilization is due to the attainment of dominance by species that, by their reactions, inhibit new invaders rather than facilitate them. Such mature communities may also be viewed as being characterized by species and life forms that make the most efficient use of available resources under prevailing environmental conditions. Thus, a relatively mature ('climax') community might be envisaged that is self-perpetuating and in a steady-state equilibrium under prevailing environmental conditions (cf. sections 4.4 & 6.2.1).

Many successions inferred from glacier foreland chronosequences are consistent with this concept of relative stabilization. Pioneer and early successional stages tend to be of relatively short duration and are replaced on older terrain by more persistent communities. However, the timescale for this change varies considerably and few details are available concerning the causal processes underlying stabilization and subsequent maintenance. Furthermore, it is doubtful whether a true steady-state equilibrium is attained, even in the absence of allogenic environmental change.

Under the most favourable environmental conditions, glacier foreland succession to relatively stable forest may take < 100 years, although 100 to 200 years is more typical (see sections 5.1.1 & 5.3.3), and 200 to 300 years is required for the restoration of climax communities of pine and birch in front of Djunkuat Glacier in the Caucasus (Turmanina & Volodina, 1978). Above and beyond the tree line, persistent heath and meadow communities generally develop within 200 to 400 years where conditions are favourable. Similar climax types develop after not less than 800–900 years in the Tien Shan according to Solomina (1989), who cites 800–1000 years for communities within the tundra zone of Eurasia, 1000–1500 years for Svalbard, and 3000–3500 or even > 9000 years for harsher climates.

Clearly there are problems of comparability where changes are relatively slow in later successional stages and where estimates involve different definitions of 'maturity' or 'climax' and/or different criteria of stability. Gross physiognomy and the dominant life form may reach a relatively stable state before other attributes of the vegetation. Even in the case of such simple criteria, however, there may still be difficulties. At Glacier Bay, for example, Cooper (1923a) recognizes that there are differences between the 'young climax forest' (or 'young subclimax') on the glacier foreland and the 'old climax forest' beyond the 'Little Ice Age' glacier limits. A distinct line of demarcation separates the younger forest from the older forest. There are few floristic differences but the latter is characterized by a darker green tone, trees with weather-beaten forms, an abundance of standing dead trunks, and the presence of broad avalanche tracks. A second example is provided by Wardle (1970) from the glacier foreland of Strachan Glacier, New Zealand. He notes that although the silver beech (*Nothofagus menziesii*) colonizes within 10 years of deglaciation, it only attains a near-climax appearance on terrain dating from the early seventeenth century.

The time interval required to attain a mature state is likely to be longer in the case of species composition than in the case of vegetational physiognomy. At Storbreen, the dominant species on terrain deglaciated for about 220 years are similar to those beyond the glacier foreland boundary, but the subordinate species differ considerably in both number and abundance (Matthews, 1978a). This difference is reflected in vegetation types and gradients defined by multivariate classification and ordination techniques (Matthews, 1979a–c; see section 5.3.2). Slow successional changes are therefore still occurring on the glacier foreland after 220 years. A similar conclusion is reached by Elven (1975) who concludes that only communities of simple structure (e.g. those of extreme snowbeds and springs) have reached a mature state after 220 years of succession on moraines in the Finse area of southern Norway.

There is evidence to suggest that reaction does not always lead to stabilization, particularly when successions are viewed on a long timescale. Vegetation on the Harris Creek Moraine (HCM), estimated to have formed between 615 and 1220 years ago, shows signs of possible retrogressive succession in front of Klutlan Glacier (Birks, 1980a). On younger moraines *Picea glauca* grows increasingly dense and tall with increasing moraine age, culminating in trees up to 12 m high and 18–46 cm in diameter on moraine

K-V (moraine age about 360 years). Beneath this *Picea–Ledum* nodum, a layer of well-humified organic-rich humus of thickness 14–20 cm overlies mineral substrata. In contrast, the predominant vegetation on HCM is depauperate *Picea glauca* forest of the *Picea–Rhytidium* nodum, with widely-spaced, old stunted trees (up to 14 cm high and 68 cm in diameter) most of which have narrow rings and heart rot. Soils consist of up to 8 cm of unhumified moss-rich litter overlying 2–3 cm of organic humus with abundant wood remains, which is in turn underlain by up to 25 cm of humified forest soil (mor humus). At the base, 2–3 cm of pale white or grey sand overlies coarse gravel.

A second example is described by Burrows (1990) from the foreland of Franz Josef Glacier, New Zealand. Of several forest phases (see section 6.1.6), mature rimu–kamahi forest (dominated by *Dacrydium cupressinum*, *Weinmannia racemosa* and *Quintinia acutifolia*) characterizes terrain deglaciated for *ca* 5000 to *ca* 12 000 years. Although this might be regarded as steady-state vegetation, forest of lower stature, with understorey species that are tolerant of very low soil fertility and waterlogging, is present on nearby surfaces that are older than 12 000 years. Burrows suggests that given sufficient time and continued strong leaching, the mature rimu-kamahi forest gradually declines in parallel with a diminishing nutrient status of the soil and increased gleying.

Somewhat similar retrogressive tendencies have been detected at Glacier Bay in *Picea sitchensis* forest, which may eventually be replaced by muskeg. Although other processes may well contribute, including longer-term allogenic effects on soils, retrogression may be initiated by nutrient immobilization (Bormann & Sidle, 1990). These authors point out that as rapidly as the spruce forest develops highly productive vegetation, it begins to deteriorate. The decline in spruce productivity (Fig. 6.11a) appears to be strongly related to declining foliar nitrogen and phosphorus concentrations in the young spruce stand (Fig. 6.11b–c). They suggest that nitrogen and phosphorus are immobilized by accumulation in soil O and Bh horizons or in spruce bolewood (Fig. 6.12).

Concentrations of most macro- and micro-nutrients in the O horizon are higher at the alder stage than at the spruce stage (Bormann & Sidle, 1990). This is consistent with rapid nutrient uptake and cycling back to the O horizon at the alder stage, followed by uptake and retention, especially of phosphorus in the stems, at the spruce stage. Nutrients returned to the O horizon beneath spruce appear to be further immobilized by a lack of effective decomposers, poorly decomposable litter, microclimate, or a combination of these factors. The exception to this pattern is exchangeable iron, which is 10–16 times more concentrated in spruce O horizons

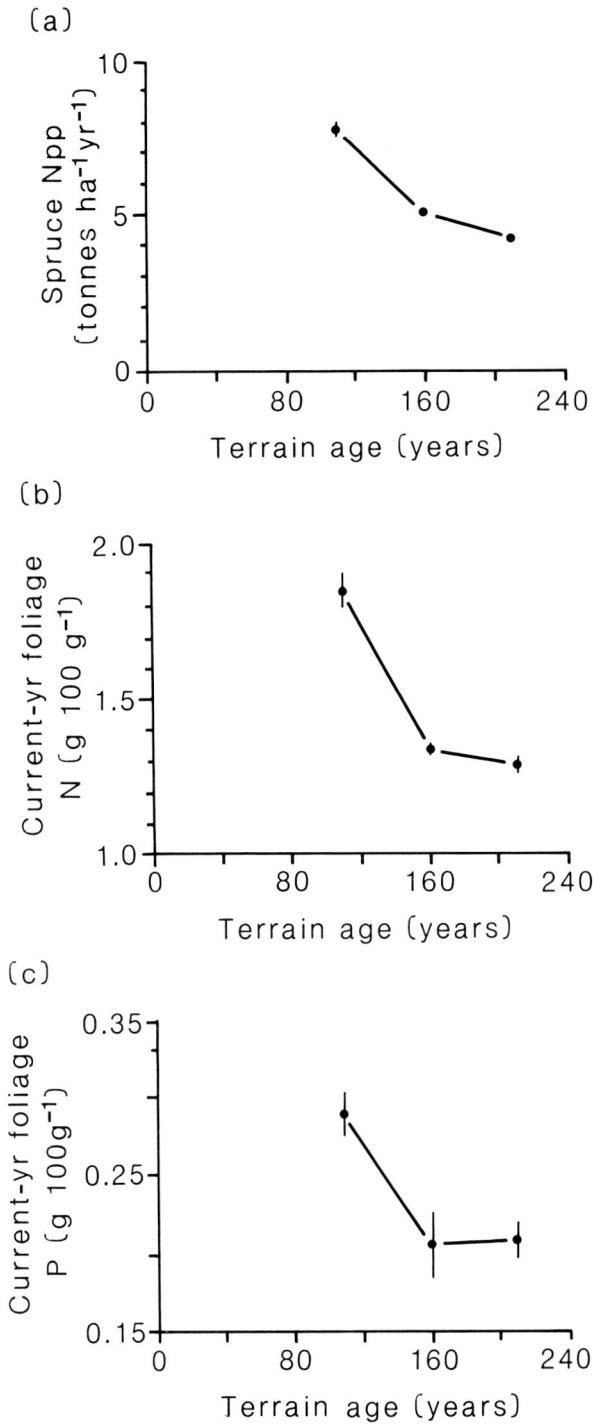

Fig. 6.11. *Picea sitchensis* productivity and foliar nutrient content in developing spruce forest, Glacier Bay, Alaska: (a) above-ground net primary production; (b, c) nitrogen (%) and phosphorus (%), respectively, in current-year foliage. Standard error bars are shown (from Bormann & Sidle, 1990).

(a)

(b)

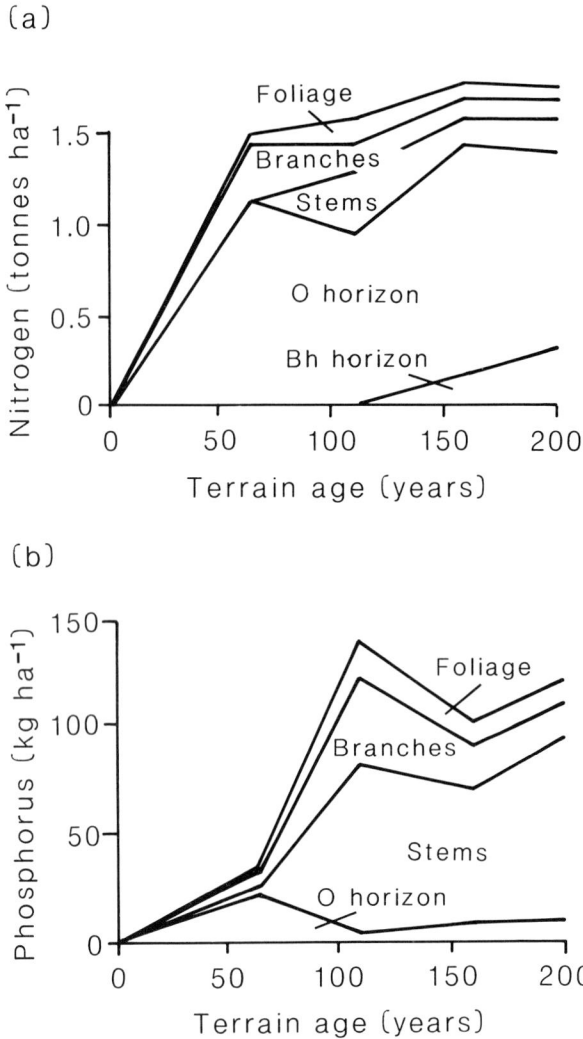

Fig. 6.12. Distribution of (a) total nitrogen in above-ground vegetation and soil horizons, and (b) total phosphorus in above-ground vegetation and available phosphorus in the O horizon, Glacier Bay, Alaska (from Bormann & Sidle, 1990).

compared with alder O horizons. Together with magnesium, potassium, zinc and total nitrogen, exchangeable iron also increases in the developing Bh horizon, a pattern consistent with acidification and podzolization.

In the longer term, soil disturbance by windthrow or other processes may increase or maintain nutrient availability (see section 5.4.2). Alternatively, further deterioration may occur. That the final stage of plant succession in southeastern Alaska is muskeg (peatland) has been proposed by many workers (e.g. Zach, 1950; Lawrence, 1958). However, as pointed out by Bormann & Sidle (1990), universal mechanisms of paludification have not been established. Mechanisms have been discussed by Noble, Lawrence & Streveler (1984), who recognize that muskeg→forest transitions are occurring or have occurred in southeastern Alaska as well as forest→muskeg transitions. In the context of recently-deglaciated terrain, they take the view that muskegs tend slowly to replace forests at suitable sites but that further transitions occur in response to changing climate, fire regime, drainage and local catastrophes. They conclude that the development of muskeg through purely autogenic processes, such as the invasion and subsequent reaction of *Sphagnum* spp., may take from 800 to several thousand years, possibly much less if allogenic processes intervene. On marine terraces near Lituya Bay (about 80 km to the west of the mouth of Glacier Bay), Ugolini & Mann (1979) describe a chronosequence in which they attribute the formation of muskeg from spruce forest to a 'pedogenic' process. They envisage a deterioration of the internal drainage of the soil as an iron-pan (placic horizon) develops during podzolization over a period of about 500 years. Anaerobic conditions induced in the overlying horizons lower the decomposition rate of organic matter in the forest floor. This leads to an increasingly thick litter layer, which is eventually able independently to retain water permanently. At the same time, organic acids lower the pH, allowing the establishment of bog vegetation.

Allogenic rather than autogenic processes appear to be responsible for long term disequilibrium on terrain deglaciated during the Holocene in the Vestfold Hills of Antarctica (Seppelt *et al.*, 1988). With increasing distance from the Sørsdal Glacier and adjacent ice sheet, species richness and community complexity at first increase as lichen assemblages give way to moss beds with epiphytes and generally richer lichen communities. According to Seppelt *et al.*, this corresponds with the classical model of succession. However, beyond 5–10 km from the ice margin, on terrain deglaciated for at least 3000–5000 years, species richness and abundance rapidly decline to form simple lichen assemblages of only one or two species. This they describe as 'climax collapse', which is attributable to the accumulation of wind-blown salt and loose sand to detrimental or toxic levels.

6.2   Models

In this section, the approach adopted is partly historical (beginning with the earliest models) and partly systematic, progressing from formal models that emphasize the pattern of succession, to mechanistic models that are more concerned with underlying processes. With specific reference to primary succession on recently-deglaciated terrain, the limitations of these models are made explicit. It should be pointed out from the outset that most mechanistic models are based largely on the range of biological processes considered in section 6.1.

6.2.1   Monoclimax and polyclimax

The idea that succession terminates in a stable, integrated, self-maintaining climax community dominates most early models of succession (Selleck, 1960; Shimwell, 1971). The most extreme model of this type is that of monoclimax, as advocated by Clements (1928, 1936). According to the monoclimax model, all successions (seres) in an area eventually converge on the same climax community. Given sufficient time, community reactions modify the environment to such an extent that habitat differences are nullified and the characteristics of the climax are determined by regional climate. All other communities within an area are therefore viewed as subordinate to this climatic climax which, at least in theory, is of regional extent.

Although not a necessary part of monoclimax theory (Whittaker, 1974), Clements' interpretation was strongly conditioned by his explicit use of an organismic analogy for the climax. He viewed the climax as a complex organism, that arises, grows, matures and reproduces itself. The developmental process of succession is therefore seen as deterministic, repeatable and always progressive; retrogression is not allowed as part of this scheme. Somewhat similar holistic concepts appear to underly the approach adopted by Margalef (1968) and Odum (1969) to ecosystem development (cf. Drury & Nisbet, 1973; McIntosh, 1980, 1981).

Even without its organismic trappings, the monoclimax model has many weaknesses. It has been widely criticized for its basic assumption that all successions in an area must converge to the same climax community; for the degree of integration and stability attributed to the climax state; for an inability to accommodate other stable communities in a satisfactory way, for an overemphasis on autogenesis and particularly on the reaction mechanism during succession; for the emphasis given to climatic control; for a failure to take due account of climatic change; and for the unacceptably long time required (often on a geological rather than a successional timescale) for convergence to unity even were it possible (see for example, perceptive early critiques by Gleason, 1917; Domin, 1923; Bourne, 1934; and Cain, 1939; and the modern reviews in Mueller-Dombois & Ellenberg, 1974; Whittaker, 1974; and White, 1979).

Most of the available evidence from glacier forelands relates to short chronosequences, typically no longer than a few hundred years. The characteristics of climax vegetation must therefore be inferred either from the trajectory of successional pathways, or from the characteristics of the vegetation beyond the foreland boundary. Bearing in mind these limitations, the evidence from recently-deglaciated terrain in general upholds the criticisms listed above of the monoclimax model. Nevertheless, there is support for some aspects of the model, in the form of the convergent pathways to woodland or forest recognized in boreal and sub-alpine zones (see section 5.3.3; Figs. 5.37–5.39). There, the effects of differences in parent materials and microtopography appear to be nullified as succession proceeds towards a relatively uniform and relatively stable mature vegetation that is climatically determined. Furthermore, the concepts of stabilization and of a potential climax state (cf. Langford & Buell, 1969) remain useful, even though the ideal, steady-state equilibrium may rarely if ever be attained (see section 6.1.7).

The polyclimax model, which is generally attributed to Tansley (1920, 1929, 1935), differs from that of monoclimax in several respects (Whittaker, 1974). During succession, there is only partial convergence of seres, leading to more than one climax community. In the landscape, these form a vegetation mosaic, which is determined by a mosaic of habitats. One of the phases of the mosaic, which may be most widespread and directly controlled by climate, comprises the climatic climax; other phases may be controlled by the parent material (edaphic climax), topography (topoclimax), or other environmental factors. The concept of a polyclimatic climax (Tüxen, 1933; Mueller-Dombois & Ellenberg, 1974; Meeker & Merkel, 1984), which recognizes more than one climatic climax in a macroclimatic region (each corresponding to a mature soil type), is best regarded as a variant of the polyclimax model.

The emphasis given in the polyclimax model to a small number of climax community types, is supported to a limited extent by the absence of uniformity in the structure and composition of mature communities on glacier forelands and beyond their boundaries. This model is also in agreement with the recognition of factors other than climate as effective influences in mature communities. However, as in the case of monoclimax, the polyclimax model is dominated by the necessity for vegetation to attain a climax state as part of a mature ecosystem and a stable landscape. It is also unable to account for the full variety of communities present in the mature vegetation landscape, or for the strongly divergent successional pathways recognized from glacier forelands in polar and alpine regions.

6.2.2   Climax pattern and site climax

According to the climax pattern model (Whittaker, 1951, 1953, 1967), the environments of a landscape form a pattern of complex environmental gradients that involve many individual factor gradients. Succession at any point in the landscape and the resulting climax community depend on the precise combination of environmental factors at that point. The climax state is defined as a steady-state equilibrium of plant species populations in relation to habitat factors, without any dependence on the attainment of soil maturity or long-term geomorphological evolution. Continuous environmental gradients imply a continuum of intergrading climax communities, rather than a community mosaic. Plant species are distributed individualistically with respect to environmental gradients and to the climax pattern of communities (cf. Gleason, 1926, 1927, 1937). Any mature community types that are recognized tend to be arbitrary subdivisions of an intergrading climax continuum (cf. McIntosh, 1967). Successional types or stages in the vegetation are correspondingly diffuse.

The site climax model, as used in range management studies (Dyksterhuis, 1949, 1958; Meeker & Merkel, 1984), resembles the climax pattern model in defining the climax state as an existing or potential plant community phenomenon in relation to existing site conditions. However, each site climax is viewed as a community type to permit classification and mapping (cf. the 'potential natural vegetation' of Küchler, 1988). The concept of site climax is regarded as less realistic than the concept of climax pattern, which views climax vegetation as a complex of vegetational gradients.

The climax pattern model goes a considerable way towards matching the full complexity of vegetation and environmental interactions that exist in late successional and 'climax' landscapes on glacier forelands. It is in agreement with the individualistic distribution of species (e.g. Fig. 5.30), accounts for a high degree of within-type variability (see section 5.3.2), and is certainly compatible with the recognition of major vegetation gradients that correspond with major environmental factor complexes (see section 5.4.3). It provides the most realistic climax model, and allows the widest range of successional patterns within a single framework. No reference is made to divergent succession, however, and little consideration is given to the mechanisms of successional change. The absence of a direct contribution to the explanation of successional pathways is perhaps the major shortcoming of this model.

6.2.3 Relay floristics and IFC

Two models that address both the form and the mechanism of succession within particular pathways have been respectively labelled 'relay floristics' and 'initial floristic composition' (IFC) (Egler, 1954, McCormick, 1968). Relay floristics refers to the successive appearance and disappearance of groups of species at a site through time (Fig. 6.13a). Each group of species invades the site at a certain stage of vegetation and site development. The species groups supposedly react on the site and produce conditions suited to the invasion of the next group. The IFC model envisages that all species are present or are able to colonize from near the start. Development unfolds from this initial flora, without additional species invasions. New groups of species assume predominance as each successive group is replaced (Fig. 6.13b). Growth rate of the various groups (typically in the sequence grasses and forbs→shrubs→trees) determines the successional sequence rather than reactions. Although this model was originally developed in the context of secondary succession, particularly involving old fields, it is also applicable to cases of primary succession where there are few limitations on migration and ecesis (Finegan, 1984). The essential point is that life-cycle differences rather than reaction mechanisms control the succession according to the IFC model.

Drury & Nisbet (1973), Miles (1979), and McIntosh (1980) all point to similarities between the IFC model and primary succession at Glacier Bay. It is noted, in particular, that the tree dominants of the later stages are present at the pioneer stage. Species' patterns that are close to the IFC model have also been noted in front of Jostedalsbreen outlet glaciers in southern Norway, where nearly all species are present on the youngest surfaces (H. J. B. Birks, pers. comm. 1990). Data supporting the early recruitment of later successional species are available from the small Nooksack cirque in northern Washington (Adams & Dale, 1987). Below Nooksack Glacier, *Abies amabilis, Tsuga heterophylla, T. mertensiana, T. plicata, Pinus contorta, P. monticola, Chamaecyparis nootkatensis* and *Alnus sinuata* all occupy sites within a few years of glacier recession. In an area exposed by the glacier for up to 37 years, the age of these trees ranged from 1 to 37 years, with a maximum height of 23 m and a density of 682 overstorey stems ha^{-1}. Adams & Dale suggest that the early establishment of these initial colonizers may determine the direction taken by plant succession and the equilibrium reached (see also Oliver & Adams, 1979; Oliver, Adams & Zasoski, 1985). Although the same three species may be

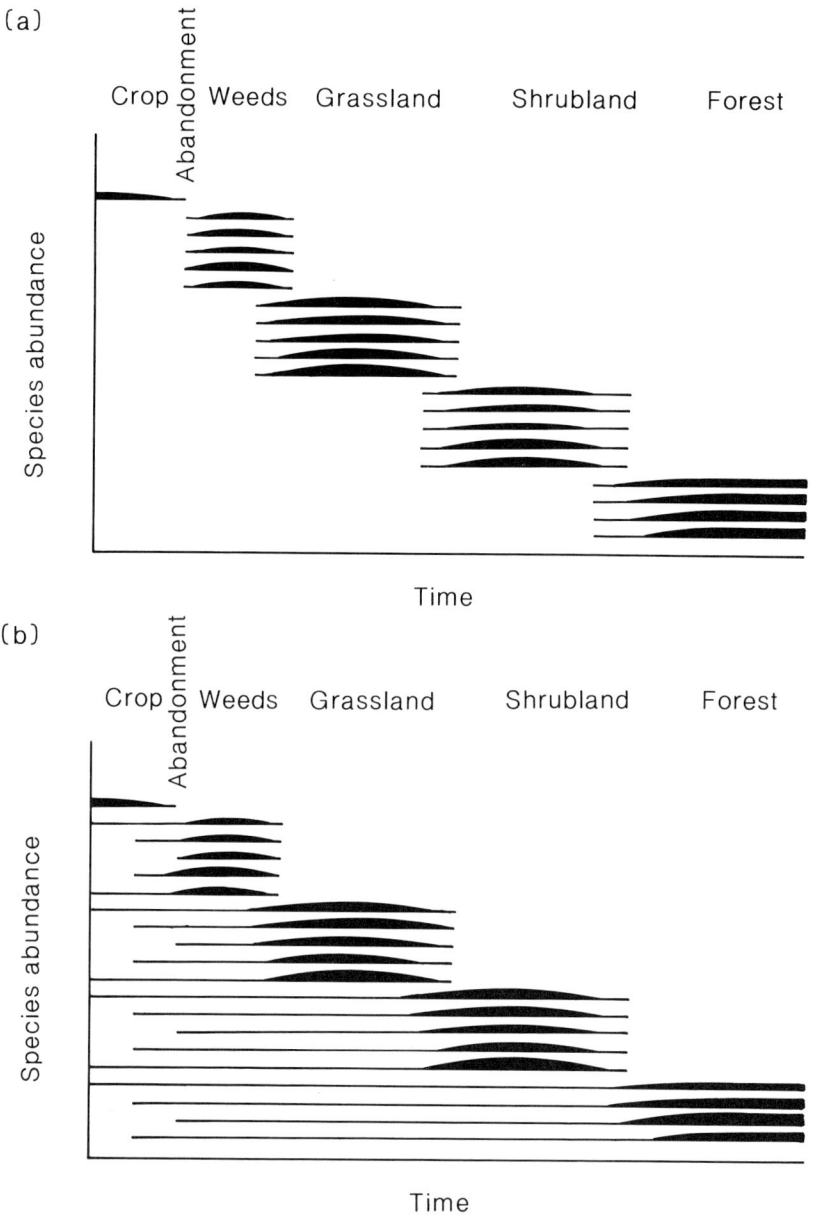

Fig. 6.13. Models of succession originating from studies of abandoned fields: (a) relay floristics; (b) initial floristic composition (IFC) (from Egler, 1954).

Table 6.5. *Species composition of overstorey stems on recently-deglaciated terrain and on terrain deglaciated for > 800 years, Nooksack cirque, Washington, U.S.A. (after Adams & Dale, 1987).*

Site	Terrain age (yr)	*Tsuga heterophylla* (%)	*Tsuga mertensiana* (%)	*Abies amabilis* (%)	other (%)
Foreland	< 37	42	37	19	2
Canyon	> 800	39	3	67	1

the major dominants on terrain deglaciated for > 800 years in Nooksack Canyon, their relative importance is different on the older terrain (Table 6.5), the number of successful colonizers is much less (22 overstorey stems ha^{-1}), and patterns exhibited by other species show evidence of other processes (e.g. competitive inhibition) (see sections 5.4.2 & 6.1.5).

Miles (1979: 53) goes as far as to say that 'the process of succession [at Glacier Bay] seems identical to that of secondary succession as explained by Egler'. However, such comparisons do not distinguish the form from the mechanism of succession and they deny the fact that the two mechanisms (i.e. reaction and life-cycle differences) are not mutually exclusive (Egler, 1954; Finegan 1984). Cooper (1923b) is convinced that the pioneering individuals of *Picea sitchensis* at Glacier Bay are permanently stunted and are not the plants that grow to become dominants when conditions improve later in the succession. Symptoms of poor growth and nutrient deficiency characteristic of such pioneer trees (Lawrence 1958, 1979) also indicate that life-cycle differences are not the only mechanisms of change (see section 6.1.4). Perhaps the best evidence that relay floristics, rather than IFC, is the more appropriate model at Glacier Bay, is the species lists in Reiners, Worley & Lawrence (1971).

It would appear that most other successional sequences inferred from glacier foreland chronosequences are closer in form to the relay floristics model than to the IFC model. There is generally an appreciable lag between colonization by pioneer species and the subsequent colonization of successive loose groupings of later colonizers (see, for example, Figs. 5.24–5.26 & 5.30). Although most of the species involved in successional sequences may occur in the propagule rain (see section 6.1.1), which may be supplemented by the release of propagules temporarily stored in the ice bank (see section 6.1.2), differential patterns of establishment clearly limit the applicability of the IFC model. On the other hand, a major limitation of the relay floristics model as depicted by Egler in Fig. 6.13a is that it does not represent in a realistic way the lack of discreteness of the species groups.

6.2.4   Non-selective and selective autosuccession

Another set of models derived from studies of secondary succession has been proposed by Muller (1940, 1952). In the context of the recolonization of disturbed sites, 'autosuccession' describes the form of the successional sequence where the recolonizing species are the same as the species present in the surrounding, undisturbed (climax) vegetation. 'Non-selective autosuccession' refers to a change from bare ground to mature vegetation in which the species colonize in no specific order. 'Selective autosuccession' describes the situation where one element of the colonizing species preceeds the establishment of another element. Both autosuccession models differ from 'true secondary succession', in which one or more species foreign to the surrounding vegetation are involved in the recolonization (i.e. there is at least a partial relay). Autosuccession is similar to the 'direct succession' model proposed by Whittaker & Levin (1977) and might also be viewed as a compact form of almost any other model (cf. Noble, 1981).

According to Muller (1952), true secondary succession is characteristic of sub- and low-alpine vegetation in northern Sweden; only autosuccession, sometimes selective, occurs in the mid-alpine belt, whilst non-selective autosuccession is characteristic of the high-alpine belt. Selective species colonization and relays of species are interpreted by Muller (1952) as evidence for interactions between species, which do not exist in the severe environments characterized by non-selective autosuccession.

These models have been applied to primary succession on the Storbreen glacier foreland (Matthews 1979b). The simplest form of primary succession at Storbreen is described as autosuccession with some selection. This occurs at the highest altitudes investigated, where many of the pioneer species also occur in the mature (climax) communities. As most dominant species of these mature communities (e.g. *Luzula arcuata* and *Salix herbacea*) are only occasionally present in the pioneer assemblages, succession cannot be termed non-selective. At even higher altitudes, at sites typical of the high-alpine belt, non-selective autosuccession may be the rule. With decreasing altitude, selection increases during succession involving the snowbed species of the mid-alpine belt. In the low-alpine belt, there occurs a relay succession from relatively distinct pioneer to heath species groups.

Quite similar models have been proposed by Svoboda & Henry (1987) who recognize 'directional–nonreplacement succession' in high resistance environments and 'nondirectional–nonreplacement succession' in extre-

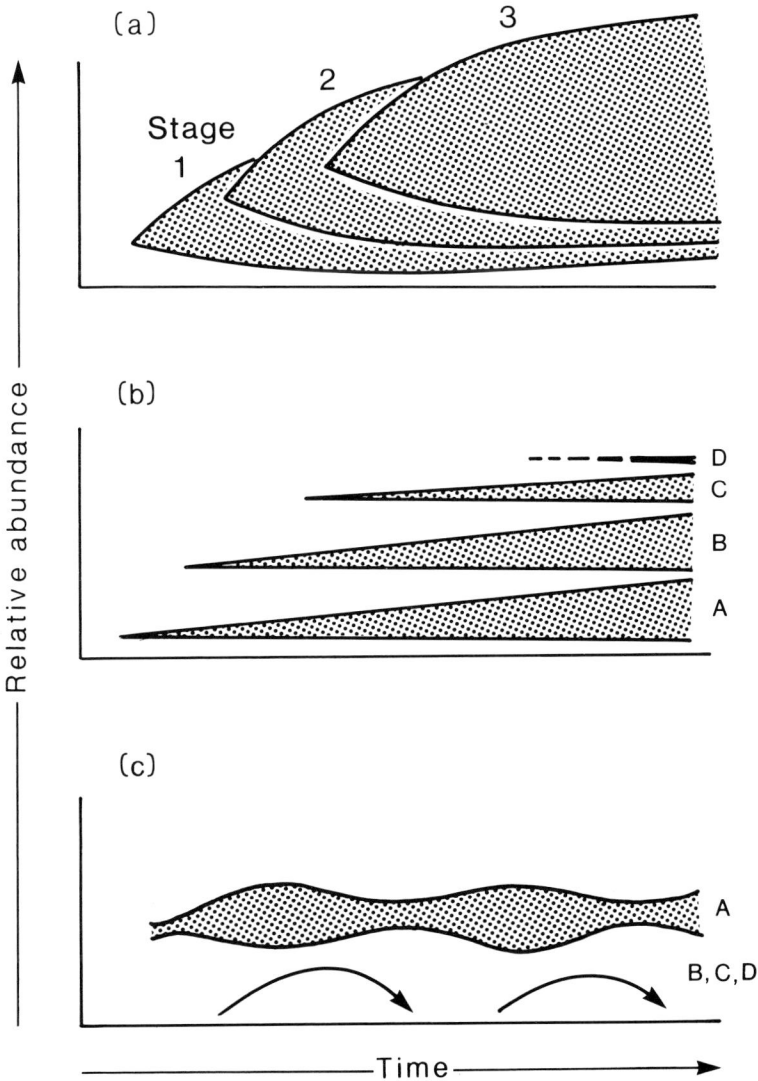

Fig. 6.14. Models of succession originating from the Canadian Arctic: (a) classi-cal directional-replacement succession in low resistance environments; (b) direc-tional-nonreplacement succession in high resistance environments;(c) nondirec-tional-nonreplacement succession in extremely resistant environments (from Svoboda & Henry, 1987).

mely resistant environments. These contrast with the classical 'directional-replacement' model of succession in low resistance (i.e. environmentally-favourable) environments in which species replacement takes place due to modification of habitat by reaction (Fig. 6.14). Directional–nonreplacement and nondirectional–nonreplacement succession are seen as characteristic of polar semidesert and polar deserts, respectively. In the former, invading species succeed in slow expansion without eliminating or replacing each other, whereas in the latter very few species succeed in establishing. In both cases free space is generally available for futher expansion. Species that do establish in polar deserts comprise populations that exhibit fluctuating cover and standing crop. Repetitive invasions of other species fail in permanent establishment.

Although these models are concerned largely with the form rather than the mechanisms of succession, they contribute towards an explanation of the range of successional patterns inferred from recently-deglaciated terrain in different environments. Such autosuccession models appear applicable to the simplest glacier foreland successions that characterize the most severe polar and alpine environments with low temperatures, low precipitation, and sometimes with high levels of disturbance from cryogenic processes.

6.2.5   Facilitation, tolerance and inhibition

Three influential models, which emphasize alternative mechanisms for species replacements during succession, have been proposed by Connell & Slatyer (1977). In all three models, the earlier species cannot invade and grow once the site is fully occupied. The most important differences between the models therefore relate to how new species appear later in the sequence (Fig. 6.15). The first model (facilitation) is essentially the traditional explanation for primary succession and has much in common with Egler's model of relay floristics (see section 6.2.3). It assumes that vegetation changes are driven by the reaction of early successional species on their environment. These species thereby facilitate the recruitment of later successional species. They 'prepare the way' for later colonizers either by reducing environmental stress or by improving resource availability to competitors. In time, the early species are eliminated through competition for resources with the late successional species.

In both the tolerance and inhibition models, all species present in the successional sequence are capable of colonizing at the earliest stage (cf. the IFC model, section 6.2.3). In the tolerance model, early colonizers have little or no effect on the recruitment of later successional species. The later appearance of certain species is attributed to their life-cycle characteristics (passive tolerance, *sensu* Pickett, Collins & Armesto, 1987a) and/or an ability to persist because they tolerate lower resource levels (active tolerance). Similar lines of reasoning account for the elimination of early

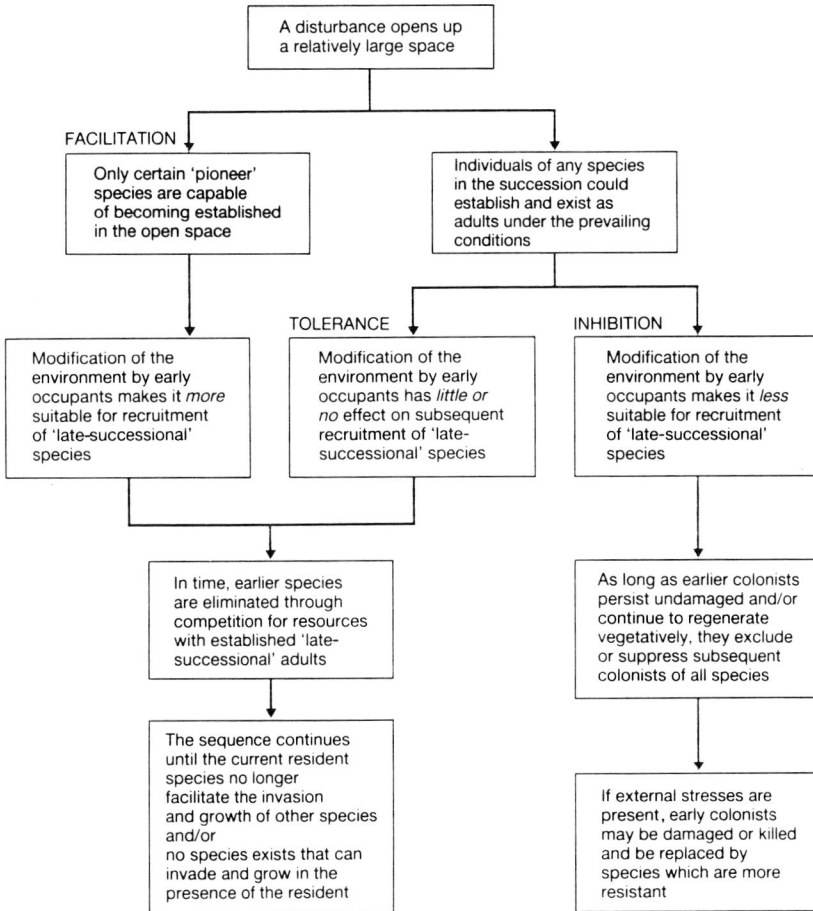

Fig. 6.15. Facilitation, tolerance and inhibition models (from Connell & Slatyer, 1977, as modified by Begon, Harper & Townsend, 1990).

successional species, which are inferior competitors at, for example, lower levels of light, moisture or nutrients.

In the inhibition model, all species exclude or inhibit competitors. The first occupants pre-empt the space and resist the invasion of subsequent colonists or suppress the growth of those already present. The latter invade or grow only when the existing dominants are damaged or killed by disturbances. In the absence of such disturbances, which might involve physical processes, herbivores or pathogens, stable successional communities may persist. Only when space is created and/or resources are released, can the later successional species reach maturity. Provided disturbance is not too extensive or intense, long-lived species tend to accumulate in the later stages of succession because they are replaced less often. Thus, early successional species may be replaced, but this is not the result of competition with the later successional species.

Although Connell & Slatyer (1977) note the absence of critical experimentation, they conclude that the facilitation model is likely to apply to primary succession, and specifically refer to glacier foreland succession in this context. They reason that there are major constraints on plant growth that require amelioration before succession can proceed. However, several other authors draw parallels between succession at Glacier Bay and the IFC model, which is closer to the tolerance and inhibition models than to the facilitation model (see section 6.2.3). The solution to this paradox lies in the fact that particular species replacements may be explained by a particular mechanism, whereas whole seres cannot be explained by a particular model. A similar conclusion has been reached by Burrows (1990) in relation to the vegetation sequence on the foreland of Franz Josef Glacier, New Zealand, where the three mechanisms are all represented, often at the same successional stage (see section 6.1.5). Unfortunately, therefore, no one model in its entirety, and to the exclusion of other models, provides a satisfactory explanation of any particular glacier foreland sere. In short, these three models do not describe a comprehensive set of alternative testable hypotheses (cf. Finegan, 1984; Pickett, Collins & Armesto, 1987a; Walker & Chapin, 1987; Connell, Noble & Slatyer, 1987).

6.2.6   Chronic disturbance, competitive hierarchy and resource ratio

Horn (1976, 1987) suggests several simple models based largely on studies of plant replacements during secondary forest succession. No discussion is required of his model of 'obligatory succession' as this corresponds to the facilitation model. Horn's second model of 'chronic patchy disturbance' is essentially a statement that all conceivable replacements are possible. Although this model has been criticized by Noble (1981) as not representing succession (because of a lack of directionality or predictability), it is useful for insights into the role of disturbance in succession. Horn's third model of a 'competitive hierarchy' corresponds roughly to the tolerance model, although it includes elements of competitive inhibition (Horn, 1987). It is also useful as a framework for consideration of the more specific 'resource ratio' model of Tilman (1985, 1988).

Chronic patchy disturbance refers to frequent, localized, physical disturbance that is more-or-less uniformly distributed in time and space, and sufficient to result in the deaths of individual plants and a few neighbours. This asynchronous process is seen as distinct from larger-scale disturbance, which simultaneously ends the life of all trees in a stand (synchronous, large-scale chronic disturbance). As any species is capable of colonizing the space created if propagules are available, the chronic patchy disturbance model resembles non-selective autosuccession. It may well be a dominant element of vegetation change on glacier forelands subject to high levels of cryogenic disturbance, not only in very severe polar and alpine climates but also in susceptible microenvironments elsewhere. A large element of stochasticity with little or no directional trend has often been considered a feature of severe environments and has led to the long debate over the applicability of concepts of succession and climax to tundra vegetation (Griggs, 1934; Churchill & Hanson, 1958; White, 1979).

A competitive hierarchy is a list of species in order of their competitive ability. Each species is able to outcompete all above it in the list, to invade open sites, and to resist invasion by those same species. Succession is therefore viewed as a uni-directional sorting process, which gradually converges on a stationary community consisting of the competitive dominant. Interactions between *Salix*, *Alnus* and *Picea* at Glacier Bay have been interpreted in this way, with particular reference to light (see section 6.1.5). Although Horn (1971) has developed his model only in terms of the tolerance of tree species to light, it is in principle much more widely applicable.

In the 'resource ratio' hypothesis of succession, Tilman (1985, 1988) provides a model that is more specific in two respects. First, the competitive hierarchy is determined relative to two resources, a limiting soil resource (usually nitrogen), and light. Second, the relative competitive abilities of the species are regulated through time by the changing ratio between the two resources. The availability or supply rate of the limiting soil resource is not necessarily determined by the plants, but can involve atmospheric nutrient inputs, inputs of litter, and modifications by decomposers. Succession is therefore seen as the response of plant species to a composite, temporal soil resource : light gradient (Fig. 6.16). Tilman suggests that the predictions of the resource-ratio model are consistent with the pattern of dominant species in the chronosequence at Glacier Bay and hence that changing soil nitrogen and light levels can explain at least the broad pattern of succession observed there (see also Tilman, 1986). However, as the actual requirements of the species in terms of these and other resources are poorly understood, the application of Tilman's model remains to be tested in the context of glacier forelands (cf. Thompson, 1987; Tilman, 1987).

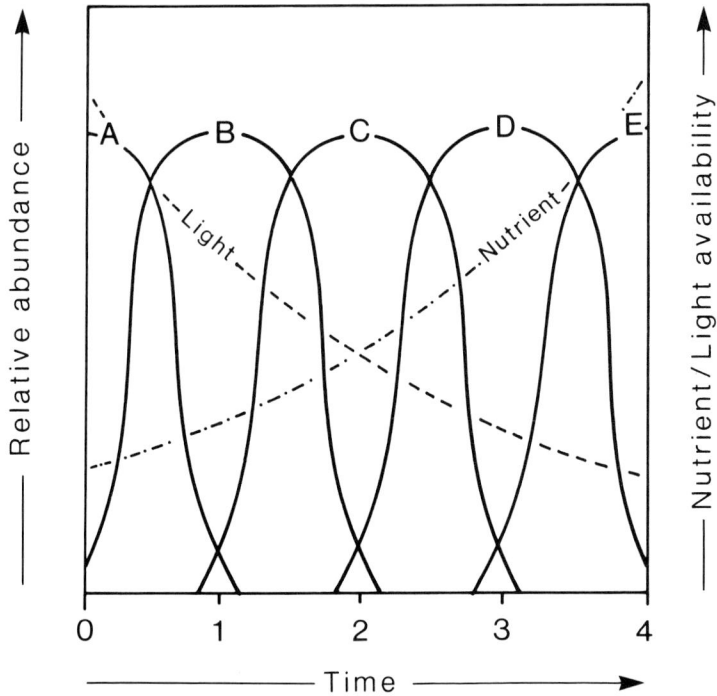

Fig. 6.16. The resource-ratio hypothesis of succession (from Tilman, 1988).

6.2.7   Evolutionary strategies

The concept of succession as a gradient in time and of species having evolved strategies that pre-adapt them for survival along different parts of this gradient, has been recently used to provide an explanation for succession (Colinvaux, 1973; Drury & Nisbet, 1973; Pickett, 1976). In its simplest form, this type of model predicts that there are opportunistic-generalist species (r-strategists) and equilibrium-specialist species (K-strategists). A population cannot be both generalist and specialist, but both kinds of strategists generally co-exist in a landscape, the r-strategists in disturbed sites, and the K-strategists in mature patches. As the former species have evolved life-cycle and physiological characteristics appropriate to dispersal and growth in disturbed conditions, they rapidly colonize unoccupied sites in the early stages of succession only to be replaced during the later stages by the K-strategists, which are slower growing and utilize resources more for sustained existence at a steady-state equilibrium (see, for example, Bazzaz, 1979; Finegan, 1984; Colinvaux, 1986).

Table 6.6. *Three primary plant stategies in relation to varying levels of disturbance and stress (after Grime, 1979).*

Intensity of disturbance	Intensity of stress	
	Low	High
Low	Competitor (C)	Stress-tolerator (S)
High	Ruderal (R)	None

Grime (1977, 1979) suggests the existence of three primary strategies, rather than two, and that these have evolved as adaptations to varying degrees of disturbance and stress (Table 6.6). Stress is here defined as phenomena that restrict photosynthetic production (such as shortages of light, water and nutrients, or sub-optimal temperatures). Disturbance results in the destruction of plant biomass by physical or biological processes. Grime views the ruderal and stress-tolerant strategies as corresponding respectively, with a r–K continuum, whilst competitors are intermediate. The major difference between the R–C–S and the r–K schemes is the recognition of a distinct stress-tolerant strategy enabling survival in conditions of limited productivity, which may arise in unproductive habitats or as a result of the reaction of the plants themselves. As high stress and high disturbance prevent the establishment of vegetation, there is no viable fourth strategy (but see Grubb, 1985; Kautsky, 1988).

Because particular taxa and life forms are seen as evolving strategies that reflect to varying degrees the three primary strategies, their positions can be plotted on a triangular diagram (Fig. 6.17a). Species exhibiting so-called secondary strategies (for example, competitive ruderals and stress-tolerant competitors) can also be recognized. During succession, species with an essentially ruderal strategy are seen as being replaced by species with increasing stress tolerance. More productive sites follow trajectories involving competitive strategies in intermediate successional stages (Fig. 6.17b).

Grime's model is useful in providing an evolutionary basis for the variety of successional pathways identified from glacier forelands. In the most favourable environments, at Glacier Bay for example, succession appears to follow a moderate productivity trajectory (P_2 in Fig. 6.17b). During succession towards spruce forest, ruderals or competitive ruderals are rapidly replaced by tree species with increasingly stress-tolerant competitive strategies. The low productivity trajectory (P_3) represents the invasion

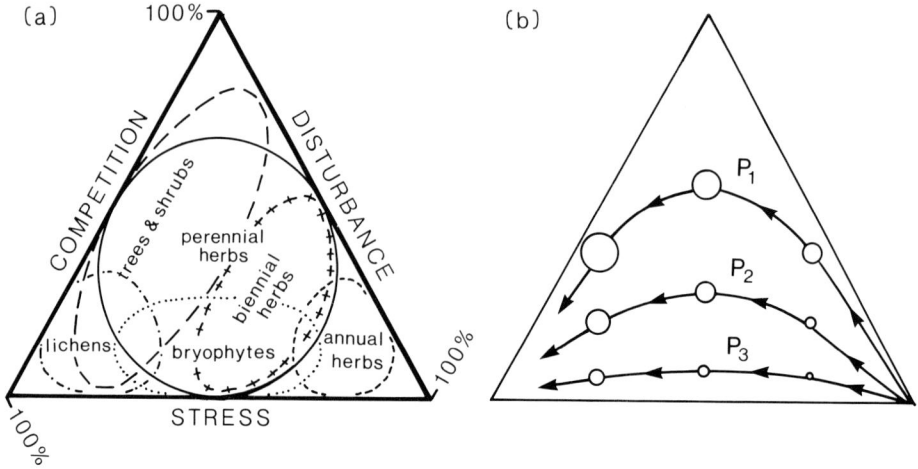

Fig. 6.17. Succession and evolutionary strategies: (a) life-forms characterized in terms of three primary strategies; (b) successional pathways in relation to strategies under conditions of high (P_1), intermediate (P_2) and low (P_3) potential productivity. In (b) circle size represents biomass (from Grime, 1979; [a] as modified by MacMahon, 1980).

of pioneer bryophyte communities by lichens in extreme high-alpine and polar environments. In less severe low-alpine and low-arctic conditions, succession involving tundra vegetation types probably follows a trajectory intermediate between P_2 and P_3.

The C–S–R scheme, and other related schemes (e.g. Southwood, 1977, 1988; Taylor, Aarssen & Loehle, 1990) must nevertheless be regarded as a gross simplification of a very wide range of possible strategies. This range is reflected in the wide variety of morphological and physiological characteristics exhibited by plants. As a considerable part of this range is involved in succession on recently-deglaciated terrain, the C–S–R scheme provides, at best, an incomplete model.

6.2.8   Vital attributes, process interactions and a causal hierarchy

All of the models discussed above are incomplete representations of the complex reality of glacier foreland succession. Indeed, many of these models are extremely simple, being based on one dominant concept or a single process. Noble (1981) and Miles (1987) have summarized how some of the models are related (Fig. 6.18). The complexity of successional phenomena in relation to the simplicity of available models has led to the suggestion that the quest for unifying, generalized theories of succession

	Clements (1916)	Egler (1954)	Horn (1976)	Connell & Slatyer (1977)	Whittaker & Levin (1977)
(a) ►A → B, D ← C	Primary succession	Relay floristics	Obligatory succession	Facilitation model	Replacement succession
(b) ►A ↺					Direct succession
(c) ►A ⇄ B$_{CD}$ (BCD), D ← C$_D$	Secondary succession	IFC			
(d) ►A → B◄, ✕, ►D ← C◄				Competitive hierarchy	Tolerance model
(e) ►A ⟶/⟶ B◄	Subclimax			Inhibition model	Plateau succession
(f) ►A ↔ B◄, ✕, ►D ↔ C◄			Chronic disturbance		

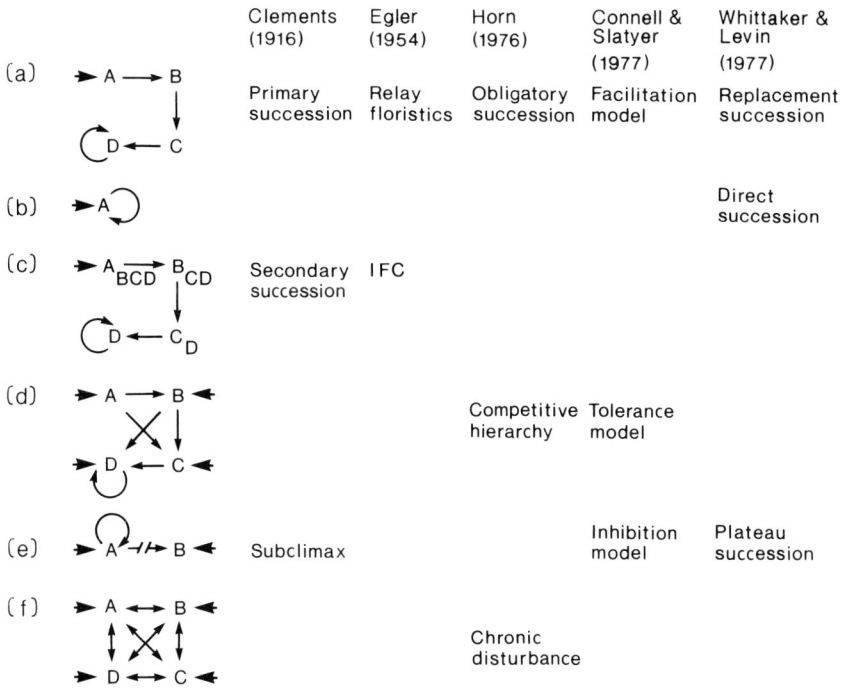

Fig. 6.18. Schematic comparison of selected successional models as replacement sequences (small arrows) of vegetation types or dominant species (A–D). Large arrows represent alternative starting points after disturbance. Subscript letters in (c) indicate species present as subordinates or propagules (omitted from d–f) (from Noble, 1981, as modified by Miles, 1987).

should be abandoned (Miles, 1987). Others do not agree (Finegan, 1984; Anderson, 1986; Matthews, 1989). Although such a theory has yet to appear, there have been at least three attempts to produce comprehensive frameworks for understanding succession. These concentrate on the idea that succession is the resultant of process interactions.

One approach is based on the notion of 'vital attributes' (Noble & Slatyer, 1977, 1980; Noble, 1981). A vital attribute is defined as 'any attribute of a species which is vital in determining its role in vegetation replacement sequences' (Noble & Slatyer, 1980: 21). The vital attributes considered by Noble & Slatyer relate to: (1) methods of recovery from disturbance (such as vegetational spread, seedling pulse from a seed bank, seedling pulse from the surrounding vegetation); (2) the ability to establish in the face of competition (such as only immediately after disturbance, for an indefinite period thereafter, or only after specific requirements have been fulfilled) and (3) the time taken to reach critical life-history stages. Although the vital attributes are based on the outcome of mechanisms, rather than the mechanisms themselves, the approach attempts to provide a qualitative statement about the major processes involved in vegetation sequences. It has been used to predict regenerative sequences in secondary succession but has only limited applicability to primary successions. As in the case of simulation modelling (Botkin, 1981; Shugart, West & Emanuel, 1981; Shugart, 1984), it may be possible to adapt this approach to investigate glacier foreland successions.

Of more direct relevance to succession on recently-deglaciated terrain is the concept, most clearly stated by Walker & Chapin (1987), that succession is the resultant of multiple process interactions. With particular reference to the models of Connell & Slatyer (1977), Walker & Chapin (1987) point out that the processes involved may be combined in other ways and with other processes, and that particular processes may vary in their relative importance in different environments and at different stages in succession. Furthermore, particular species replacements as well as the overall vegetation change may be viewed as a net response to a range of processes. For example, alder may increase soil nitrogen content and reduce the degree of nutrient limitation of spruce growth (suggesting facilitation) yet its overall net effect may be inhibitory. This has been demonstrated by them on Alaskan floodplains (Walker & Chapin, 1986; Walker, Zasada & Chapin, 1986) and suggested for the Glacier Bay chronosequence (Chapin & Walker, 1990).

Walker & Chapin (1987) conclude that succession is a complex of many individual processes acting simultaneously. They propose what is in effect an individualistic process model of succession (Fig. 6.19). This predicts the relative importance of various processes at different stages of succession, in different types of succession (primary, secondary and regenerative), and in severe (i.e. low resource) or favourable environments. They suggest that succession is a continuum from early stages, where factors governing colonization are most important, to late stages, where factors governing senescence and mortality predominate (although disturbance may prevent the attainment of this senescent stage). The type of succession is important primarily in determining the mode of arrival of initial colonizers. Resource availability then controls the relative importance of other successional processes. Many features of this model are applicable to succession on glacier forelands and will be referred to in more detail in section 6.3.

Finally, the construction of a comprehensive framework for successional mechanisms has been attempted in the form of a hierarchy of

(a)

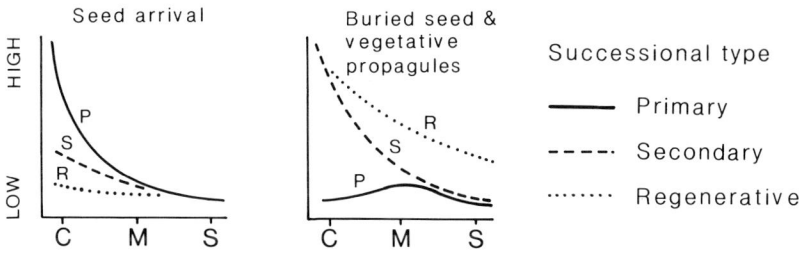

(b)

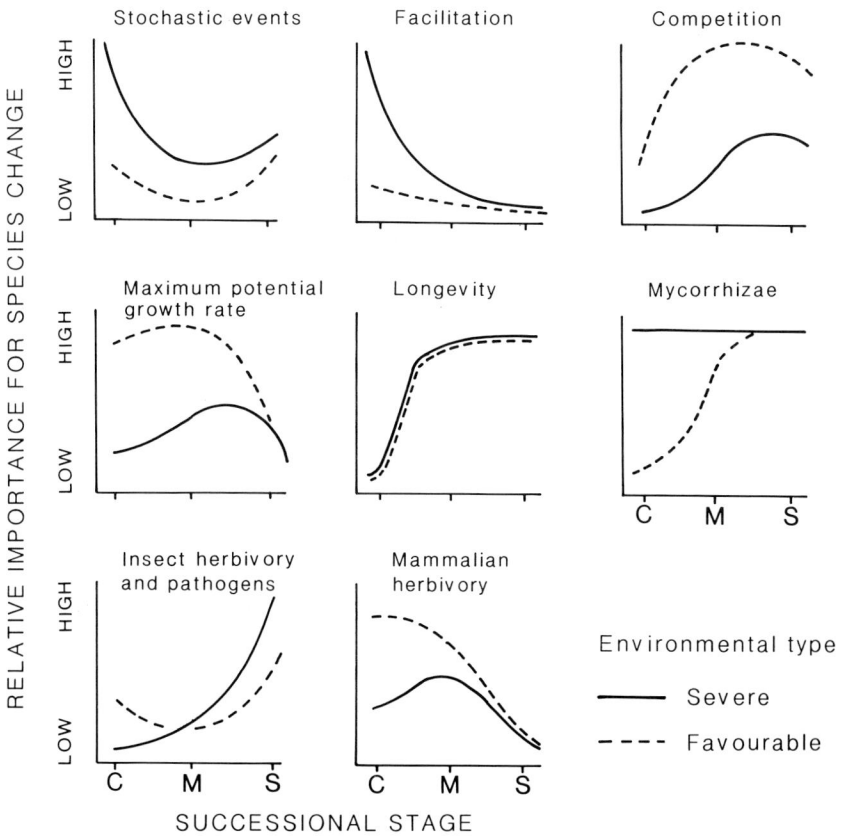

SUCCESSIONAL STAGE

Fig. 6.19. Individualistic process model of controls on species composition during succession: the influence of (a) successional type (P, primary; S, secondary; R, regenerative); and (b) environmental severity (severe or favourable) on major successional processes are shown for colonization (C), maturation (M) and senescence (S) stages of succession (from Walker & Chapin, 1987).

Table 6.7. *A hierarchy of successional causes (after Pickett, Collins &*
Armesto, 1987a,b).

General causes of succession	Contributing processes or conditions	Defining or modifying factors or behaviours
Site availability	Coarse-scale disturbance	Size, Severity, Time, Dispersion
Differential species availability	Dispersal	Landscape configuration Dispersal agents
	Propagule pool	Time since disturbance Land use
Differential species performance	Resource availability	Soil conditions Topography Microclimate Site history
	Ecophysiology	Germination requirements Assimilation rates Growth rates Population differentiation
	Life history strategy	Allocation pattern Reproductive timing Reproductive mode
	Environmental stress	Climate cycles Site history Prior occupants
	Competition	Presence of competitors Identity of competitors Fine-scale disturbance Predators and herbivores Resource base
	Allelopathy	Soil characteristics Microbes Neighbouring plants
	Herbivory, disease and predation	Climate cycles Consumer cycles Plant vigour Plant defenses Community composition Patchiness

successional causes (Pickett, Collins & Armesto, 1987a,b; Pickett &
McDonnell, 1989). The hierarchy has three levels, with specific mechanisms
nested within more general causes. (Table 6.7). The first level of the
hierarchy represents minimal defining phenomena: (1) sites become available; (2) species are differentially available at a site; and (3) species have
different evolved or enforced capacities for dealing with the site and one
another. The second level of the hierarchy comprises the broad processes.

Interactions or conditions that must be considered to understand each case of succession. The third and most detailed level encompasses site-specific factors or behaviours that act to cause or limit change in the second level. Understanding at each level is achieved by a knowledge of the processes or phenomena at the next lower level. Although the hierarchy of causes may provide a context for studies at specific sites, a scheme for formulating general and testable hypotheses, and the initial tentative outline of a complete successional theory (Pickett, Collins & Armesto, 1987a), it is in fact little more than an exhaustive check-list or causal repertoire (Pickett & McDonnell, 1989). Perhaps its greatest failing is that, unlike the model of Walker & Chapin (1987), it reveals little about process interactions.

All three models discussed in this section are flawed by the emphasis given to biological processes. Consequently, there is an unsatisfactory treatment of physical environmental processes. Both the nature and scale of environmental influences on resource availability are neglected. Allogenesis is not adequately integrated with autogenic processes. As in most discussions of succession, physical environmental processes tend to be discussed only in relation to disturbance. This will be remedied in the next section in the context of a geoecological model.

6.3   A geoecological model

It is clear that most of the general models discussed above have only limited applicability in the context of succession on recently-deglaciated terrain. Some may be more appropriate for secondary succession. Others tend either to overemphasize a particular concept or process, or to provide a large range of possible contributory processes without directing attention to those actually important on glacier forelands. What is required is a model based on those processes that are both necessary and sufficient to account for the available evidence.

In this section, the outline of a geoecological model is presented. It possesses two distinctive features. First, physical environmental processes as well as biological processes form an integral part of the model (coupling of physical and biological processes). Second, spatial variation in temporal dynamics is emphasized, both at the within- and between-foreland scales. That is, attention is paid to succession as a landscape process, which varies according to site position and the environmental characteristics of the habitat (spatio-temporal dynamics). These two distinctive features respectively represent the vertical (ecological) and horizontal (geographical) aspects of geoecology (see section 1.4).

6.3.1   Coupling of physical and biological processes

The underlying justification for including physical and biological processes in a successional model is that both occur at any site and hence that both are likely to contribute to succession. As demonstrated throughout the text, it is inadequate to describe physical environmental factors solely in terms of an unchanging set of initial site conditions. Because of the existence of physical environmental factors behaving as influx variables (see, in particular, section 5.4.2), it is invalid to regard succession as a purely biological phenomenon.

Separate treatment of physical and biological processes seems to have originated from the recognition of a distinction between allogenic (externally-driven) and autogenic (internally- or community-driven) succession (Gams, 1918; Cooper, 1926; Tansley, 1935). Although interdependence of succession and soil development is widely acknowledged, allogenic processes have tended to be regarded as somehow separate from succession. Thus, communities have been described as exhibiting passive 'adaptive change' as a result of alterations in non-soil factors of the environment, whilst they exhibit succession *sensu stricto* in response to autogenic processes associated with vegetation and soil development (Moravec, 1969). There are many difficulties with this view. In reality, allogenic and autogenic processes interact during succession at any site and are often inseparable.

As recognized in Fig. 6.19, the relative importance of various processes vary with successional stage and environmental severity. For example, in the pioneer stage of primary succession, Walker & Chapin (1987) follow Viereck (1966) and others in stressing the importance of dispersal mechanisms that promote seed arrival from distant sources (see also, Finegan, 1984; Davis, *et al.*, 1985; del Moral & Clampitt, 1985; Wood & del Moral, 1987). Colonization from buried seed and vegetative propagules are depicted as being of low importance, even in mid-succession.

Several lines of evidence from glacier forelands suggest that the dependence of early successional change on differential seed arrival has been overemphasized. Glacier forelands seem to be well supplied with diaspores from the full range of species characteristic of the available flora (Ryvarden, 1971, 1975). Even on recently-deglaciated nunataks (which are relatively distant from potential seed sources) a large diaspore number appears to be more important than specialized dispersal adaptations (Elven, 1980). In addition, pseudo-viviparous tumbling (a means of vegetative propagation) appears to be particularly important amongst the earliest pioneers (see section 6.1.1). In these respects glacier foreland succession may be closer to secondary succession than is implied by Walker & Chapin (1987). The long-held view of the absence of life on newly-deglaciated substrates must also be revised in the light of the possibility of ice propagule banks (section 6.1.2) and sub-glacial microbial colonisers (section 6.1.4).

The evidence suggests that ecesis is more important than migration in determining species composition on newly-deglaciated terrain. Why do only certain species successfully establish and come to characterize the pioneer stage of succession? This is a separate question from why are the pioneers subsequently replaced by later colonizers? The answer to the first question probably lies as much in the physical environmental characteristics of the site and, especially, in the physico-chemical characteristics of the substrate, as in the biological characteristics of the plants. This is in agreement with the suggestion by Lawrence (1979) that the pioneers at Glacier Bay require some special physical, chemical or microbial conditions as well as a lack of competition.

The distinctive nature of freshly-deposited till has been neglected in previous models of primary succession. This has too often been characterized as an unfavourable, nutrient-poor medium, when nitrogen may be the only macronutrient that is deficient (Fitter, 1984). Even nitrogen is present in small quantities from such sources as precipitation or the infall of plant and animal remains. The availability of other nutrients, albeit in small quantities, in a substrate characterized by a high pH and a high base saturation (see section 4.2.4), appear to favour members of the pioneer species group that do not have a high nitrogen requirement. Conditions seem to be particularly favourable for basiphilous and calcicolous species, which often comprise a distinctive element of the pioneer flora (see section 5.4.2).

Other characteristics of newly-deglaciated terrain and the immediate proglacial area, such as the texture of the sediment and high moisture availability, can also be regarded as generally favourable for plant colonization, although exposure to glacier winds (section 3.3.2) and substrate instability of the paraglacial zone (section 3.4) are generally unfavourable. The latter contribute to the high rates of turnover in pioneer populations. High diversity levels in the late pioneer stage (section 6.1.5) may be considered as resulting from the availability of open space and low levels of competition, but only within limits set by the number of species able to colonize.

Replacement of pioneer species by later colonizers is also a multiprocess phenomenon. Biological processes, involving life cycle traits, facilitation and competition, are undoubtedly important (much as proposed by Walker & Chapin in Fig. 6.19). But so are gradual, allogenic changes in the physical environment, which parallel plant community change and soil development, and are causal influences on succession (see especially section 5.4.2). Included are such processes as the downwashing of fine particles (pervection) from the substrate, the leaching of cations and the corresponding acidification of the substrate, changes in the intensity of frost-heave and frost-sorting, deflation and aeolian deposition, and the climatic and hydrological changes induced by retreat of the glacier snout. Such predictable changes differ from the disturbances termed 'stochastic events' by Walker & Chapin (1987), and from the unpredictable 'environmental stress' of Pickett, Collins & Armesto (1987), although they could be accommodated as factors modifying 'resource availability' in the causal hierachy of Pickett and others (Table 6.7).

Disturbance (abrupt, disruptive, allogenic change) is increasingly recognized as interacting with biological processes during succession (Walker & Chapin, 1987; cf. Sousa, 1984; Breitburg, 1985; Walker, Zasada & Chapin, 1986; Whittaker, 1989), and as possessing endogenous as well as exogenous aspects (White, 1979; Pickett & White, 1985). Oliver, Adams & Zasoski (1985) estimate that of the total area deglaciated by Nooksack Glacier since A.D. 1800, about 63% has been subjected to secondary disturbances. A given area may be affected by more than one type of disturbance, which include, in decreasing order of area affected: avalanches (29%); rockslides (22%); intermittent snowfields (i.e. late-lying snow in cool summers or in summers following winters of heavy snow) (19%); creeping snowfields (15%); and glacio-fluvial streams (5%) (see also section 5.4.2 & Fig. 5.44). Whittaker (1991) concludes that disturbance slows, interrupts and often reverses progressive vegetation development on the Storbreen glacier foreland. Disturbances of various types may also enhance, redirect, or prevent successional change, depending on their type, scale, frequency and intensity. Whereas it is legitimate to regard large-scale, intense disturbance as nudation, small-scale disturbance of lower intensity but greater frequency (e.g. cryoturbation) can be an integral part of succession (Connell & Slatyer, 1977; White, 1979). Such stochastic events may be a strong influence, particularly on early primary succession (Fig. 6.19; cf. Shure & Ragsdale, 1977; van der Maarel, de Cock & de Wildt, 1985).

Decline of pioneer populations is not necessarily causally related to the rise of species characteristic of the next successional stage. For example, at least an element of the pioneer flora appears to decline in response to the physico-chemical changes associated with leaching. This process is accentuated by inputs of organic matter. However, it is likely to be effective on unvegetated or sparsely colonized terrain characterized by a low cation exchange capacity (see sections 4.2.4, 4.3.1 & 4.3.3). Precise mechanisms have not been investigated, but some or all of the following may be involved: (1) nutrient depletion; (2) changes in the balance of available nutrients; or (3) toxicity effects associated with pH changes.

Some of the gradual allogenic changes in the physical environment, such as the formation of stone pavement, may disadvantage later colonizers as well as pioneers. Others appear to favour certain later colonizers, such as acidophilous, calcifuge and heliophyte species. There is a parallel here with biological successional processes, some of which favour and some of which inhibit later colonizers (facilitation and competition, respectively). However, the likelihood of successional changes being driven by such allogenic processes probably follows a pattern close to that indicated for facilitation in Fig. 6.19.

Walker and Chapin (1987) follow Lawrence *et al.* (1967), Connell & Slatyer (1977) and Finegan (1984) in suggesting that facilitation is important primarily in the early stages of community development (see also Wood & del Moral, 1987). Reactions that ameliorate a severe environment are likely to improve conditions for other species, but for facilitation to be an effective mechanism of succession, the early colonizers must be disadvantaged or at least benefit less than the later colonizers from the changed conditions. Despite the frequency with which vegetation change in primary succession has been attributed to facilitation, particularly with respect to nitrogen fixation, it has yet to be demonstrated conclusively on glacier forelands (see section 6.1.4). There is also widespread support for competition as an important process in many successional sequences, especially in association with secondary succession (e.g. Parrish & Bazzaz, 1982; Connell, 1983; Schoener, 1983; Tilman, 1985, 1987b). It is likely that the importance of competition increases during primary succession (Fig. 6.19). However, as is the case with facilitation, its existence on glacier forelands lacks experimental proof (see section 6.1.5).

According to Walker & Chapin (1987) life-history traits are important throughout primary succession, although different traits become important at different successional stages (Fig. 6.19). In addition to mode of arrival, maximum potential growth rate affects changes in species composition in early to mid-succession in all situations, but especially in favourable environments where abundant resources are available to support rapid growth (cf. Hosner & Minckler, 1963; Uhl & Jordan, 1984). Differential

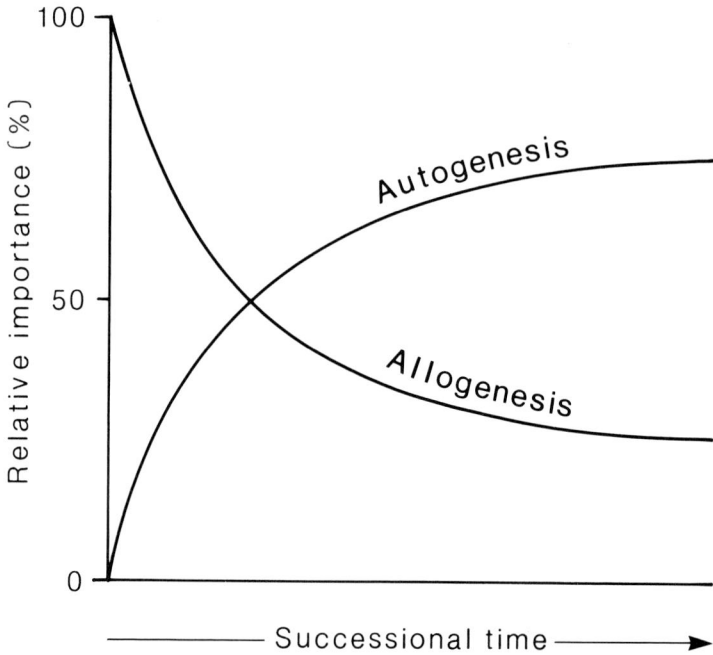

Fig. 6.20. Schematic representation of the relative importance of autogenic and allogenic processes during succession.

longevity, on the other hand, may explain changes in species dominance in mid- to late-succession in many communities (cf. van der Valk, 1981; Hibbs, 1983). These life-history traits would appear to be less important on those glacier forelands characterized by seres of long duration. In such situations successional stages, particularly in mid- to late-succession, can be much longer than the life-cycle lengths of the constituent species. However, little attention has been given to relationships between life-cycle traits (or species interactions) on the one hand, and allogenic environmental processes on the other.

The overall decline through time in the importance of allogenesis as a cause of directional vegetation change results in part from a reduction in the intensity of certain processes, including self-terminating processes (such as leaching of cations and the pervection of fines from surface horizons) and some types of disturbance, particularly paraglacial disturbance (see section 4.4). Allogenesis also declines relative to autogenesis as vegetation cover, biomass and plant reactions increase. This concept is summarized in Fig. 6.20, in which allogenic processes decline rapidly from 100% importance as autogenic processes are initiated and intensify. It should be noted, however, that allogenesis remains of significance in the later stages of succession (cf. Oliver, Adams & Zasoski, 1985). The relative importance of autogenesis and allogenesis at any stage, and hence the slopes of the lines in Fig. 6.20, depends on environmental severity. This idea is elaborated in the following section.

6.3.2   Spatio-temporal dynamics

Given that physical and biological processes are coupled, the second aspect of the geoecological model involves providing an explanatory framework for variations in succession both within and between glacier forelands. Details of the successional patterns at local and regional scales have been considered at length in section 5.3. Here, the underlying theme is that the relative importance of physical (allogenic) and biological (autogenic) causes varies in a predictable way that is dependent on the severity of the physical environment.

Suggested relationships between environmental severity and the relative importance of allogenic and autogenic processes during succession are shown in Fig. 6.21. In favourable (high resource, low stress) environments, steep curves are indicative of a rapid increase in the importance of autogenic processes through time. This reflects in particular the rapid accumulation of biomass due to such factors as a favourable climate, a relatively nutrient-rich substrate or an undisturbed habitat. In increasingly severe environments, greater constraints are placed on biological production, allogenic processes are relatively more important at any stage, and the phase controlled primarily by allogenic processes becomes longer. In the most severe environments, autogenic processes may be so weak that allogenesis remains more important than autogenesis indefinitely.

Fig. 6.21 can be applied to environmental gradients at various scales. For example, along macro-scale latitudinal, altitudinal and continentality gradients, major differences in environmental stresses are controlled largely by climate. At this scale the three sets of curves approximately correspond with a latitudinal gradient from boreal forest (e.g. Glacier Bay) to tundra and polar desert. Stress may result primarily from low air and soil temperatures, although other 'stressors' present include temperature ranges and rhythms, growing-season length, day length, wind strength, flooding and waterlogging, drought, snow lie, nutrient deficiencies, cryo-turbation and permafrost (cf. Billings, 1987). Similarly, the three sets of curves provide a good representation of differences on an altitudinal gradient of increasing severity from sub-alpine to low-alpine and high-

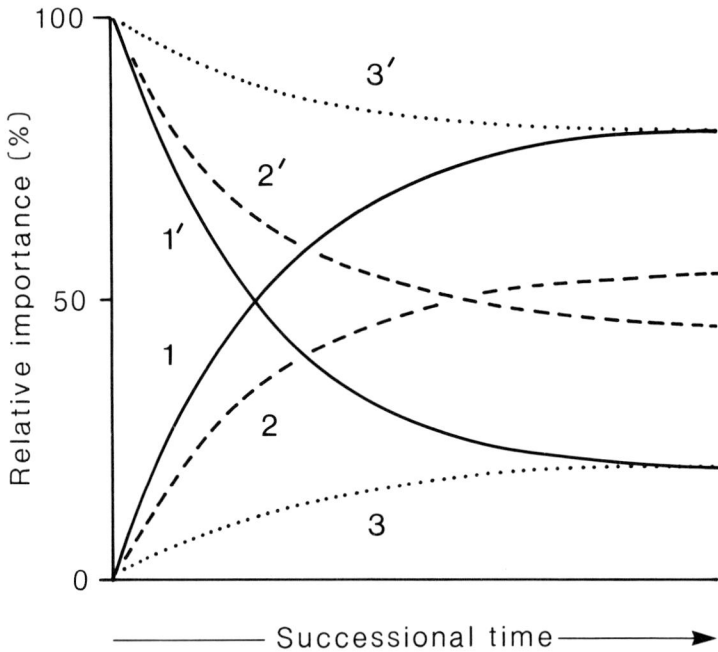

Fig. 6.21. Schematic representation of the effect of increasing environmental severity (solid, broken and dotted curves, respectively) on the relative importance of autogenic (1→2→3) and allogenic (1'→2'→3') processes during succession.

alpine belts. At the within-foreland scale, differences in environmental severity associated with, for example, microtopographic gradients, can be accommodated. Thus, within the low-alpine belt of southern Norway, moraine-ridge crests represent the most severe microenvironment, at least in terms of stresses imposed by exposure and drought.

This concept bears some similarity to the notion, developed by Svoboda & Henry (1987), that succession rate is determined by a 'biological driving force' (BDF) in relation to 'environmental resistance' (ER). BDF is defined as the sum of the biological (intrinsic) forces driving succession, whereas ER is the sum of the extrinsic abiotic and biotic factors resisting invasion, establishment, population expansion and soil development. For directional change in the vegetation, BDF must exceed ER. The main limitation of this idea is that the extrinsic abiotic factors are conceptualized only in terms of resistance to intrinsic biological forces. One of the main points of the present geoecological approach is that physical allogenic processes can reinforce or deflect, as well as resist, vegetation changes driven by biological processes. In short, succession involves environmental driving forces as well as biological ones.

Many aspects of the form and complexity of succession can be related to environmental severity and the relative importance of autogenic and allogenic processes. These include: (1) the number of species and variety of growth forms taking part in succession as well as the biomass; (2) the degree to which pioneer and mature communities differ in terms of species composition; (3) the number of successional stages that can be recognized; (4) the extent of divergence and/or convergence in successional pathways; (5) the rate of succession and the time required to attain a mature condition.

Species richness and the variety of growth forms involved in succession tend to be less in severe as opposed to favourable environments (see section 5.2.5). On the latitudinal gradient, for example, species richness falls as the number of morphological, physiological and reproductive options open to vascular plants for establishment and survival is reduced poleward (Billings, 1987). In severe environments, the small available flora, together with its limited biomass and relatively weak autogenic processes, lead to relatively small differences in the environment for plants between early and late successional stages (MacMahon, 1980, 1981). Any differences that exist are primarily due to allogenic processes. Where these processes produce strongly directional changes in the environment, succession may occur; however, in the absence of such directional allogenic changes, the various vegetation stages remain similar and succession is not apparent. Thus, non-selective autosuccession may be regarded as characteristic of the most severe polar and high-alpine environments. These environments are not only severe in terms of low temperatures, short growing seasons, etc., but cryogenic disturbance often constrains any directional change in the substrate that might arise due to allogenic processes. In less severe environments, greater autogenic changes combine with allogenic changes to promote larger differences between early and late stages and permit a larger number of successional stages. Relay succession, with successive species' replacements, is therefore increasingly apparent in increasingly favourable environments.

Whether successional pathways converge or diverge depends on the mix of autogenic and allogenic processes. In principle, processes of either type can alter successional pathways to produce convergence or divergence. Successional pathways will tend to converge where there are unifying processes creating similar environments. This may account for the varied opinions that have been expressed regarding the existence of convergence or divergence (see, for example, Olson, 1958; Monte, 1973; Matthews, 1979c,d; Glenn-Lewin, 1980; Pineda, *et al.*, 1981; Christensen & Peet, 1984; McCune & Allen, 1985; Matthews & Whittaker, 1987; Inouye & Tilman, 1988; Rydin & Borgegård, 1988; Facelli & D'Angela, 1990). Although there is no simple relationship, convergence on glacier forelands tends in practice to be associated with autogenesis in relatively favourable environments and in later successional stages (see section 5.3.3).

In boreal and sub-alpine zones, trees may rapidly extend strong biotic control over extensive areas of the landscape; hence succession may be strongly convergent. This is often accentuated by the presence in the later stages of only a single dominant species (e.g. *Picea sitchensis* or *Betula pubescens*). In the tundra and in the alpine zone, succession is slower and appears to be strongly divergent. A small number of dominant species each becomes confined to a limited range of habitats (e.g. heath and snowbed dominants). Strong environmental gradients persist largely unmodified during succession. Biotic modification of the habitat eventually produces, at most, only partial convergence along restricted parts of microtopographic gradients. In extreme polar and high-alpine environments, with low levels of autogenesis, overwhelmingly abiotic control of vegetation patterns, and simple non-selective autosuccessions, concepts of convergence and divergence have little meaning.

With the continued existence and interaction of autogenic and allogenic processes, succession tends towards a climax landscape of relative stability (see section 6.1.7). A steady-state gradient-rich landscape (Whittaker, 1953, 1974) is a useful abstract concept but such an equilibrium is rarely, if ever, attained on recently-deglaciated terrain. An apparent steady-state condition is most likely to be approached where successional change due to autogenic processes is rapid, and hence where environmental conditions are least severe (e.g. in boreal and sub-alpine climates or on productive substrates). However, gradual allogenic changes as well as disturbances generally prevent the attainment of a steady-state and, over several generations, the effects of one or a few dominant organisms can give rise to soil changes leading to retrogressive change. Thus, glacier foreland landscapes in general and those characteristic of severe environments in particular (e.g. extreme polar and high alpine environments) are best described as being in a permanent state of disequilibrium.

7

The ecological significance of recently-deglaciated terrain

Although it has been of major importance in the development of ideas on succession and soil development, the ecology of recently-deglaciated terrain has never been the subject of a comprehensive review. The first aim of this book has therefore been to provide the first review of this type. The scattered literature on ecological investigations of glacier forelands world-wide has been brought together and the present state of knowledge has been summarized with particular reference to succession and related concepts. The second aim has been, through the geoecological approach, to provide an integrated and interdisciplinary theoretical framework for this research. This final chapter contains a brief overview of both the research and the approach, and points to some possible future directions for study. Despite a relatively long history of investigation, the full potential of glacier forelands for ecological research has yet to be realized.

7.1   Chronosequences

By analysing chronosequences, a large quantity of information has accumulated on primary succession and soil development. In effect, glacier forelands have been viewed as large-scale natural ecological experiments operating on timescales of decades and centuries. Although much of the available information is descriptive and sometimes anecdotal, and has often been collected during short summer field seasons, it nevertheless represents an extremely important data source. This data source has been and can be used further to develop concepts and theory and to test those originating elsewhere. Opportunities for the study of chronosequences are rare, and there are few alternative approaches available for investigating long-term ecological change (Likens, 1989).

Much of the theory of succession has been based on secondary succession (see section 6.2) and, at present, this theory tends to be of site-specific rather than universal applicability (West, Shugart & Botkin, 1981). The same can be said about primary succession, although slower change makes it more difficult to gather information and to generalize. Data from glacier forelands are of importance not only in contributing to the theory of primary succession (cf. Adams & Dale, 1987; Walker, 1991) but also in clarifying the similarities and differences in nature and rate between primary and secondary succession. Suggestions that the development of a general theory of succession may be a futile objective (West, Shugart & Botkin, 1981) or that the search for a unifying, generalized theory of succession should be abandoned (Miles, 1987) are premature. Such a theory must, however, be capable of encompassing succession as a variable and complex phenomenon (cf. Huston & Smith, 1987; Burrows, 1990). Thus, it is hoped that the variability and complexity of glacier foreland chronosequences described in this book is a useful contribution towards an attainable, general theory of succession.

The greatest weakness of studies of chronosequences on glacier fore-lands is that plant succession and soil development are only inferred. There are surprisingly few cases where such inferences have actually been tested (see sections 5.1.2 & 5.1.3). Further tests involving both direct observation and retrospective analysis are urgently required. It is particularly important that existing permanent quadrats are maintained and recorded on a regular basis, and that new ones are established at strategically-located sites. Tests based on retrospective analysis are likely to provide greater immediate returns, although they cannot provide the same resolution as direct observation. Considerable ingenuity will be necessary to design appropriate and sophisticated analytical procedures capable of filtering out the effects of particular factors and clarifying their interactions.

Chronosequence studies increasingly involve quantification but have rarely applied rigorous sampling procedures. Small sample sizes at so-called optimal sites that are supposedly representative of particular terrain ages, tend to be the rule, partly due to logistical considerations. This is a characteristic feature of most phytosociological studies of plant communities and also of investigations of pedogenesis. Greater attention to rigorous sampling designs, with due regard for natural variability and replication, is necessary in order to obtain full benefit from future chronosequence research. Good models in this respect include those of Sondheim & Standish (1983) in their study of soil development in front of Robson Glacier, British Columbia, and the successional studies of the author at Storbreen, Jotunheimen (see section 5.3.1).

Increased attention has been paid in recent years to the investigation of successional processes in chronosequences (see especially section 6.1). This development has led to major advances in our understanding of succession. Many more such studies are required and must be integrated in order to understand succession as a phenomenon of process interactions (Walker & Chapin, 1987). Lack of understanding of plant population processes during glacier foreland succession is slowly being remedied. However, a more concentrated effort is required to gain essential information on autecology and species' interactions, particularly in relation to the dominant species. Manipulative experiments provide a most powerful method for testing the importance of successional processes generally (cf. Hairston, 1989; Aarssen & Epp, 1990). To date, few experiments of this type have been attempted on glacier forelands. Those at Glacier Bay, which were loosely designed to test facilitation (section 6.1.4), are being extended by L. R. Walker and F. S. Chapin (pers. comm. 1989). Another set of manipulative experiments were begun at Storbreen, Jotunheimen, by R. J. Whittaker, J. R. Petch and the author in 1987.

Relatively few studies have involved a detailed consideration of interactions between vegetation and soil. This is surprising given the close and interrelated nature of succession and soil development. Further understanding of both plant succession and soil development necessitates a synthesis of ideas from these two fields, and an integrated methodology. The beginnings of such a methodology can be seen in the investigation of nutrients in soils and biomass (e.g. Jacobson & Birks, 1980; Fitter & Parsons, 1987; Bormann & Sidle, 1990; cf. Walker, 1989). It is also a fundamental part of the geoecological approach, which is discussed in the section below.

7.2   The geoecological approach

A geoecological approach has been adumbrated in this book not as a well-articulated model but more as a conceptual framework. This is clear from section 6.3, which has summarized its two key aspects: (1) the concept that biological and physical systems interact within the evolving glacier foreland landscape; and (2) the concept that glacier foreland landscapes are spatial systems within which the characteristics of succession and soil development vary broadly in relation to environmental severity.

The importance of the dynamics of the physical system to the ecology of recently-deglaciated terrain has long been recognized but in most investigations there remains a reluctance fully to embrace its implications. All studies recognize that initial physical conditions constrain succession and soil development. Most recognize that allogenic factors are effective, even if they are usually regarded as external to the biological system. However, almost all view change, and the processes responsible for change, within a primarily autogenic framework. The geoecological approach allows for physical and biological systems to interact and for allogenic and autogenic processes to control the trajectory of change, depending on time and place.

This approach has points of contact with several other approaches to vegetation dynamics. For example, the 'dynamical systems perspective' of Roberts (1987) treats vegetation and environment as being coupled by the physiological requirements of the component species and their ability to modify the environment. It is also related to the 'individual-based model' of Huston & Smith (1987), which depicts succession as a non-equilibrium process producing steady-state communities, the properties of which depend on abiotic conditions and on the type and frequency of disturbances. A somewhat similar view is taken by Burrows (1990) who summarizes the 'kinetic concept' of vegetation change as recognising disturbance and continually changing vegetation as normal, and self-maintaining vegetation as rather unusual (cf. Raup, 1957, 1981; Drury & Nisbet, 1973; Veblen & Ashton, 1978; White, 1979; Veblen, Schlegel & Escobar, 1980; Oliver, 1981; Whitmore, 1982; Pickett & White, 1985).

In this book, the physical system has been treated largely analytically. The components of the physical environment (including climate) were first examined in turn (see chapter 3). Many physico-chemical processes have thereby been identified as producing changes in the physical environment. These processes should be regarded as at least potential driving forces in succession and soil development. However, there have been few attempts to study in detail actual interactions between physical and biological processes on glacier forelands. In some cases correlations have been observed between physical environmental changes and changes in the vegetation or soil system (see especially sections 4.4.1, 5.4.2 & 5.4.3). Such circumstantial evidence is of a similar type to much of the basis for belief in biological processes as the driving forces in succession and soil development.

The present situation is well summarized by the diagrammatic representation of the succession concept (Fig. 7.1) that was compiled by H. S. Horn and J. F. Franklin (West, Shugart & Botkin, 1981). This shows elements of the 'classical view of succession', in which primary succession is initiated by geophysical processes but thereafter is largely a deterministic biological process unless interrupted by environmental change. Additional elements of a 'modern concept of succession' include the recognition of physical disturbance as a normal feature of the environmental regime, a greater

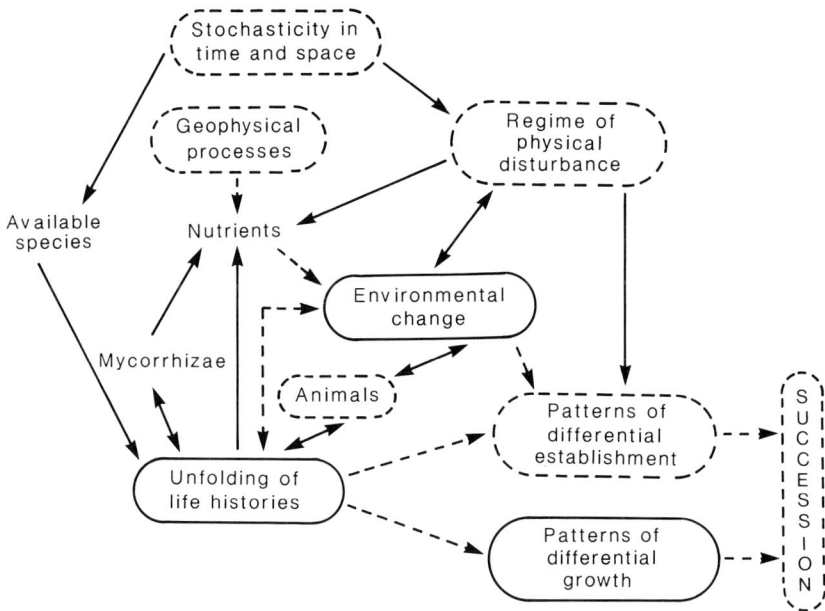

Fig. 7.1. Schematic representation of the succession concept indicating classical (broken arrows) and modern (solid arrows) elements, areas where working models are available (solid compartments), and areas where models are presently only in a suggestive phase of development (broken compartments) (after H. S. Horn & J. F. Franklin, from West, Shugart & Botkin, 1981).

degree of uncertainty, and a wider range of possible causal factors determining successional pathways. Few of the subsystems or links are well understood. On glacier forelands, the disturbance regime, other aspects of the physical environmental system, and the roles of animals, mycorrhizae and microbial components, are particularly poorly known. Additional complications are introduced by what might be termed the multifarious and multidimensional nature of vegetation dynamics in space and time. Cyclic processes (cf. Watt, 1947) may occur alongside directional processes, various directional processes may have different durations (cf. the short-term, secular and geohistorical changes of Jenik, 1986) and processes may interact at different scales, some in a hierarchical fashion (cf. Allen & Starr, 1982; O'Neill *et al.*, 1986). Such complexities have hardly begun to be addressed on glacier forelands, despite the relative simplicity of these landscapes.

A proper synthesis of the physical and biological systems has yet to be achieved on any glacier foreland. It should be recognized that the phytosociological vegetation maps produced in the Alps and Norway (see section

5.3.1) represent a step in this direction. The more objective mapping of numerically-defined community types and community gradients at Storbreen has extended this approach (see section 5.3.2). A further important step has been the approach developed at Storbreen by Whittaker (1987, 1989), who recognizes environmental factor complexes that are related to vegetational gradients (see section 5.4.3). Although powerful numerical methods now permit the combined analysis of vegetation and environmental data sets, available field measurements of environmental variables are rarely comparable in quality or quantity to those available for the vegetation.

At the between-foreland scale, generalizations have been attempted in this book (see sections 5.3.4 & 6.3.2). These are, however, qualitative and there are major deficiencies in the data base, particularly outside of Europe and North America. On a world-wide basis, the most significant gap in knowledge exists with respect to tropical alpine glacier forelands, where even the patterns of plant succession have yet to be described in detail. Based on the fragmentary studies available from Mount Kenya (e.g. Coe, 1964, 1967; Spence, 1989), isolated individuals of many species appear to colonize early. However, in this apparently hostile environment, it seems to take > 1000 years for the vegetation to achieve a cover and floristic composition similar to the mature Afroalpine vegetation complex present on terrain deglaciated for *ca* 12 500 years (Mahaney, 1990). From a geoecological point of view, it is important to explore the effects of the distinctive tropical alpine environment on successional processes and soil development. A project of this type has been proposed for Mount Jaya, Irian Jaya, as part of a larger integrated mountain research programme for Indonesia (Kilmaskossu & Hope, 1985).

Although all of these approaches have made contributions to our understanding of vegetation-environment interrelationships, with the exception of the small number of studies on nutrient cycling noted above (section 7.1), they do not directly tackle functional interactions between the physical and biological systems. A detailed understanding of functional relationships requires a concerted effort in the fields of monitoring, experiment and modelling. This might best be attempted at a small number of forelands where the uncoupled biological and physical systems are best understood. However, even at sites such as Glacier Bay or Storbreen, a fundamental lack of knowledge of the physical subsystem would appear to be a major stumbling block to the implementation of a truly functional geoecological approach.

7.3   Some broader implications

The main contribution of studies of recently-deglaciated terrain to ecology has undoubtedly resulted from viewing these areas as field laboratories, or microcosms, within which natural experiments in ecosystem and landscape development are unfolding. This is encapsulated within the chronosequence concept. Whether emphasis is given to plant succession, soil development, whole ecosystems or landscapes, all studies of glacier foreland chronosequences take advantage of the unique opportunities offered for detailed study of long-term, time-dependent processes, which are otherwise largely inaccessible to ecological investigation. In this last section, brief comments are made on five broader areas within which ecological studies of glacier forelands are of at least potential relevance. These are, perhaps surprisingly, very wide ranging: (1) resolving conflicts between holistic and reductionist approaches to ecology; (2) providing modern analogues for palaeoenvironmental reconstruction; (3) developing biogeographical theory; (4) applying succession theory to land restoration; and (5) understanding cultural landscapes and human environmental impacts.

On recently-deglaciated terrain, the complementarity of holistic and reductionist approaches to ecology is apparent. Individuals and populations do not exist in isolation; they occur together in communities and ecosystems and as part of the vegetation cover in landscapes. It is as incomplete to reduce the ecology of glacier forelands merely to individuals and populations as it is to ignore these components of the landscape. Although the geoecological approach begins from a holistic perspective, it requires a consideration of individuals and populations if processes and mechanisms are to be understood at particular sites. In turn, studies that focus on these building blocks of ecology need to be placed in a broader perspective of interactions with other ecosystem components and of larger-scale and longer-term phenomena. It is clear, therefore, that both holistic and reductionist approaches are necessary for a complete ecology. Studies of glacier foreland ecology may well contribute to the forthcoming synthesis.

Second, some attempts have been made to use landscape development on recently-deglaciated terrain as an analogue of deglaciation in the past. The interdisciplinary project to investigate the glacier foreland and surge moraines of Klutlan Glacier, Yukon Territory, was launched explicitly to improve understanding of the history of landscape evolution and to aid interpretation of the Late Wisconsin stratigraphic record in Minnesota

(Wright, 1980). Studies of ice wastage, lake formation, soil development and revegetation associated with the stagnating ice-cored moraines of Klutlan Glacier seem to provide a satisfactory analogue, even though there are many differences in the details of how the two landscapes evolved. In particular, there is strong support for the hypothesis that most Minnesota lakes originated as shallow depressions in a fully-vegetated supraglacial till mantle.

Although perfect analogues are unlikely to be found, there is considerable scope for this type of study elsewhere. Parallels between sediment and soil changes in the Late Glacial and soil changes in chronosequences have been discussed by Berglund & Malmer (1971). These authors point out, however, that present-day glacier forelands deviate from the regions exposed by ice recession during Late Glacial times in terms of their small area and being surrounded by vegetation. Similarly, despite very different time scales, Lüdi (1958) has compared vegetation succession and soil development in Central Europe at the beginning of the Allerød with chronosequences in front of present-day glaciers in the central Alps. At a more detailed level, several workers have used the presence of particular pollen taxa or pollen assemblages near the base of pollen profiles as evidence of the early stages of succession following deglaciation. For example, based on the presence of *Oxyria* on recently-deglaciated terrain in Norway, Macpherson (1982) has used high *Oxyria* pollen percentages near the base of a pollen profile from Newfoundland (see Birks, 1973, and Tipping, 1988, for similar inferences based in Scotland). One way in which progress can be made is to relate the present vegetation in recently-deglaciated areas to the present pollen rain, thereby clarifying the interpretation of fossil pollen assemblages (Jochimsen, 1972, 1979, 1986; Birks, 1980b; Caseldine, 1989).

A third area of potential broader relevance for the ecology of recently-deglaciated terrain is in the development of biogeographical theory *sensu* MacArthur & Wilson (1967). MacArthur & Wilson's equilibrium theory of island biogeography considers the species richness of islands as a function of rates of immigration and extinction, which are in turn affected by island size and distance from source areas. Many researchers have since interpreted 'islands' in a wider sense, including not only true islands but also hills surrounded by plains, isolated patches of vegetation, nature reserves within agricultural landscapes, and gaps in otherwise vegetated areas (e.g. van der Maarel, 1988). Glacier forelands of varying sizes represent another type of 'island' situation and could be profitably investigated in such terms. Indeed, Elven (1980) in an original and wide-ranging paper on plant immigration and succession on southern Norwegian nunataks (islands within a sea of ice) clearly envisages this development. Investigations of glacier forelands from this viewpoint would seem particularly welcome given recent criticisms of island biogeographic theory on the basis of an inadequate data base and insufficient validation (Gilbert, 1980).

Fourth, succession theory derived from glacier forelands may find direct applications to the restoration of land. This field has already been mentioned as a justification for the investigation of succession in general and primary succession in particular (see section 1.2). Matthews & Whittaker (1987) discuss several specific implications of the theory of primary succession as derived from glacier forelands for vegetation restoration. The complexity of successional patterns on glacier forelands suggests that a particular course of action may not be appropriate for all parts of a landscape being restored; it may therefore be necessary to adopt modified or different treatments in different areas. By use of appropriate treatments to simulate successional divergence, it may be possible to divert succession along a different trajectory towards an alternative 'stable' state. By altering the balance within a controlling factor complex, it may be possible to find alternative treatments capable of bringing about the same desired successional effect. Attention could also be given to the reduction or prevention of those features of the system leading to retrogression (e.g. disturbance) as well as or instead of encouraging progressive succession. It should be noted that not all natural successions are equally suitable as a source of theoretical guidance for restoration programmes. Glacier forelands would appear to be particularly suitable as a source of information relevant to the restoration of sites in open, disturbed arctic and alpine environments, for which observational or experimental data from closed, undisturbed tundra may be misleading.

Lastly, ecological studies of recently-deglaciated terrain can contribute both directly and indirectly to an understanding of cultural landscapes and human impacts. A direct contribution may be forthcoming where human activities impinge on glacier forelands. This ranges from the impact of tourists and scientists at Glacier Bay (Lawrence, 1979) and in the Antarctic (Smith, 1982), to grazing animals, cultivation and industry in Norway (Fægri, 1933, 1986; Alexander, 1982), the Alps (Lüdi, 1958; Jochimsen, 1970; Schubiger-Bossard, 1988) and the U.S.S.R. (Turmanina & Volodina, 1979; Solomina, 1989). There is considerable scope for historical investigations of the cultural landscape, *sensu* Birks *et al.* (1988), on glacier forelands located close to agricultural communities. Particularly in the European sub-alpine and low-alpine belts, practices such as grazing and cutting for fodder or fuel are likely to have been much more extensive in past centuries, with large residual effects on the present vegetation.

Indirectly, glacier forelands may provide data on the natural background to processes that become affected by human activities. This has been recognized by Bormann & Sidle (1990) in the context of forest-resource issues, such as long-term site productivity, forest decline, acid rain and climatic change. All these issues require a better understanding of likely natural long-term trends upon which the human impact is being or will be superimposed. Where largely undisturbed by human activities, glacier forelands provide a window of opportunity in this area.

A timely conclusion is provided by experiments that have been carried out on Signy Island in the Antarctic by Smith (1991; see also Smith, 1990). The experiments using cloches, with the resulting germination and establishment of plants from propagule pools (see section 6.1.2), have been interpreted as giving an indication of the likely impact on a glacier foreland of global warming due to the greenhouse effect. It should of course be recognized that greenhouse warming is probably already increasing the rate of glacier retreat and hence increasing the area of glacier forelands world-wide!

References

Aarssen, L. W. & Epp, G. A. (1990). Neighbour manipulations in natural vegetation: a review. *Journal of Vegetation Science*, **1**, 13–30.

Adams, A. B. & Dale, V. H. (1987). Vegetation succession following glacial and volcanic disturbances in the Cascade Mountain Range of Washington, U.S.A. In *Mount St Helens 1980. Botanical Consequences of the Explosive Eruptions*, ed. D. E. Bilderback, pp. 70–147. Berkeley: University of California Press.

Aellen, M. (1981). Recent fluctuations of glaciers. In *Switzerland and Her Glaciers*, eds. P. Kasser & W. Haeberli, pp. 70–89. Berne: Kümmerly & Frey.

Ahlmann, H. W. (1953a). *Glacier Variations and Climatic Fluctuations*. New York: American Geographical Society. (Bowman Memorial Lectures, Series 3.)

Ahlmann, H. W. (1953b) Glaciärer och klimat i Norden under de senaste tusentalen år. *Norsk Geografisk Tidsskrift*, **13**, 56–75.

Åkerman, J. (1980). Studies on periglacial geomorphology in West Spitzbergen. *Meddelanden från Lunds Universitets Geografiska Institution Avhandlingar*, **89**, 1–297.

Alestalo, J. (1971). Dendrochronological interpretation of geomorphic processes. *Fennia*, **105**, 1–140.

Alexander, M. J. (1982). Soil development at Engabredal, Holandsfjord, north Norway. *Okstindan Research Project Preliminary Report for 1980 and 1981*, ed. J. Rose, pp. 1–23. London: University of London, Birkbeck College.

Alexander, M. J. (1986). Micro-scale soil variability along a short moraine ridge at Okstindan, northern Norway. *Geoderma*, **37**, 341–60.

Alexandrova, V. D. (1988). *Vegetation of the Soviet Polar Deserts* Cambridge: Cambridge University Press.

Alexandrovsky, A. L. (1989). Soils as an indicator of the age of mountain moraines. In *Moraines as a Source of Glaciological Information*, eds. L. R. Serebryanny, A. V. Orlov & O. N. Solomina, pp. 168–72. Moscow: Nauka. [In Russian]

Allen, T. F. H. & Starr, T. B. (1982). *Hierarchy: Perspectives for Ecological Complexity*. Chicago: Chicago University Press.

Ammann, K. (1975). Pollenanalytische und Vegetationskundliche Untersuchungen im Vorfeld. *Zeitschrift für Gletscherkunde und Glazialgeologie*, **11**, 63–73.

Ammann, K. (1979). Gletschernahe vegetation in der Oberaar Einst und Jetzt. In *Werden und Vergehen von Pflanzengesellschaften*, eds. O. Wilmanns & R. Tüxen, pp. 227–51. Vaduz: Cramer.

Andersen, J. L. & Sollid, J. L. (1971). Glacial chronology and glacial geomorphology in the marginal zones of the glaciers Midtdalsbreen and Nigardsbreen, South Norway. *Norsk Geografisk Tidsskrift*, **25**, 1–38.

Anderson, D. J. (1967). Studies on structure in plant communities. III. Data on pattern in colonizing species. *Journal of Ecology*, **55**, 397–404.

Anderson, D. J. (1986). Ecological succession. In *Community Ecology*, eds. J. Kikkawa & D. J. Anderson, pp. 269–85. Oxford: Blackwell Scientific Publications.

Anderson, H. A., Berrow, M. L., Farmer, V. C., Hepburn, A., Russell, J. D. & Walker, A. D. (1982). A reassessment of podzol formation processes. *Journal of Soil Science* **33**, 125–36.

Andrews, J. T. (1975). *Glacial Systems: An Approach to Glaciers and their Environments*. North Scituate, Massachusetts: Duxbury Press.

Archer, A. C. (1973). Plant succession in relation to a sequence of hydromorphic soils formed on glacio-fluvial sediments in the alpine zone of the Ben Ohau Range, New Zealand. *New Zealand Journal of Botany*, **11**, 331–48.

Archer, A. C., Simpson, M. A. J. & Macmillan, B. H. (1973). Soils and vegetation of the lateral moraine at Malte Brun, Mount Cook Region, New Zealand. *New Zealand Journal of Botany*, **11**, 23–48.

Arnold, R. W. (1983). Concepts of soils and pedology. In *Pedogenesis and Soil Taxonomy. I. Concepts and Interactions*, eds. L. P. Wilding, N. E. Smeck & G. F. Hall, pp. 1–21. Amsterdam: Elsevier.

Ballantyne, C. K. (1978). Variations in the size of coarse clastic particles over the surface of a small sandur, Ellesmere Island, N.W.T., Canada. *Sedimentology*, **25**, 141–7.

Ballantyne, C. K. & Matthews, J. A. (1982). The development of sorted circles on recently-deglaciated terrain, Jotunheimen, Norway. *Arctic and Alpine Research*, **14**, 341–54.

Ballantyne, C. K. & Matthews, J. A. (1983). Desiccation cracking and sorted polygon development, Jotunheimen, Norway. *Arctic and Alpine Research*, **15**, 339–49.

Baranowski, S. (1977). Results of dating the fossil tundra in the forefield of Werenskioldbreen. *Acta Universitatis Wratislaviensis*, **387**, 31–6.

Baranowski, S. & Karlén, W. (1976). Remnants of Viking Age tundra in Spitzbergen and northern Scandinavia. *Geografiska Annaler*, **58(A)**, 35–40.

Baranowski, S. & Pękala, K. (1982). Nival-eolian processes in the tundra areas and in the nunatak zone of the Hans and Werenskiold Glaciers (SW Spitsbergen). *Acta Universitatis Wratislaviensis*, **525**, 11–27.

Bardo, K. S. (1980). *Soil Development on Neoglacial Deposits in the Nooksack Cirque Area*. M.S. Thesis: University of Washington, Seattle. [not seen]

Barry, R. G. (1981). *Mountain Weather and Climate*. London: Methuen.

Barry, R. G. & Ives, J. D. Introduction. In *Arctic and Alpine Environments*, eds. J. D. Ives & R. G. Barry, pp. 1–13. London: Methuen.

Baxter, D. V. & Middleton, J. T. (1961). Geofungi in forest successions following retreat of Alaskan Glaciers. *Recent Advances in Botany*, **2**, 1514–17. [Lectures and Symposia presented to the 9th International Botanical Congress, Montreal, 1959. Toronto: University of Toronto Press.]

Bazzaz, F. A. (1979). The physiological ecology of plant succession. *Annual Reviews of Ecology and Systematics*, **10**, 351–71.

Beck, E., Mägdefrau, K. & Senser, M. (1986). Globular mosses. *Flora oder Allemeine Botanische Zeitung (Jena)*, **178**, 73–83.

Becking, J. H. (1979). Nitrogen fixation by *Rubus ellipicus* J. E. Smith. *Plant and Soil*, **53**, 541–54.

Becking, J. H. (1984). Identification of the endophyte of *Dryas* and *Rubus* (Rosaceae). *Plant and Soil*, **78**, 105–28.

Begét, J. E. (1983). Radiocarbon-dated evidence of worldwide early Holocene climatic change. *Geology*, **11**, 389–93.

Benedict, J. B. (1976). Frost creep and gelifluction features: a review. *Quaternary Research*, **6**, 55–76.

Benedict, J. B. (1989). The use of *Silene acaulis* for dating: the relationship of cushion diameter to age. *Arctic and Alpine Research*, **21**, 91–6.

Benninghoff, W. S. (1952). Interaction of vegetation and soil frost phenomena. *Arctic*, **5**, 34–44.

Berglund, B. E. & Malmer, N. (1971). Soil conditions and late-glacial stratigraphy. *Geologiska Föreningens i Stockholm Förhandlingar*, **93**, 575–86.

Bergsma, B. M., Svoboda, J. & Freedman, B. (1984). Entombed plant communities released by a retreating glacier at central Ellesmere Island, Canada. *Arctic*, **37**, 49–52.

Bergström, E. (1955). British Ruwenzori Expedition, 1952, glaciological observations – preliminary report. *Journal of Glaciology*, **2**, 468–76.

Berry, M. E. (1987). Morphological and chemical characteristics of soil catenas on Pinedale and Bull Lake moraine slopes in the Salmon River Mountains, Idaho. *Quaternary Research*, **28**, 210–25.

Beschel, R. E. (1950). Flechten als Altersmasstab rezenter Moränen. *Zeitschrift für Gletscherkunde und Glazialgeologie*, **1**, 152–61.

Beschel, R. E. (1957). Lichenometrie im Gletschervorfeld. *Jahrbuch, Verein zum Schutz der Alpenpflanzen und -tierre (München)*, **22**, 164–85.

Beschel, R. E. (1958). Lichenometrical studies in West Greenland. *Arctic*, **11**, 254.

Beschel, R. E. (1961). Dating rock surfaces by lichen growth and its application to glaciology and physiography (lichenometry). In *Geology of the Arctic, Volume 2*, ed. G. O. Raasch, pp. 1044–62. Toronto: University of Toronto Press.

Beschel, R. E. (1963). Growth ring studies on Arctic willows. In *Jacobsen-McGill Arctic Research Expedition 1959–62. Preliminary Report 1961–1962*, ed. F. Müller, pp. 189–98. Montreal: McGill University. (Axel Heiberg Island Research Reports).

Beschel, R. E. & Weidick, A. (1973). Geobotanical and geomorphological reconnaissance in West Greenland, 1961. *Arctic and Alpine Research* **5**, 311–19.

Bezinge, A. (1976). Troncs fossiles morainiques et climat de la periode Holocene en Europe. *Bulletin de la Murithienne*, **93**, 93–111.

Bickerton, R. (1978). *Colonisation of Pioneer Species on a Recently-exposed Glacier Foreland at Styggedalsbreen in Jotunheimen, Southern Norway*. B.Sc. Thesis. University of Sheffield, Department of Geography.

Billings, W. D. (1974). Arctic and alpine vegetation: plant adaptations to cold summer climates. In *Arctic and Alpine Environments*, eds. J. D. Ives & R. G. Barry, pp. 403–43. London: Methuen.

Billings, W. D. (1987). Constraints to plant growth, reproduction, and establishment in Arctic environments. *Arctic and Alpine Research*, **19**, 357–65.

Billings, W. D. & Bliss, L. C. (1959). An alpine snowbank environment and its effects on vegetation, plant development, and productivity. *Ecology*, **30**, 388–97.

Billings, W. D. & Mooney, H. A. (1968). The ecology of arctic and alpine plants. *Biological Reviews*, **43**, 481–529.

Birkeland, P. W. (1974). *Pedology, Weathering and Geomorphological Research*, Oxford: Oxford University Press.

Birkeland, P. W. (1978). Soil development as an indicator of relative age of Quaternary deposits, Baffin Island, N.W.T., Canada. *Arctic and Alpine Research*, **10**, 733–47.

Birkeland, P. W. & Burke, R. M. (1988). Soil catena chronosequences on eastern Sierra Nevada moraines, California, U.S.A., *Arctic and Alpine Research*, **20**, 473–84.

Birkeland, P. W., Burke, R. M. & Benedict, J. B. (1989). Pedogenic gradients for iron and aluminium accumulation and phosphorus depletion in arctic and alpine soils as a function of time and climate. *Quaternary Research*, **32**, 193–204.

Birks, H. H., Birks, H. J. B., Kaland, P. & Moe, D. eds. (1988). *The Cultural Landscape – Past, Present and Future*. Cambridge: Cambridge University Press.

Birks, H. J. B. (1973). *Past and Present Vegetation of the Isle of Skye: A Palaeoecological Study*. Cambridge: Cambridge University Press.

Birks, H. J. B. (1980a). The present flora and vegetation of the moraines of the Klutlan

Glacier, Yukon Territory, Canada. *Quaternary Research*, **14**, 60–86.

Birks, H. J. B. (1980b). Modern pollen assemblages and vegetational history of the moraines of the Klutlan Glacier, and its surroundings, Yukon Territory, Canada. *Quaternary Research*, **14**, 101–29.

Birks, H. J. B., Birks, H. H., Lamb, H. F. & Wright, H. E. Jr (in prep.). Floristic changes on glacial moraines in Jostedal between 1930 and 1990.

Birnie, R. V. (1977). A snow-bank push mechanism for the formation of some "annual" moraine ridges. *Journal of Glaciology*, **18**, 77–85.

Blachut, S. P. & Ballantyne, C. K. (1976). Ice-dammed lakes: a critical view of their nature and behaviour. *McMaster University Department of Geography, Research Discussion Paper*, **6**, 1–90.

Black R. F. (1981). Late Quaternary climate changes in the Aleutian Islands, Alaska. In *Quaternary Paleoclimate*, ed. W. C. Mahaney, pp. 47–62. Norwich: GeoAbstracts.

Bliss, L. C. (1971). Arctic and alpine plant life cycles. *Annual Review of Ecology and Systematics*, **2**, 405–38.

Bluck, B. J. (1974). Structure and directional properties of some valley sandur deposits in southern Iceland. *Sedimentology*, **21**, 533–54.

Blundon, D. J. (1989). *Primary Succession Following Retreat of the Robson Glacier, British Columbia*. Ph.D. Thesis. University of Alberta.

Blundon, D. J. & Dale, M. R. T. (1990). Dinitrogen fixation (acetylene reduction) in primary succession near Mount Robson, British Columbia, Canada. *Arctic and Alpine Research*, **22**, 255–63.

Blundon, D. J. & Russell, W. (in prep.) Primary succession on the Mount Robson moraines, British Columbia, Canada.

Bockheim, J. G. (1980). Solution and use of chronofunctions in studying soil development. *Geoderma*, **24**, 71–85.

Bockheim, J. G. (1990). Soil development rates in the Transantarctic Mountains. *Geoderma*, **47**, 59–77.

Bohn, H. L., McNeal, B. L. & O'Conner, G. A. (1979). *Soil Chemistry*, New York: Wiley.

Bond, G. (1956). Evidence for fixation of nitrogen by root nodules of alder (*Alnus*) under field conditions. *New Phytologist*, **55**, 147–53.

Bond, G. (1976). The results of the IBP survey of root nodule formation in non-leguminous angiosperms. In *Symbiotic Nitrogen Fixation in Plants*, ed. P. S. Nutman, 443–74. Cambridge: Cambridge University Press.

Bonde, E. K. (1969). Plant disseminules in wind-blown debris from a glacier in Colorado. *Arctic and Alpine Research*, **1**, 135–40.

Boothroyd, J. C. & Ashley, G. M. (1975). Processes, bar morphology and sedimentary structures on braided outwash fans, northeastern Gulf of Alaska. In *Glaciofluvial and Glaciolacustrine Sedimentation.*, eds. A. V. Jopling & B. C. McDonald, 193–222. Tulsa, Oklahoma: Society of Economic Palaeontologists and Mineralogists. (SEPM Special Publication No. 23)

Boothroyd, J. C. & Nummedal, D. (1978). Proglacial braided outwash: a model for humid alluvial fan deposits. In *Fluvial Sedimentology*, ed. A. D. Miall, pp. 641–68. Calgary: Canadian Society of Petroleum Geologists. (CSPG Memoir No. 5)

Bormann, B. T. & Sidle, R. C. (1990). Changes in productivity and distribution of nutrients in a chronosequence at Glacier Bay National Park, Alaska. *Journal of Ecology*, **78**, 561–78.

Bormann, B. T. & Tarrant, R. F. (1989). Rain and cloud water at remote sites in Alaska and Oregon. *Journal of Environmental Quality*, **18**, 149–52.

Bormann, F. H. & Likens, G. E. (1979). *Pattern and Process in a Forested Ecosystem*. New York: Springer Verlag.

Bothamley, I. K. (1987). *The Nature and Origin of Neoglacial Terminal Moraines:*

Jotunheimen and Jostedalsbreen, Southern Norway. Ph.D. Thesis, University of Wales (University College Cardiff).

Botkin, D. B. (1981). Causality and succession. In *Forest Succession: Concepts and Applications*, eds. D. C. West, H. H. Shugart & D. D. Botkin, 36–55. New York: Springer Verlag.

Boulton, G. S. (1970). The deposition of subglacial and melt-out tills at the margins of certain Svalbard glaciers. *Journal of Glaciology*, **9**, 231–45.

Boulton, G. S. (1976a). A genetic classification of tills and criteria for distinguishing tills of different origin. *Uniwersytet im. Adam Mickiewicza W Poznaniu, Seria Geografia*, **12**, 65–80.

Boulton, G. S. (1976b). The origin of glacially fluted surfaces: observations and theory. *Journal of Glaciology*, **17**, 287–301.

Boulton, G. S. (1978). Boulder shapes and grain-size distributions of debris as indicators of transport paths through a glacier and till genesis. *Sedimentology*, **25**, 773–99.

Boulton, G. S. & Dent, D. L. (1974). The nature and rates of post-depositional changes in recently deposited till from south-east Iceland. *Geografiska Annaler*, **56(A)**, 121–34.

Boulton, G. S., Dent, D. L. & Morris, E. M. (1974). Subglacial shearing and crushing, and the role of water pressure in tills from south-east Iceland. *Geografiska Annaler*, **56(A)**, 135–45.

Boulton, G. S. & Eyles, N. (1979). Sedimentation by valley glaciers: a model and genetic classification. In *Moraines and Varves: Origin, Genesis, Classification*. Ed. Ch. Schlüchter, pp. 11–23. Rotterdam: Balkema.

Boulton, G. S. & Paul, M. A. (1976). The influence of genetic processes on some geotechnical properties of glacial tills. *Quarterly Journal of Engineering Geology*, **9**, 159–94.

Bourne, R. (1934). Some ecological conceptions. *Empire Forestry Journal*, **13**, 15–30.

Bradshaw, A. D. & Chadwick, M. J. (1980). *The Restoration of Land: the Ecology and Reclamation of Derelict and Degraded Land.* Oxford: Blackwell Scientific Publications.

Brandani, A. A. (1983). Glacial processes and disturbance in vegetation richness. In *Tills and Related Deposits*, eds. E. B. Evenson, Ch. Schlüchter & J. Rabassa, 403–10. Rotterdam: Balkema.

Braun-Blanquet, J. (1932). *Plant Sociology: The Study of Plant Communities.* New York: Hafner Publishing Co. (English translation of '*Pflanzensoziologie*')

Braun-Blanquet, J. & Jenny, H. (1926). Vegetations-entwicklung und Bodenbildung in der alpinen Stufe der Zentralalpen. *Ergebnisse der wissenschaftlichen Untersuchung des schweizer Nationalparks*, **63**(2), 1–208.

Bray, J. R. & Struik, G. J. (1963). Forest growth and glacial chronology in eastern British Columbia, and their relation to recent climatic trends. *Canadian Journal of Botany*, **41**, 1245–71.

Brázdil, R. (1988). Variation of air temperatures and atmospheric precipitation in the region of Svalbard and of Jan Mayen. In *Recent Climatic Change*, ed. S. Gregory, 53–68. London: Belhaven Press.

Breitburg, D. L. (1985). Development of a subtidal epibenthic community: factors affecting species composition and mechanisms of succession. *Oecologia*, **65**, 173–84.

Bridge-Cook, Wm. & Lawrence, D. B. (1959). Soil mould fungi isolated from recently glaciated soils in south eastern Alaska. *Journal of Ecology*, **47**, 529–49.

Brink, V. C. (1964). Plant establishment in the high snowfall alpine and subalpine regions of British Columbia. *Ecology*, **45**, 431–8.

Brodzikowski, K. & Van Loon, A. J. (1987). A systematic classification of glacial and periglacial environments, facies and deposits. *Earth Science Reviews*, **24**, 297–381.

Brugman, M. M. & Post, A. (1981). Effects of volcanism on the glaciers of Mount St Helens. *United States Geological Survey Circular*, 850-D, 1–11.

Brunner, K. (1987). Glacier mapping in the Alps (with 3 map sheets). *Mountain Research and Development*, **7**, 375–85.

Bryant, I. D. (1982). Loess deposits in Lower Adventdalen, Spitzbergen. *Polar Research*, **2**, 93–103. (Norsk Polarinstitutt).

Budd, W. F., Keage, P. L. & Blundy, N. A. (1979). Empirical studies of ice sliding. *Journal of Glaciology*, **23**, 157–70.

Büdel, J. (1982) *Climatic Geomorphology*. Princeton: Princeton University Press.

Bunting, B. T. (1965). *The Geography of Soil*. London: Hutchinson.

Burdon, J. J. (1987). *Diseases and Plant Biology*. Cambridge: Cambridge University Press.

Burga, C. A. (1987). *Gletscher-und Vegetationsgeschichte der Südrätischen Alpen seit der Späteiszeit*. Basel: Birkhäuser.

Burger, R. & Franz, H. (1969). Die Bodenbildung in der Pasterzenlandschaft. *Wissenschaftliche Alpenvereinsheft* (München), **21**, 253–64.

Burke, R. M. & Birkeland, P. W. (1979). Reevaluation of multiparameter relative dating techniques and their application to the glacial sequence along the eastern escarpment of the Sierra Nevada, California. *Quaternary Research*, **11**, 21–51.

Burke, R. M. & Birkeland, P. W. (1984). Holocene glaciation in the mountain ranges of the western United States. In *Late-Quaternary Environments of the United States, Volume 2, The Holocene*, ed. H. E. Wright Jnr, pp. 3–11. London: Longman.

Burrows, C. J. (1990). *Processes of Vegetation Change*. Boston: Unwin Hyman.

Burrows, C. J. & Gellatly, A. F. (1982). Holocene glacier activity in New Zealand. *Striae*, **18**, 41–7.

Burtscher, M. (1982). *Zur Vegetation und Flora Zweier Gletschervorfelder im Venedigergebiet*. Dissertation, University of Innsbruck. [not seen]

Bushnell, T. M. (1942). Some aspects of the soil catena concept. *Soil Science Society of America, Proceedings*, **7**, 466–76.

Butters, F. K. (1914). Some peculiar causes of plant distribution in the Selkirk Mountains, British Columbia. *Minnesota Botanical Studies*, **4**, 313–31.

Cain, S. A. (1939). The climax and its complexities. *American Midland Naturalist*, **21**, 146–81.

Cairns, J. Jr ed. (1980). *The Recovery Process in Damaged Ecosystems*. Ann Arbor, Michigan: Ann Arbor Science Publishers.

Campbell, I. B. & Claridge, G. G. C. (1987). *Antarctica: Soils, Weathering Processes and Environment*. Amsterdam: Elsevier.

Canada Soil Survey Committee (1978). *The Canadian System of Soil Classification*. Ottawa: Supply & Service Canada. (Canada Department of Agriculture, Publication No. 1646)

Carroll, T. (1974). Relative age dating techniques and a Late Quaternary chronology, Arikaree Cirque, Colorado. *Geology*, **2**, 321–5.

Caseldine, C. J. (1985). The extent of some glaciers in northern Iceland during the Little Ice Age and the nature of recent deglaciation. *Geographical Journal*, **151**, 215–27.

Caseldine, C. J. (1987). Neoglacial glacier variations in northern Iceland: examples from the Eyjafjördur area. *Arctic and Alpine Research*, **19**, 296–304.

Caseldine, C. J. (1989). Pollen assemblage – plant community relationships on the Storbreen glacier foreland, Jotunheimen Mountains, southern Norway. *New Phytologist*, **111**, 105–18.

Caseldine, C. J. & Cullingford, R. A. (1981). Recent mapping of Gljúfurárjökull and Gljúfurárdalur. *Jökull*, **31**, 11–22.

Caseldine, C. J. & Matthews, J. A. (1987). Podzol development, vegetation change and glacier variations at Haugabreen, southern Norway. *Boreas*, **16**, 215–30.

Chapin III, F. S. (1991). Physiological controls over plant establishment in primary succession. In *Primary Succession on Land*, eds. J. Miles & D. H. Walton, Oxford: Blackwell Scientific Publications (in press).

Chapin III, F. S. & Walker, L. R. (1990). The importance of Glacier Bay to tests of current theories of plant succession. In *Proceedings 2nd Glacier Bay Science Symposium*, eds. A. M. Milner & J. D. Wood, 136–9. Anchorage, Alaska: United States Department of the Interior, National Parks Service.

Charlesworth, J. K. (1957). *The Quaternary Era*. London: Edward Arnold.

Chernov, Yu. I. (1985). *The Living Tundra*. Cambridge: Cambridge University Press.

Cherrett, J. M. (1989). Key concepts: the results of a survey of our members' opinions. In *Ecological Concepts: the Contribution of Ecology to an Understanding of the Natural World*, ed. J. M. Cherrett, pp. 1–16. Oxford: Blackwell Scientific Publications.

Chesworth, W. (1973). The parent rock effect in the genesis of soil. *Geoderma*, **10**, 215–25.

Chesworth, W. (1976a). Conceptual models in pedogenesis: a rejoinder. *Geoderma*, **16**, 257–60.

Chesworth, W. (1976b). Conceptual models in pedogenesis: a further rejoinder. *Geoderma*, **16**, 265–6.

Childs, C. W., Parfitt, R. L. & Lee, R. (1983). Movement of aluminium as an inorganic complex in some podzolised soils, New Zealand. *Geoderma*, **29**, 139–55.

Chinn, T. J. H. (1981). Use of rock weathering-rind thickness for Holocene absolute-age dating in New Zealand. *Arctic and Alpine Research*, **13**, 33–45.

Chorley, R. J. & Kennedy, B. A. (1971). *Physical Geography: A Systems Approach*. London: Prentice Hall.

Christensen, N. L. & Peet, R. K. (1984). Convergence during secondary forest succession. *Journal of Ecology*, **72**, 25–36.

Church, M. A. (1972). Baffin Island sandurs: a study of Arctic fluvial processes. *Geological Survey of Canada Bulletin*, **216**, 1–208.

Church, M. A. & Gilbert, R. (1975). Proglacial fluvial and lacustrine environments. In *Glaciofluvial and Glaciolacustrine Sedimentation*, eds. A. V. Joplin & B. C. McDonald, pp. 22–100. Tulsa, Oklahoma: Society of Economic Palaeontologists and Mineralogists. (SEPM Special Publication No. 23)

Church, M. A. & Ryder, J. (1972). Paraglacial sedimentation: a consideration of fluvial processes conditioned by glaciation. *Geological Society of America Bulletin*, **83**, 3059–72.

Churchill, E. D. & Hanson, H. C. (1958). The concept of climax in arctic and alpine vegetation. *Botanical Review*, **24**, 127–91.

Claridge, G. G. C. & Campbell, I. B. (1977). The salts in Antarctic soils, their distribution and relationship to soil processes. *Soil Science*, **123**, 377–84.

Clark, M. J. (1988). *Advances in Periglacial Geomorphology*. Chichester: Wiley.

Clayton, L. & Moran, S. R. (1974). A glacial process-form model. In *Glacial Geomorphology*, ed. D. R. Coates, pp. 89–119. Binghampton, New York: State University of New York.

Clements, F. E. (1904). *Development and Structure of Vegetation*. Lincoln, Nebraska: University Publishing Company. (Report of the Botanical Survey of Nebraska, No. 7)

Clements, F. E. (1928). *Plant Succession and Indicators: A Definitive Edition of Plant Succession and Plant Indicators*. New York: H. W. Wilson. [New York: Hafner, 1963]

Clements, F. E. (1936). Nature and structure of the climax. *Journal of Ecology*, **24**, 252–84.

Coaz, J. (1887). Erste Ansiedelung phanerog. Pflanzen auf von Gletschern verlassenem Boden. *Mitteilungen der Naturforschenden Gesellschaft in Bern: aus dem Jahre 1886*, Nos 1143–68, 3–12.

Coe, M. J. (1964). Colonisation in the nival zone of Mount Kenya. *Proceedings 1st Symposium of the East African Academy, Kampala*, 137–40.

Coe, M. J. (1967). *The Ecology of the Alpine Zone on Mount Kenya*. The Hague: Junk.

Cole, D. W. & Rapp, M. (1981). Elemental cycling in forest ecosystems. In *Dynamic Principles of Forest Ecosystems*, ed. D. E. Reichle, 341–409. Cambridge: Cambridge University Press.

Colinvaux, P. A. (1973). *Introduction to Ecology*. New York: Wiley.

Colinvaux, P. A. (1986). *Ecology*. New York: Wiley.

Collins, N. J. (1976). The development of moss-peat banks in relation to changing climate and ice cover on Signy Island in the Maritime Antarctic. *British Antarctic Survey Bulletin*, **43**, 85–102.

Colman, S. M. (1981). Rock-weathering rates as a function of time. *Quaternary Research*, **15**, 250–64.

Colman, S. M. & Dethier, D. P. (1986). *Rates of Chemical Weathering of Rocks and Minerals*. Orlando, Florida: Academic Press.

Colman, S. M., Pierce, K. L. & Birkeland, P. W. (1987). Suggested terminology for Quaternary dating methods. *Quaternary Research*, **28**, 314–9.

Connell, J. H. (1983). On the prevalence and relative importance of interspecific competition: evidence from field experiments. *American Naturalist*, **122**, 661–96.

Connell, J. H., Noble, I. R. & Slatyer, R. O. (1987). On the mechanisms producing successional change. *Oikos*, **50**, 136–7.

Connell, J. H. & Slatyer, R. O. (1977). Mechanisms of succession in natural communities and their role in community stability and organization. *American Naturalist*, **111**, 1119–44.

Cooper, W. S. (1916). Plant succession in the Mount Robson region, British Columbia. *The Plant World*, **19**, 211–38.

Cooper, W. S. (1923a). The recent ecological history of Glacier Bay, Alaska: I. The interglacial forests of Glacier Bay. *Ecology*, **4**, 93–128.

Cooper, W. S. (1923b). The recent ecological history of Glacier Bay, Alaska: II. The present vegetation cycle. *Ecology*, **4**, 223–46.

Cooper, W. S. (1923c). The recent ecological history of Glacier Bay, Alaska: III. Permanent quadrats at Glacier Bay: An initial report upon a long-period study. *Ecology*, **4**, 355–65.

Cooper, W. S. (1926). The fundamentals of vegetation change. *Ecology*, **7**, 391–413.

Cooper, W. S. (1931). A third expedition to Glacier Bay, Alaska. *Ecology*, **12**, 61–95.

Cooper, W. S. (1939). A fourth expedition to Glacier Bay, Alaska. *Ecology*, **20**, 130–55.

Corner, R. W. M. & Smith, R. I. L. (1973). Botanical evidence of ice recession in the Argentine Islands. *British Antarctic Survey Bulletin*, **35**, 83–6.

Corte, A. (1976). Rock glaciers. *Biuletyn Peryglacjalny*, **26**, 175–97.

Costin, A. B., Jennings, J. N., Bautovich, B. C. & Wimbush, D. J. (1973). Forces developed by snowpatch action, Mt. Twynam, Snowy Mountains, Australia. *Arctic and Alpine Research*, **5**, 121–6.

Crandell, D. R. & Miller, R. D. (1974). Quaternary stratigraphy and extent of glaciation in the Mount Rainier region, Washington. *United States Geological Survey, Professional Paper*, **847**, 1–59.

Crawford, R. M. M. & Wishart, D. (1967). A rapid multivariate method for the detection and classification of groups of ecologically related species. *Journal of Ecology*, **55**, 505–24.

Crocker, R. L. (1952). Soil genesis and the pedogenic factors. *Quarterly Review of Biology*, **27**, 139–68.

Crocker, R. L. (1959). The plant factor in soil formation. *Australian Journal of Science*, **21**, 180–93.

Crocker, R. L. & Dickson, B. A. (1957). Soil development on the recessional moraines of the Herbert and Mendenhall Glaciers, south-eastern Alaska. *Journal of Ecology*, **45**, 169–85.

Crocker, R. L. & Major, J. (1955). Soil development in relation to vegetation and surface age at Glacier Bay, Alaska. *Journal of Ecology*, **43**, 427–48.

Crouch, H. (1991). Plant distribution patterns and primary succession on a glacier foreland: a comparative study of cryptogams and higher plants. In *Primary Succession on Land*, eds. J. Miles & D. H. Walton, Oxford: Blackwell Scientific Publications (in press).

Cruickshank, J. G. (1972). *Soil Geography*. Newton Abbot: David & Charles.

Dahl, E. (1956). Rondane: mountain vegetation in south Norway and its relation to environment. *Skrifter utgitt av det Norske Videnskaps Academie i Oslo. Matematisk-naturvidenskapelig klasse*, **3**, 1–374.

Dahl, E. (1986). Zonation in arctic and alpine tundra and fjellfield ecobiomes. In *Ecosystem Theory and Applications*, ed. N. Polunin, pp. 35–62. Chichester: Wiley.

Dale, M. B. & Blundon, D. J. (1990). Quadrat variance analysis and pattern development during primary succession. *Journal of Vegetation Science*, **1**, 153–64.

Dale, M. B. & Blundon, D. J. (1991). Quadrat covariance analysis and the scales of interspecific association during primary succession. *Journal of Vegetation Science*, **2**, 103–12.

Dale, M. B. & MacIsaac, D. A. (1989). New methods for the analysis of spatial pattern in vegetation. *Journal of Ecology*, **77**, 78–91.

Darmody, R. G., Thorn, C. E. & Rissing, J. M. (1987). Chemical weathering of fine debris from a series of Holocene moraines: Storbreen, Jotunheimen, southern Norway. *Geografiska Annaler*, **69**(A), 405–13.

Davis, B. N. K., Lakhani, K. H., Brown, M. C. & Park, D. G. (1985). Early seral communities in a limestone quarry – an experimental study of treatment effects on cover and richness of vegetation. *Journal of Applied Ecology*, **22**, 473–90.

Davy de Virville, A. (1929). La flore de deux glaciers inférieurs des Pyrénnées. *Revue générale de botanique*, **14**, 1–23.

Dearman, W. R. & Baynes, F. J. (1979). Etch-pit weathering of feldspars. *Proceedings of the Ussher Society*, **4**, 390–401.

Decker, H. F. (1966). Plants. In *Soil Development and Ecological Succession in a Deglaciated Area of Muir Inlet, Southeast Alaska*, ed. A. Mirsky, pp. 73–95. Columbus, Ohio: Ohio State University Research Foundation. (Institute of Polar Studies, Report No. 20)

Delibrias, G., Le Roy Ladurie, M. & Le Roy Ladurie, E. (1975). La forêt fossile de Grindelwald: nouvelles datations. *Annales économies sociétés civilisations*, **1**, 137–47.

DeLong, D. M. (1966). Insects. In *Soil Development and Ecological Succession in a Deglaciated Area of Muir Inlet, Southeast Alaska*, ed. A. Mirsky, pp. 97–120. Columbus, Ohio: Ohio State University Research Foundation. (Institute of Polar Studies, Report No. 20)

del Moral, R. & Clampitt, C. A. C. (1985). Growth of native plant species on recent volcanic substrates from Mount St Helens. *American Midland Naturalist*, **114**, 374–83.

Denton, G. H. & Karlén, W. (1973). Holocene climatic variations – their pattern and possible cause. *Quaternary Research*, **3**, 155–205.

Denton, G. H. & Karlén, W. (1977). Holocene glacial and tree-line variations in the White River valley and Skolai Pass, Alaska and Yukon Territory. *Quaternary Research*, **7**, 63–111.

Derbyshire, E. & Blackmore, R. (1975). Meteorological observations at Blaisen,Hardangerjökulen ice cap, Norway, 26th August–9th September 1972. *Journal of Meteorology*, **1**, 59–66.

Detwyler, T. R. (1974). Vegetation – snow cover relations in an alpine pass, Alaska. In *Icefield Ranges Research Project, Scientific Results, Volume 4*, eds. V. C. Bushnell & M. G. Marcus, pp. 355–60. New York & Montreal: American Geographical Society & Arctic Institute of North America.

Domin, K. (1923). Is the evolution of the earth's vegetation tending towards a small number of climatic formations? *Acta Botanica Bohemica*, **2**, 54–60.

Dowdeswell, J. A. & Morris, S. E. (1983). Multivariate statistical approaches to the analysis of late Quaternary relative age data. *Progress in Physical Geography*, **7**, 157–76.

Dreimanis, A. (1988). Tills: their genetic terminology and classification. In *Genetic Classification of Glacigenic Deposits*, eds. R. P. Goldthwait & C. L. Matsch, pp. 17–83. Rotterdam: Balkema.

Dreimanis, A. & Vagners, U. J. (1971). Bimodal distribution of rock and mineral fragments in basal tills. In *Tills: A Symposium*, ed. R. P. Goldthwait, pp. 237–50. Columbus, Ohio: Ohio State University Press.

Drewry, D. (1986). *Glacial Geologic Processes*. London: Edward Arnold.

Drewry, D. & Liestøl, O. (1985). Glaciological investigations of surging ice caps in Nordaustlandet, Svalbard, 1983. *Polar Record*, **22**, 359–78.

Driscoll, F. G. Jr (1980). Wastage of the Klutlan ice-cored moraines, Yukon Territory, Canada. *Quaternary Research*, **14**, 31–49.

Drury, W. H. & Nisbet, I. C. T. (1973). Succession. *Journal of the Arnold Arboretum*, **54**, 331–68.

Duchaufour, P. (1982). *Pedology*. London: George Allen & Unwin.

Dugmore, A. J. (1989). Tephrochronological studies of Holocene glacier fluctuations in south Iceland. In *Glacier Fluctuations and Climatic Change*, ed. J. Oerlemans, pp. 37–55. Dordrecht: Kluwer.

Dyksterhuis, E. J. (1949). Condition and management of range land based on quantitative ecology. *Journal of Range Management*, **2**, 104–15.

Dyksterhuis, E. J. (1958). Ecological principles in range evaluation. *Botanical Review*, **24**, 253–72.

Dzierzek, J. & Nitychoruk, J. (1987). Types of rock glaciers in northwestern Wedel Jarlsberg Land, Spitzbergen. *Polish Polar Research*, **8**, 231–41.

Edwards, P. J. & Gillman, M. P. (1987). Herbivores and plant succession. In *Colonization, Succession and Stability*, eds. A. J. Gray, M. J. Crawley & P. J. Edwards, pp. 295–314. Oxford: Blackwell Scientific Publications.

Egler, F. E. (1954). Vegetation science concepts I. Initial floristic composition, a factor in old-field vegetation development. *Vegetatio*, **4**, 412–7.

Eglitis, A. (1984). The spruce beetle in Glacier Bay National Park and Preserve. In *Proceedings 1st Glacier Bay Science Symposium*, eds. J. D. Wood, M. Gladziszewski, I. A. Worley & C. Veqvist, pp. 30–1. Atlanta, Georgia: United States Department of the Interior, National Parks Service.

Einarsson, E. (1970). Plant ecology and succession in some nunataks in the Vatnajökull glacier in south-east Iceland. *Proceedings Helsinki Symposium: Ecology of Subarctic Regions*, pp. 247–56. Paris: UNESCO.

Einarsson, E. (1971). Appendix C: Botanical investigations. In *Red Rock Ice Cliff, Nunatarssuaq, Greenland*, by R. P. Goldthwait, pp. 23–7. (United States Army Cold Regions Research and Engineering Laboratory, Technical Report, No. 224)

Ellenberg, H. (1988). *Vegetation Ecology of Central Europe*. Cambridge: Cambridge University Press.

Elliot. J. K. (1989). An investigation of the change in surface roughness over time on the foreland of Austre Okstindbreen. *Okstindan Preliminary Report for 1984*, pp. 4–20. Reading: University of Reading, Department of Geography. (Geographical Papers, University of Reading, No. 104)

Ellis, J. M. & Calkin, P. E. (1984). Chronology of Holocene glaciation, central Brooks Range, Alaska. *Geological Society of America Bulletin*, **95**, 897–912.

Ellis, S. (1975). A study of plant colonization and succession in relation to the deglacierization of Oksfjellbreen. In *Okstindan Research Project preliminary report for 1973*, ed. R. B. Parry & P. Worsley, pp. 37–49. Reading: University of Reading, Department of Geography.

Ellis, S. (1979). The identification of some Norwegian mountain soil types. *Norsk Geografisk Tidsskrift*, **33**, 205–12.

Ellis, S. (1980a). Soil-environmental relationships in the Okstindan Mountains, north Norway. *Norsk Geografisk Tidsskrift*, **34**, 167–76.

Ellis, S. (1980b). Physical and chemical characteristics of a podzolic soil formed in Neoglacial till, Okstindan, northern Norway. *Arctic and Alpine Research*, **12**, 65–72.

Ellis, S. (1980c). An investigation of weathering in some arctic-alpine soils on the northeast flank of Okskolten, north Norway. *Journal of Soil Science*, **31**, 371–85.

Ellis, S. (1983). Micromorphological aspects of arctic-alpine pedogenesis in the Okstindan Mountains, Norway. *Catena*, **10**, 133–48.

Elven, R. (1974). *Artsinnvandring og Vegetasjonsutvikling på Resente Morener i Finseområdet*. Cand. Real Thesis, University of Oslo.

Elven, R. (1975). Plant communities on recently deglaciated moraines at Finse, southern Norway. In *IBP in Norway, Methods and Results Section PT–UM Grazing Project, Hardangervidda Botanical Investigations. Appendix I*, eds. Norwegian National IBP Committee, 381–467. Oslo: Norwegian National IBP Committee. (Norwegian National IBP Committee, Annual Report for 1974)

Elven, R. (1978a). Subglacial plant remains from the Omnsbreen glacier area, southern Norway. *Boreas*, **7**, 83–9.

Elven, R. (1978b). Association analysis of moraine vegetation at the glacier Hardangerjøkulen, Finse, South Norway. *Norwegian Journal of Botany*, **25**, 171–91.

Elven, R. (1978c). Vegetasjonen ved Flatisen og Osterdalsisen, Rana, Nordland, med vegetationskart over Vesterdalen i 1:15,000. *Det Kgl. Norske Videnskabers Selskab, Museet Rapport, Botanisk Serie*, 1978–1, pp. 1–83.

Elven, R. (1980). The Omnsbreen glacier nunataks – a case study of plant immigration. *Norwegian Journal of Botany*, **27**, 1–16.

Elven, R. & Aarhus, A. (1984). A study of *Draba cacuminum* (Brassicaceae). *Nordic Journal of Botany*, **4**, 425–41.

Elven, R. & Ryvarden, L. (1975). Dispersal and primary establishment of vegetation. In *Fennoscandian Tundra Ecosystems, Part I*, ed. F. E. Wielgolaski, pp. 81–5. Berlin: Springer Verlag.

Embleton, C. (1979). Nival processes. In *Process in Geomorphology*, eds. C. Embleton & J. Thornes, pp. 307–24. London: Edward Arnold.

Embleton, C. & King, C. A. M. (1975a). *Glacial Geomorphology*. London: Edward Arnold.

Embleton, C. & King, C. A. M. (1975b). *Periglacial Geomorphology*. London: Edward Arnold.

Erikstad, L. & Sollid, J. L. (1986). Neoglaciation in South Norway using lichenometric methods. *Norsk Geografisk Tidsskrift*, **40**, 85–105.

Everett, K. R. (1971). Soils of the Meserve Glacier area, Wright Valley, Southern Victoria Land, Antarctica. *Soil Science*, **112**, 425–38.

Evers, W. (1951). Gletscherwinde am Nigardsbre (Südnorwegen). In *Landschaft und Land*, ed. K. Kayser, pp. 123–35. Remegen: Verlag des Amtes für Landeskunde. (Festschrift Erich Obst)

Eyles, N. (1979). Facies of supraglacial sedimentation on Icelandic and Alpine temperate glaciers. *Canadian Journal of Earth Science*, **16**, 1341–61.

Eyles, N. (1983a). Glacial geology: a landsystem approach. In *Glacial Geology*, ed. N. Eyles, pp. 1–18. Oxford: Pergamon Press.

Eyles, N. (1983b). The glaciated valley landsystem. In *Glacial Geology*, ed. N. Eyles, pp. 91–110. Oxford: Pergamon Press.

Eyles, N. & Menzies, J. (1983). The subglacial landsystem. In *Glacial Geology*, ed. N. Eyles, pp. 19–70. Oxford: Pergamon Press.

Facelli, J. M. & D'Angela, E. (1990). Directionality, convergence, and rate of change during early succession in the Inland Pampa, Argentina. *Journal of Vegetation Science*, **1**, 255–60.

Fægri, K. (1933). Über die Längenvariationen einiger Gletscher des Jostedalsbre und die dadurch bedingten Pflanzensukzessionen. *Bergens Museums Årbok 1933, Naturvidenskapelig rekke*, **7**, 1–255.

Fægri, K. (1950). On the variations of western Norwegian glaciers during the last 200 years. *International Association for Scientific Hydrology, General Assembly Oslo*, **2**, 293–303.

Fægri, K. (1986). Plant succession at Nigardsbreen. In *The Cultural Landscape – Past, Present and Future, Excursion Guide*, ed. H. J. B. Birks, pp. 99–103. Bergen: University of Bergen, Botanical Institute. (Botanisk Institutt Report No. 42)

Fahnestock, R. K. (1963). Morphology and hydrology of a glacial stream: White River, Mount Rainier, Washington. *United States Geological Survey Professional Paper*, **422A**, 1–70.

Falconer, G. (1966). Preservation of vegetation and patterned ground under a thin ice body in northern Baffin Island, N.W.T. *Geographical Bulletin* (Ottawa), **8**, 194–200.

Fastie, C. (1990). Inference and verification in chronosequence studies at Glacier Bay. In *Proceedings 2nd Glacier Bay Science Symposium*, eds. A. M. Milner & J. D. Wood, 147–9. Anchorage, Alaska: United States Department of the Interior, National Park Service.

Fenn, C. R. & Gurnell, A. M. (1987). Proglacial channel processes. In *Glacio-fluvial Sediment Transfer: An Alpine Perspective*, eds. A. M. Gurnell & M. J. Clark, 423–72. Chichester: Wiley.

Fenton, J. H. C. (1982). Vegetation re-exposed after burial by ice and its relationship to changing climate in the South Orkney Islands. *British Antarctic Survey Bulletin*, **51**, 247–55.

Finegan, B. (1984). Forest succession. *Nature*, **312**, 109–14.

Fitter, A. (1984). Succession theory is too reductionist. *Nature*, **312**, 705. [Comment on Finegan, 1984]

Fitter, A. H. & Hay, R. K. M. (1987). *Environmental Physiology of Plants*. London: Academic Press.

Fitter, A. H. & Parsons, W. F. J. (1987). Changes in phosphorus and nitrogen availability on recessional moraines of the Athabasca Glacier, Alberta. *Canadian Journal of Botany*, **65**, 210–3.

Fitze, P. F. (1980). Zur Bodenentwicklung auf Moränen in den Alpen. *Geographica Helvetica*, **35**, 97–106.

Fitze, P. F. (1982). Zur relativdatierung von moränen aus der sicht der bodenentwicklung in den kristallinen Zentralalpen. *Catena*, **9**, 265–306.

Flint, R. F. (1971). *Glacial and Quaternary Geology*. New York: Wiley.

Flohn, H. (1969). Local wind systems. In *World Survey of Climatology, Volume 2*, ed. H. Flohn, pp. 139–71. Amsterdam: Elsevier.

Florin, M. B. & Wright, H. E. Jnr (1969). Diatom evidence for the persistence of stagnant glacial ice in Minnesota. *Geological Society of America Bulletin*, **50**, 695–704.

Forman, R. T. T. & Godron, M. (1986). *Landscape Ecology*. New York: Wiley.

Fraser, B. E. C. (1970). *Vegetation Development on Recent Alpine Glacier Forelands in Garabaldi Park, British Columbia*. Ph.D. Thesis, University of British Columbia.

Frederiksen, S. (1971). The flora of some nunataks in Frederikshåb District, West Greenland. *Botanisk Tidsskrift*, **66**, 60–8.

French, D. D. & Smith, V. R. (1985). A comparison between Northern and Southern Hemisphere tundras and related ecosystems. *Polar Biology*, **5**, 5–21.

French, H. M. (1976). *The Periglacial Environment*. London: Longman.

French, H. M. (1988). Active layer processes. In *Advances in Periglacial Geomorphology*, ed. M. J. Clark, pp. 151–77. Chichester: Wiley.

Frey, E. (1922). Die Vegetationsverhältnisse der Grimselgegend im Gebiet der zukünftigen Stauseen. *Mitteilungen der Naturforschenden Gesellschaft in Bern*, **6**, 1–196.

Friedel, H. (1934). Boden- und Vegetationsentwicklung am Pasterzenufer. *Carinthia*, **2**, 29–41.

Friedel, H. (1936). Wirkungen der Gletscherwinde auf die Ufervegetation der Pasterze. *Bioklimatische Beiblåtter der Meteorologischen Zeitschrift* (Braunschweig), **3**, 21–5.

Friedel, H. (1937). Boden- und Vegetations-entwicklung im Vorfelde des Rhonegletschers. Vorläufiger Bericht. *Bericht über das Geobotanische Forschungsinstitutt Rübel in Zurich*, for 1937, 65–76.

Friedel, H. (1938). Die Pflanzenbesiedlung im Vorfeld des Hintereisferners. *Zeitschrift für Gletscherkunde*, **26**, 215–39.

Friedel, H. (1952). Gesetze der Niederschlagsverteilung im Hochgebirge. *Wetter und Leben: Zeitschrift für Praktische Bioclimatologie* (Vienna), **4**, 73–86.

Friedel, H. (1956). Die Alpine Vegetation des Oberstulen Mölltales (Hohe Tauern). *Wissenschaftliche Alpenvereinshefte*, **16**, 1–155. (Universitätsverlag Wagner, Innsbruck)

Furley, P. A. (1968). Soil formation and slope development: 2. The relationship between soil formation and gradient angle in the Oxford area. *Zeitschrift für Geomorphologie, N.F.*, **12**, 25–42.

Furley, P. A. (1971). Relationships between slope form and soil properties developed over chalk parent materials. In *Slopes, Form and Process*, ed. D. Brunsden, pp. 141–64. (Institute of British Geographers, Special Publication No. 3)

Furrer, G. & Holzhauser, H. (1984). Gletscher- und klimageschichtliche Auswertung fossiler Hölzer. *Zeitschrift für Geomorphologie, N.F., Supplement Band*, **50**, 117–36.

Galon, R. (1973). A synthetic description of deposits and landforms observed on the proglacial area of Skeidarárjökull. Conclusions with regard to the age of the deposits and the way in which deglaciation is proceeding. *Geographica Polonica*, **26**, 139–50.

Gams, H. (1918). Principienfragen der Vegetationsforschung. *Vierteljahrsschrift der Naturforschenden Gesellschaft in Zürich*, **67**, 132–56.

Geiger, R. (1969). Topoclimates. In *World Survey of Climatology, Volume 2*, ed. H. Flohn, pp. 105–38. Amsterdam: Elsevier.

Geiger, R. (1971). *The Climate Near the Ground*. Cambridge, Massachusetts: Harvard University Press.

Gellatly, A. F. (1984). The use of rock weathering–rind thickness to redate moraines in Mount Cook National Park, New Zealand. *Arctic and Alpine Research*, **16**, 225–32.

Gellatly, A. F. (1985). Phosphate retention: relative dating of Holocene soil development. *Catena*, **12**, 227–40.

Gellatly, A. F. (1987). Establishment of soil covers on tills of variable texture and implications for interpreting palaeosols – a discussion. In *International Geomorphology 1986, Part II*, ed. V. Gardiner, pp. 775–84. London: Wiley.

Gellatly, A. F., Whalley, W. B. & Gordon, J. E. (1986). Topographic control over recent glacier changes in southern Lyngen Peninsula, North Norway. *Norsk Geografisk Tidsskrift*, **40**, 211–18.

Gennadiyev, A. N. (1979). Study of soil formation by the chronosequence method (as exemplified by the soils of the Elbruz region). *Soviet Soil Science*, **10**, 707–16. (Translated from *Pochvovedeniye*, **12**, 33–43, 1978)

Gerrard, A. J. (1981). *Soils and Landforms*. London: George Allen & Unwin.

Geyh, M. A., Röthlisberger, F. & Gellatly, A. (1985). Reliability tests and interpretation of ^{14}C dates from palaeosols in glacier environments. *Zeitschrift für Gletscherkunde und Glazialgeologie*, **21**, 275–81.

Giardino, J. R., Shroder, J. F. Jnr, & Vitek, J. D. (1987). *Rock Glaciers*. Boston: Allen & Unwin.

Gibbons, A. B., Megeath, J. D. & Pierce, K. L. (1984). Probability of moraine survival in a succession of glacial advances. *Geology*, **12**, 327–30.

Gilbert, F. S. (1980). The equilibrium theory of island biogeography. *Journal of Biogeography*, **7**, 209–35.

Gillberg, G. (1977). Redeposition: a process of till formation. *Geologiska Föreningens i*

Stockholm Förhandlingar, **99**, 246–53.

Given, D. R. & Soper, J. H. (1975). Pioneer vegetation on moraines near Clachnacudainn Snowfield, British Columbia. *Syesis*, **8**, 349–54.

Gjaerevoll, O. (1956). The plant communities of the Scandinavian alpine snow-beds. *Det Kongelige Norske Videnskabers Selskabs Skrifter for 1956*, **1**, 1–406.

Gjaerevoll, O. & Ryvarden, L. (1977). Botanical investigations of J. A. D. Jensens Nunatakker in Greenland. *Det Kongelige Norske Videnskabers Selskab Skrifter*, **4**, 1–40.

Gleason, H. A. (1917). The structure and development of the plant association. *Bulletin of the Torrey Botanical Club*, **44**, 463–81.

Gleason, H. A. (1926). The individualistic concept of the plant association. *Bulletin of the Torrey Botanical Club*, **53**, 7–26.

Gleason, H. A. (1927). Further views on the succession concept. *Ecology*, **8**, 299–326.

Gleason, H. A. (1937). The individualistic concept of the plant association. *American Midland Naturalist*, **21**, 92–110.

Glenn-Lewin, D. C. (1980). The individualistic nature of plant community development. *Vegetatio*, **43**, 141–6.

Goldthwait, R. P. (1960). Study of ice cliff in Nunatarssuaq, Greenland. *United States Army Cold Regions Research and Engineering Laboratory Technical Report*, **39**, 1–108.

Goldthwait, R. P. (1963). Dating the Little Ice Age in Glacier Bay, Alaska. *Proceedings of the 12th International Geological Congress (Norden, 1960)*, part 27, 37–46.

Goldthwait, R. P. (1966). Glacial history. In *Soil Development and Ecological Succession in a Deglaciated Area of Muir Inlet, Southeast Alaska*, ed. A. Mirsky, pp. 1–17. Columbus, Ohio: Ohio State University Research Foundation. (Institute of Polar Studies Report No. 20)

Goldthwait, R. P. (1974). Rates of formation of glacial features in Glacier Bay. Alaska. In *Glacial Geomorphology*, ed. D. R. Coates, pp. 163–85. London: George Allen & Unwin.

Goldthwait, R. P. (1976). Frost sorted patterned ground: a review. *Quaternary Research*, **6**, 27–35.

Goldthwait, R. P. (1988). Classification of glacial morphologic features. In *Genetic Classification of Glacigenic Deposits*, eds. R. P. Goldthwait & C. L. Matsch, pp. 267–77. Rotterdam: Balkema.

Goldthwait, R. P. & Matsch, C. L. eds. (1988). *Genetic Classification of Glacigenic Deposits*. Rotterdam: Balkema.

Golubev, G. N. & Kotyakov, V. M. (1978). Glacial landscapes and their spatial variability in the temperate and subpolar latitudes. *Arctic and Alpine Research*, **10**, 277–82.

Gomez, B. & Small, R. J. (1985). Medial moraines of the Haut Glacier d'Arolla, Valais, Switzerland: debris supply and implications for moraine formation. *Journal of Glaciology*, **31**, 303–7.

Good, E. E. (1966). Mammals. In *Soil Development and Ecological Succession in a Deglaciated Area of Muir Inlet, Southeast Alaska*, ed. A. Mirsky, pp. 145–55. Columbus, Ohio: Ohio State University Research Foundation. (Institute of Polar Studies Report No. 20)

Goodwin, R. G. (1988). Holocene glaciolacustrine sedimentation in Muir Inlet and ice advance at Glacier Bay, Alaska, U.S.A. *Arctic and Alpine Research*, **20**, 55–69.

Gorbunov, A. P. (1978). Permafrost investigations in high mountain regions. *Arctic and Alpine Research*, **10**, 283–94.

Greig-Smith, P. (1961a). Data on pattern within plant communities. I. The analysis of pattern. *Journal of Ecology*, **49**, 695–702.

Greig-Smith, P. (1964). *Quantitative Plant Ecology*. London: Butterworth.

Griffey, N. J. (1976). Stratigraphical evidence for an early Neoglacial glacier maximum of Steikvassbreen, Okstindan, north Norway. *Norsk Geologisk Tidsskrift*, **56**, 187–94.

Griffey, N. J. (1978). Lichen growth on supraglacial debris and its implications for lichenometric studies. *Journal of Glaciology*, **20**, 163–72.

Griffey, N. J. & Matthews, J. A. (1978). Major Neoglacial glacier expansion episodes in southern Norway: evidences from moraine ridge stratigraphy with ^{14}C dates on buried palaeosols and moss layers. *Geografiska Annaler*, **60(A)**. 73–90.

Griffey, N. J. & Worsley, P. (1978). The pattern of Neoglacial glacier variations in the Okstindan region of northern Norway during the last three millennia. *Boreas*, **7**, 1–17.

Griggs, R. F. (1933). The colonization of the Katmai ash, a new and inorganic "soil". *American Journal of Botany*, **20**, 92–113.

Griggs, R. F. (1934). The problem of arctic vegetation. *Journal of the Washington Academy of Sciences*, **25**, 153–75.

Griggs, R. F. & Ready, D. (1934). Growth of liverworts from Katmai in nitrogen-free media. *American Journal of Botany*, **21**, 265–77.

Grime, J. P. (1977). Evidence for the existence of three primary strategies in plants and its relevance to ecological and evolutionary theory. *The American Naturalist*, **111**, 1169–94.

Grime, J. P. (1979). *Plant Strategies and Vegetation Processes*. New York: Wiley.

Grove, J. M. (1979). The glacial history of the Holocene. *Progress in Physical Geography*, **3**, 1–54.

Grove, J. M. (1985). The timing of the Little Ice Age in Scandinavia. In *The Climatic Scene*, ed. M. J. Tooley & G. M. Sheail, pp. 132–53. London: George Allen & Unwin.

Grove, J. M. (1988). *The Little Ice Age*. London: Methuen.

Grubb. P. J. (1985). Plant populations and vegetation in relation to habitat, disturbance and competition. In *The Population Structure of Vegetation*, ed. J. White, pp. 595–622. Dordrecht: Junk.

Grubb, P. J. (1986). The ecology of establishment. In *Ecology and Design in Landscape*, eds. A. D. Bradshaw, D. A. Goode & E. Thorp, pp. 83–97. Oxford: Blackwell Scientific Publications.

Grubb, P. J. (1987). Some generalizing ideas about colonization and succession in green plants and fungi. In *Colonization, Succession and Stability*, eds. A. J. Gray, M. J. Crawley & P. J. Edwards, pp. 81–102. Oxford: Blackwell Scientific Publications.

Haeberli, W., King, L. & Flotron, A. (1979). Surface movement and lichen-cover studies at the active rock glacier near Grubengletscher, Wallis, Swiss Alps. *Arctic and Alpine Research*, **11**, 421–41.

Haeberli, W., Müller, P., Alean, P. & Bösch, H. (1989). Glacier changes following the Little Ice Age – a survey of the international data base and its perspectives. In *Glacier Fluctuations and Climatic Change*, ed. J. Oerlemans, pp. 77–101. Dordrecht: Kluwer.

Hagen, J. O., Wold, B., Liestøl, O., Østrem, G. & Sollid, J. L. (1983). Subglacial processes at Bondhusbreen, Norway: preliminary results. *Annals of Glaciology*, **4**, 91–8.

Haines-Young, R. H. (1983). Size variation of *Rhizocarpon* on moraine slopes in southern Norway. *Arctic and Alpine Research*, **15**, 295–305.

Hairston, N. G. Snr (1989). *Ecological Experiments: Purpose, Design, and Execution*. Cambridge: Cambridge University Press.

Haldorsen, S. (1981). Grain-size distribution of subglacial till and its relation to glacial crushing and abrasion. *Boreas*, **10**, 91–105.

Hall, G. F. (1983). Pedology and geomorphology. In *Pedogenesis and Soil Taxonomy. I. Concepts and Interactions*, eds. L. P. Wilding, N. E. Smeck & G. F. Hall, pp. 117–40. Amsterdam: Elsevier.

Hall, K. J. (1985). Some observations on ground temperatures and transport processes at a nivation site in northern Norway. *Norsk Geografisk Tidsskrift*, **39**, 27–37.

Hallet, B. & Anderson, R. S. (1980). Detailed glacial geomorphology of a proglacial bedrock area at Castleguard Glacier, Alberta, Canada. *Zeitschrift für Gletscherkunde und Glazialgeologie*, **16**, 171–84.

Haraldsson, H. (1981). The Markarfljót sandur area, southern Iceland: sedimentological, petrographical and stratigraphical studies. *Striae*, **15**, 3–65.

Harmsen, R., Spence, J. R. & Mahaney, W. C. (1990). Glacial-interglacial cycles and the development of the Afroalpine ecosystem on East African mountains. II. Origins and development of the biotic component. *Journal of African Earth Sciences* (in press).

Harris, C. (1981). *Periglacial Mass-wasting: A Review of Research*. Norwich: Geobooks. (BGRG Research Monograph No. 4)

Harris, C. (1982). The distribution and altitudinal zonation of periglacial landforms, Okstindan, Norway. *Zeitschrift für Geomorphologie N.F.*, **26**, 283–304.

Harris, C. (1983). Vesicles in thin sections of periglacial soils from north and south Norway. *Proceedings 4th International Permafrost Conference* (Fairbanks, Alaska), pp. 445–9. Washington D.C.: National Academy Press.

Harris, C. (1985). Geomorphological applications of soil micromorphology with particular reference to periglacial sediments and processes. In *Geomorphology and Soils*, eds. K. S. Richards, R. R. Arnett & S. Ellis, pp. 219–32. London: George Allen & Unwin.

Harris, C. & Cook, J. D. (1986). The detection of high altitude permafrost in Jotunheimen, Norway, using seismic refraction techniques: an assessment. *Arctic and Alpine Research*, **18**, 19–26.

Harris, C. & Ellis, S. (1980). Micromorphology of soils in soliflucted materials, Okstindan, northern Norway. *Geoderma*, **23**, 11–29.

Harris, C. & Matthews, J. A. (1984). Some observations on boulder-cored frost boils. *Geographical Journal*, **150**, 63–73.

Hartman, E. L. & Rottman, M. L. (1986). The vascular flora of five rock glaciers in the San Juan Mountains, Colorado. *Phytologia*, **60**, 225–35.

Hastenrath, S. (1973). Observations on the periglacial morphology of Mts Kenya and Kilimanjaro, East Africa. *Zeitschrift für Geomorphology, Supplement Band*, **16**, 161–79.

Hattersley-Smith, G. (1974). Present arctic ice cover. In *Arctic and Alpine Environments*, eds. J. D. Ives & R. G. Barry, pp. 195–223. London: Methuen.

Haworth, L. A., Calkin, P. E. & Ellis, J. M. (1986). Direct measurement of lichen growth in the Central Brooks Range, Alaska, U.S.A., and its application to lichenometric dating. *Arctic and Alpine Research*, **18**, 289–96.

Healy, M. J. R. (1965). Descriptive use of discriminant functions. In *Mathematics and Computer Science in Biology and Medicine*, ed. Medical Research Council, pp. 93–102. London: H.M.S.O.

Hedberg, O. (1964). Features of Afroalpine plant ecology. *Acta Phytogeographica Suecica* No. 49. 144 pp.

Heikkinen, O. (1984a). Dendrochronological evidence of variations of Coleman Glacier, Mount Baker, Washington, U.S.A. *Arctic and Alpine Research*, **16**, 53–64.

Heikkinen, O. (1984b). Dendrokronologian menetelmiä ja sovellutuksia. *Terra*, **96**, 1–22.

Heikkinen, O. & Fogelberg, P. (1980). Bodenentwicklung im Hochgebirge: Ein Beispiel vom Vorfeld des Steingletschers in der Schweiz. *Geographica Helvetica*, **35**, 107–12.

Heim, D. (1983). Glaziäre Entwässerung und Sanderbildung am Kötlujökull, Südisland. *Polarforschung*, **53**, 17–29.

Heusser, C. J. (1956). Postglacial environments in the Canadian Rocky Mountains. *Ecological Monographs*, **26**, 263–302.

Heusser, C. J., Schuster, R. L. & Gilkey, A. K. (1954). Geobotanical studies of the Taku Glacier anomaly. *Geographical Review*, **44**, 224–39.

Hibbs, D. E. (1983). Forty years of forest succession in central New England. *Ecology*, **64**, 1394–401.

Hoinkes, H. (1954). Beiträge zur Kenntis des Gletscherwindes. *Archiv für Meteorologie, Geophysik und Bioklimatologie, Ser. B*, **6**, 36–53.

Holzhauser, H. (1982). Neuzeitliche Gletscherschwankungen. *Geographica Helvetica*, **37**, 115–26.

Holzhauser, H. (1984a). Zur Geschichte der Aletschgletscher und des Fieschergletschers. *Physische Geographie* (Zürich), **13**, 1–448.

Holzhauser, H. (1984b). Rekonstruktion von gletscherschwankungen mit Hilfe fossiler Hölzer. *Geographica Helvetica*, **39**, 3–15.

Holzhauser, H. (1985). Neue Ergebnisse zur Gletscher- und Klimageschichte des Spätmittelalters und der Neuzeit. *Geographica Helvetica*, **40**, 168–85.

Hooke, J. M. & Kain, R. J. P. (1982). *Historical Change in the Physical Environment: A Guide to Sources and Techniques.* London: Butterworth Scientific.

Hope, G. S. (1976). Vegetation. In *The Equatorial Glaciers of New Guinea*, eds. G. S. Hope, J. A. Peterson, U. Radok & I. Allison, pp. 113–72. Rotterdam: Balkema.

Hoppe, G. & Schytt, V. (1953). Some observations on fluted moraine surfaces. *Geografiska Annaler*, **35**, 105–15.

Horak, E. (1960). Die Pilzvegetation im Gletschervorfeld (2290–2350 m) des Rotmoosferners in den Ötztaler Alpen. *Nova Hedwigia*, **2**, 487–508.

Horak, E. (1961). Floristische Untersuchungen in Gletschervorfeldern der Silvretta-, Lischana- und Sesvennagruppe. *Jahresbericht der Naturforschenden Gesellschaft Graubündens* N.F., **89** (1959/60 u. 1960/61): 112–135.

Horn, H. S. (1971). *The Adaptive Geometry of Trees.* Princeton, New Jersey: Princeton University Press.

Horn, H. S. (1974). The ecology of secondary succession. *Annual Review of Ecology and Systematics*, **5**, 25–37.

Horn, H. S. (1976). Succession. In *Theoretical Ecology: Principles and Applications*, ed. R. M. May, pp. 187–204. Oxford: Blackwell Scientific Publications.

Horn, H. S. (1987). Some causes of variety in patterns of secondary succession. In *Forest Succession: Concepts and Applications*, eds. D. C. West, H. H. Shugart & D. B. Botkin, pp. 24–35. New York: Springer Verlag.

Hosner, J. F. & Minckler, L. S. (1963). Bottomland hardwood forests in southern Illinois – regeneration and succession. *Ecology*, **44**, 29–41.

Howell, J. D. & Harris, S. A. (1978). Soil-forming factors in the Rocky Mountains of southwestern Alberta, Canada. *Arctic and Alpine Research*, **10**, 313–24.

Huggett, R. J. (1976). Conceptual models in pedogenesis – a discussion. *Geoderma*, **16**, 261–2.

Hurlbert, S. H. (1984). Pseudoreplication and the design of ecological field experiments. *Ecological Monographs*, **54**, 187–211.

Huston, M. & Smith, T. (1987). Plant succession: life history and competition. *American Naturalist*, **130**, 168–98.

Ingold, R. S. & Navarre, A. T. (1952). "Polluted" water from the leaching of igneous rocks. *Science*, **116**, 595–6.

Ingram, M. J., Underhill, D. J. & Farmer, G. (1981). The use of documentary sources for the study of past climates. In *Climate and History*, eds. T. M. L. Wigley, M. J. Ingram & G. Farmer, pp. 180–211. Cambridge: Cambridge University Press.

Innes, J. L. (1984). Relative dating of Neoglacial moraine ridges in North Norway. *Zeitschrift für Gletscherkunde und Glazialgeologie*, **20**, 53–63.

Innes, J. L. (1985a). Lichenometry. *Progress in Physical Geography*, **9**, 187–254.

Innes, J. L. (1985b). A standard *Rhizocarpon* nomenclature. *Boreas*, **14**, 83–5.

Innes, J. L. (1986a). The use of percentage cover measurements in lichenometric dating. *Arctic and Alpine Research*, **18**, 209–16.

Innes, J. L. (1986b). Influence of sampling design on lichen size-frequency distributions and its effect on derived lichenometric indices. *Arctic and Alpine Research*, **18**, 201–8.

Inouye, R. S. & Tilman, D. (1988). Convergence and divergence of oldfield plant communities along experimental nitrogen gradients. *Ecology*, **69**, 995–1004.

Jacks, G. V. (1965). The role of organisms in the early stages of soil formation. In

Experimental Pedology, eds. E. G. Hallsworth & D. B. Crawford, pp. 219–26. London: Butterworth.

Jacobson, G. L. Jnr & Birks, H. J. B. (1980). Soil development on recent end moraines of the Klutlan Glacier, Yukon Territory, Canada. *Quaternary Research*, **14**, 87–100.

Janetschek, H. (1949). Tierische successionen auf hochalpinen Neuland. *Bericht des Naturwissenschaftlich-medizinischen Vereins in Innsbruck*, **49**, 1–215. [not seen]

Jenik, J. (1986). Forest succession: theoretical concepts. In *Forest Dynamics Research in Western and Central Europe*, ed. J. Fanta, pp. 7–16. Wageningen: Centre for Publishing and Documentation (Pudoc).

Jenny, H. (1941). *Factors of Soil Formation*. New York: McGraw-Hill.

Jenny, H. (1946). Arrangement of soil series and types according to functions of soil-forming factors. *Soil Science*, **61**, 375–91.

Jenny, H. (1958). Role of the plant factor in the pedogenic functions. *Ecology*, **39**, 5–16.

Jenny, H. (1961). Derivation of state factor equations of soils and ecosystems. *Proceedings of the Soil Science Society of America*, **25**, 385–8.

Jenny, H. (1965). Bodenstickstoff und seine Abhängigkeit von Zustandfaktoren. *Zeitschrift für Pflanzenernährung, Düngung und Bodenkunde*, **109**, 97–112.

Jenny, H. (1980). *The Soil Resource: Origin and Behavior*. New York: Springer Verlag.

Jochimsen, M. (1962). Das Gletschervorfeld – keine Wüste. *Jahrbuch des Österreichischen Alpenvereins*, **87**, 135–42.

Jochimsen, M. (1963). Vegetationsentwicklung im hochalpinen Neuland. *Bericht das Naturwissenschaftlich-medizinischen Vereins in Innsbruck*, **53**, 109–23.

Jochimsen, M. (1970). Die Vegetationsentwicklung auf Moränenboden in Abhängigkeit von einigen Umweltfaktoren. *Veröffentlichungen der Universität Innsbruck*, **46**, 1–22.

Jochimsen, M. (1972). Pollenniederschlag und rezente Vegetation in Gletschervorfeldern der Alpen. *Bericht der Deutschen Botanischen Gesellschaften*, **35**, 13–27.

Jochimsen, M. (1973). Does the size of lichen thalli really constitute a valid measure for dating glacial deposits? *Arctic and Alpine Research*, **5**, 417–24.

Jochimsen, M. (1979). Pollenanalytische Kriterien zur Identifizierung Alpiner Pioniergemeinschaften. In *Werden und Vergehen von Pflanzengesellschaften*, eds. O. Wilmanns & R. Tüxen, pp. 215–25. Vaduz: J. Cramer.

Jochimsen, M. (1986). Zum Problem des Pollenfluges in den Hochalpen. *Dissertationes Botanicae*, **90**, 1–249.

Johnson, D. L., Keller, E. A. & Rockwell, T. K. (1990). Dynamic pedogenesis: new views on some key soil concepts, and a model for interpreting Quaternary soils. *Quaternary Research*, **33**, 306–19.

Johnson, D. L. & Watson-Stegner, D. (1987). Evolution model of pedogenesis. *Soil Science*, **143**, 349–66.

Johnson, P. G. (1984). Paraglacial conditions of instability and mass movement. A discussion. *Zeitschrift für Geomorphologie N.F.*, **28**, 235–50.

Johnson, P. G. (1987). Rock glaciers: glacier debris systems or high-magnitude low-frequency flows. In *Rock Glaciers*, eds. J. R. Giordano, J. F. Shroder Jnr & J. D. Vitek, pp. 175–92. Boston: Allen & Unwin.

Johnson, P. G. & Lacasse, D. (1988). Rock glaciers of the Dalton Range, Kluane Ranges, south-west Yukon Territory, Canada. *Journal of Glaciology*, **34**, 327–32.

Jordan III, W. R., Gilpin, M. E. & Aber, J. D. eds. (1987). *Restoration Ecology: A Synthetic Approach to Ecological Research*. Cambridge: Cambridge University Press.

Karlén, W. (1973). Holocene glacier and climatic variations, Kebnekaise Mountains, Swedish Lappland. *Geografiska Annaler*, **55(A)**, 29–63.

Karlén, W. (1976). Lacustrine sediments and tree-limit variations as indicators of Holocene climatic fluctuations in Lappland: Northern Sweden. *Geografiska Annaler*, **58(A)**, 1–34.

Karlén, W. (1979). Glacier variations in the Svartisen area, northern Norway. *Geografiska*

Annaler, **61(A)**, 11–28.

Karlén, W. (1981). Lacustrine sediment studies. *Geografiska Annaler*, **63(A)**, 273–81.

Karlén, W. (1982). Holocene glacier fluctuations in Scandinavia. *Striae*, **18**, 26–34.

Karlén, W. & Denton, G. H. (1976). Holocene glacial variations in Sarek National Park, northern Sweden. *Boreas*, **5**, 25–56.

Kautsky, L. (1988). Life strategies of aquatic soft bottom macrophytes. *Oikos*, **53**, 126–35.

Kelly, M. (1980). The status of the Neoglacial in western Greenland. *Grønlands Geologiske Undersøgelse Rapport*, **96**, 1–24.

Kelly, P. M., Jones, P. D., Sear, C. B., Cherry, B. S. G., & Tavakol, R. K. (1982). Variations in surface air temperatures: Part 2. Arctic regions, 1881–1980. *Monthly Weather Review*, **110**, 71–83.

Kershaw, K. A. & Looney, J. H. H. (1985). *Quantitative and Dynamic Plant Ecology*. London: Edward Arnold.

Khapayev, S. A. (1978). Dynamics of avalanche natural complexes: an example from the high-mountain Teberda State Reserve, Caucasus Mountains, U.S.S.R. *Arctic and Alpine Research*, **10**, 335–44.

Kienholz, H. (1975). Versuch einer relativen Altersbestimmung mit Hilfe von Aktivitätsgradmessungen des 'Freien Eisens'. *Zeitschrift für Gletscherkunde und Glazialgeologie*, **11**, 53–60.

Kilmaskossu, M. St E. & Hope, G. S. (1985). A mountain research programme for Indonesia. *Mountain Research and Development*, **5**, 339–48.

King, L. (1983). High mountain permafrost in Scandinavia. *Proceedings 4th International Permafrost Conference* (Fairbanks, Alaska), pp. 612–7. Washington: National Academy Press.

King, L. (1984). Permafrost in Scandinavien. Undersuchungsergebnisse aus Lappland, Jotunheimen und Dovre/Rondane. *Heidelberger Geographische Arbeiten*, **76**, 1–174.

Kinzl, H. (1929). Beiträge zur Geschichte der Gletscherschwankungen in den Ostalpen. *Zeitschrift für Gletscherkunde für Eiszeitforschung und Geschichte des Klimas* (Leipzig), **17**, 66–121.

Klimek, K. (1972). Fluvial processes on the direct glacier foreland. In *Processus Périglaciaires Études sur le Terrain*, eds. P. Macar & A. Pissart, pp. 177–85. Liège: University of Liège.

Kolishchuk, V. G. (1990). Dendroclimatological study of prostrate woody plants. In *Methods of Dendrochronology*, eds. E. R. Cook & L. A. Kairiukstis, pp. 51–5. Dordrecht: Kluwer.

Koster, E. A. (1988). Ancient and modern cold-climate aeolian sand deposition: a review. *Journal of Quaternary Science*, **3**, 69–83.

Krebs, C. J. (1985). *Ecology: The Experimental Analysis of Distribution and Abundance*. New York: Harper & Row.

Krigstrom, A. (1962). Geomorphological studies of sandur plains and their braided rivers in Iceland. *Geografiska Annaler*, **44**, 328–46.

Krüger, J. (1979). Structures and textures in till indicating subglacial deposition. *Boreas*, **8**, 323–40.

Krüger, J. & Humlum, O. (1981). The proglacial area of Mýrdalsjökull. *Folia Geographica Danica*, **15(1)**, 1–58.

Kuc, M. (1964). Deglaciation of Treskelen-Treskelodden in Horsund, Vestspitsbergen, as shown by vegetation. *Studia Geologica Polonica*, **11**, 197–205.

Kuc, M. (1970). Additions to the Arctic moss flora. V. The role of mosses in plant succession and the development of peat on Fitzwilliam Owen Island (Western Canadian Arctic). *Revue Bryologique et Lichénologique*, **37**, 931–9.

Küchler, A. W. (1988). The nature of vegetation. In *Vegetation Mapping*, eds. A. W. Küchler & I. S. Zonneveld, pp. 13–23. Dordrecht: Kluwer.

Kurimo, H. (1980). Depositional deglaciation forms as indicators of different glacial and glaciomarginal environments. *Boreas*, **9**, 179–91.

LaMarche, V. C. Jnr & Fritts, H. C. (1971). Tree rings, glacier advance and climate in the Alps. *Zeitschrift für Gletscherkunde und Glazialgeologie*, **7**, 125–31.

Lamb, H. H. (1977). *Climate: Present, Past and Future. Volume 2: Climate History and the Future.* London: Methuen.

Lamb, H. H. (1984). Climate in the last thousand years: natural climatic fluctuations and change. In *The Climate of Europe: Past, Present and Future*, eds. H. Flohn & R. Fantechi, pp. 25–64. Dordrecht: Reidel.

Lance, G. N. & Williams, W. T. (1968). Note on a new information statistic classificatory program. *Computer Journal*, **11**, 195.

Langford, A. N. & Buell, M. F. (1969). Integration, identity and stability in the plant association. *Advances in Ecological Research*, **6**, 84–135.

Lautridou, J. P. (1988). Recent advances in cryogenic weathering. In *Advances in Periglacial Geomorphology*, ed. M. J. Clark, pp. 33–47. Chichester: Wiley.

Lautridou, J. P. & Ozouf, J. C. (1982). Experimental frost shattering: 15 years of research at the Centre de Géomorphologie du CNRS. *Progress in Physical Geography*, **6**, 215–32.

Lawler, D. M. (1988). Environmental limits of needle ice: a global survey. *Arctic and Alpine Research*, **20**, 137–59.

Lawrence, D. B. (1950). Estimating dates of recent glacier advances and recession rates by studying tree growth layers. *Transactions American Geophysical Union*, **31**, 243–8.

Lawrence, D. B. (1951). Recent glacier history of Glacier Bay, Alaska, and development of vegetation on deglaciated terrain with special reference to the importance of alder in the succession. *American Philosophical Society Yearbook for 1950*, pp. 175–6.

Lawrence, D. B. (1953). Development of vegetation and soil in south-eastern Alaska with special reference to the accumulation of nitrogen. *Final Report of the United States Office of Naval Research*, Project NR160–183. [not seen]

Lawrence, D. B. (1958). Glaciers and vegetation in south-eastern Alaska. *American Scientist*, **46**, 89–122.

Lawrence, D. B. (1979). Primary versus secondary succession at Glacier Bay National Monument, southeastern Alaska. *Proceedings 1st Conference on Scientific Research in the National Parks*, pp. 213–24. United States Department of the Interior, National Parks Service, Transactions and Proceedings Series No. 5. (U.S. Government Printing Office Stock No. 024–005–00743–1)

Lawrence, D. B. (1984). Ecosystem development following Neoglacial recession since 1750 A.D.: an introduction. In *Proceedings 1st Glacier Bay Science Symposium*, eds. J. D. Wood, M. Gladziszewski, I. A. Worley & C. Veqvist, pp. 26–7. Atlanta, Georgia: United States Department of the Interior, National Parks Service.

Lawrence, D. B. & Hulbert, L. (1950). Growth stimulation of adjacent plants by lupine and alder on recent glacier deposits in southeastern Alaska. *Bulletin of the Ecological Society of America*, **31**, 58.

Lawrence, D. B., Noble, M. G. & Tilman, G. D. (in prep.) Terrestrial ecology of the Glacier Bay region, Alaska. *Advances in Ecological Research*.

Lawrence, D. B., Schoenike, R. E., Quispel, A. & Bond, G. (1967). The role of *Dryas drummondii* in vegetation development following ice recession at Glacier Bay, Alaska, with special reference to its nitrogen fixation by root nodules. *Journal of Ecology*, **55**, 793–813.

Lawson, D. E. (1979). Sedimentological analysis of the western terminus of the Matanuska Glacier, Alaska. *United States Army Corps of Engineers, Cold Regions Research and Engineering Laboratory*, Report No. 79–9, pp. 1–122.

Lawson, D. E. (1981a). Sedimentological characteristics and classification of depositional processes and deposits in the glacial environment. *United States Army Corps of*

Engineers. Cold Regions Research and Engineering Laboratory, Report No. 81–27, pp. 1–22.

Lawson, D. E. (1981b). Distinguishing characteristics of diamictons at the margin of Matanuska Glacier, Alaska. *Annals of Glaciology*, **2**, 78–84.

Lawson, D. E. (1982). Mobilization, movement and deposition of active subaerial sediment flows, Matanuska Glacier, Alaska. *Journal of Geology*, **90**, 279–300.

Lawson, D. E. (1988). Glacigenic resedimentation: classification concepts and application to mass-movement processes and deposits. In *Genetic Classification of Glacigenic Deposits*, eds. R. P. Goldthwait & C. L. Matsch, pp. 147–69. Rotterdam: Balkema.

Leonard, E. M. (1986a). Varve studies at Hector Lake, Alberta, Canada, and the relationship between glacial activity and sedimentation. *Quaternary Research*, **25**, 199–214.

Leonard, E. M. (1986b). Use of lacustrine sedimentary sequences as indicators of Holocene glacial history, Banff National Park, Alberta, Canada. *Quaternary Research*, **26**, 218–31.

Le Roy Ladurie, E. (1972). *Times of Feast, Times of Famine: A History of Climate Since the Year 1000*. London: George Allen & Unwin.

Lewkowski, A. G. (1988). Slope processes. In *Advances in Periglacial Geomorphology*, ed. M. J. Clark, pp. 325–68. Chichester: Wiley.

Liestøl, O. (1967). Storbreen glacier in Jotunheimen, Norway. *Norsk Polarinstitutt Skrifter*, **141**, 1–63.

Liestøl, O. (1978). Pingoes, springs and permafrost in Spitsbergen. *Norsk Polarinstitutt Årsbok*, **1978**, 7–29.

Likens, G. E. ed. (1989). *Long-term Studies in Ecology: Approaches and Alternatives*. New York: Springer Verlag.

Lindner, L. & Marks, L. (1985). Types of debris slope accumulations and rock glaciers in South Spitzbergen. *Boreas*, **14**, 139–53.

Lindröth, C.H. (1965). Skaftafell, Iceland: a living glacial refugium. *Oikos* Supplementum, **6**, 1–142.

Lindsay, D. C. (1971). Vegetation of the South Shetland Islands. *British Antarctic Survey Bulletin*, **25**, 59–83.

Locke III, W. W. Andrews, J. T. & Webber, P. J. (1979). A manual for lichenometry. *British Geomorphological Research Group Technical Bulletin*, **26**, 1–47.

Lockwood, J. A., Thompson, C. D., Debrey, L. D., Love, C. M., Nunamaker, R. A. & Pfadt, R. E. (1991). Preserved grasshopper fauna of Knife Point Glacier, Fremont County, Wyoming, U.S.A. *Arctic and Alpine Research*, **23**, 108–14.

Longton, R. E. (1988). *Biology of Polar Bryophytes and Lichens*. Cambridge: Cambridge University Press.

Lowrison, G. C. (1974). *Crushing and Grinding*. London: Butterworth.

Luckman, B. H. (1986). Reconstruction of Little Ice Age events in the Canadian Rocky Mountains. *Géographie Physique et Quaternaire*, **40**, 17–28.

Luckman, B. H. (1988). Dating the moraines and recession of Athabasca and Dome Glaciers, Alberta, Canada. *Arctic and Alpine Research*, **20**, 40–54.

Lüdi, W. (1921). Die Pflanzengesellschaften des Lauterbrunnentales und ihre Sukzession. *Beiträge zur Geobotanischen Landesaufnahme*, **9**, 1–364.

Lüdi, W. (1945). Besiedlung und Vegetationsentwicklung auf den jungen Seitenmoränen des Grossen Aletschgletschers mit einem Vergleich der Besiedlung im Vorfeld des Rhonegletschers und des Oberen Grindelwaldgletschers. *Bericht über das Geobotanische Forschungsinstitut Rübel in Zurich*, 1944, pp. 35–112.

Lüdi, W. (1958). Beobachtungen über die Besiedlung von Gletschervorfeldern in den Schweizeralpen. *Flora* (Jena), **146**, 386–407.

Luken, J. O. (1990). *Directing Ecological Succession*. London & New York: Chapman & Hall.

Lundqvist, J. (1981). Moraine morphology. *Geografiska Annaler*, **63(A)**, 127–38.

Lundqvist, J. (1988). Glacigenic processes, deposits and landforms. In *Genetic Classification of Glacigenic Deposits*, eds. R. P. Goldthwait & C. L. Matsch, pp. 3–16. Rotterdam: Balkema.

Lutz, H. J. (1930). Observations on the invasion of newly formed glacial moraines by trees. *Ecology*, **11**, 562–7.

MacArthur, R. H. & Wilson, E. O. (1967). *The Theory of Island Biogeography*. Princeton, New Jersey: Princeton University Press.

Macfadyen, A. (1963). *Animal Ecology: Aims and Methods*. London: Pitman.

MacMahon, J. A. (1980). Ecosystems over time: succession and other types of change. In *Forests: Fresh Perspectives from Ecosystem Analysis*, ed. R. Waring, pp. 27–58. Corvallis, Oregon: Oregon State University Press.

MacMahon, J. A. (1981). Succession processes: comparisons amongst biomes with special reference to probable roles of and influences on animals. In *Forest Succession: Concepts and Application*, eds. D. C. West, H. H. Shugart & D. B. Botkin, pp. 277–304. New York: Springer Verlag.

Macpherson, J. B. (1982). Postglacial vegetation history of the eastern Avalon Peninsula, Newfoundland, and Holocene climatic change along the eastern Canadian seaboard. *Géographie Physique et Quaternaire*, **36**, 175–96.

Magurran, A. E. (1988). *Ecological Diversity and its Measurement*. London: Croom Helm.

Mahaney, W. C. (1987). Lichen trimlines and weathering features as indicators of mass balance changes and successive retreat stages of the Mer de Glace in the Western Alps. *Zeitschrift für Geomorphologie N.F.*, **31**, 411–8.

Mahaney, W. C. (1990). *Ice on the Equator: Quaternary Geology of Mount Kenya*. Sister Bay, Wisconsin: Wm Caxton Ltd.

Mahaney, W. C. & Spence, J. R. (1990). Neoglacial chronology and floristics in the Middle Teton area, central Teton Range, Western Wyoming. *Journal of Quaternary Science*, **5**, 53–66.

Maizels, J. K. (1973). Le Glacier des Bossons. Quelques aspects caractéristiques de l'environnement proglaciaire. *Revue de Géographie Alpine*, **61**, 427–47.

Maizels, J. K. (1979). Proglacial aggradation and changes in braided channel patterns during a period of glacier advance: an Alpine example. *Geografiska Annaler*, **61(A)**, 87–101.

Maizels, J. K. & Dugmore, A. J. (1985). Lichenometric dating and tephrochronology of sandar deposits, Sólheimajökull area, southern Iceland. *Jökull*, **35**, 69–77.

Maizels, J. K. & Petch, J. R. (1985). Age determination of intermoraine areas, Austerdalen, southern Norway. *Boreas*, **14**, 51–65.

Mann, D. H. (1986). Reliability of a fjord glacier's fluctuations for palaeoclimatic reconstructions. *Quaternary Research*, **25**, 10–24.

Marangunic, C. (1972). *Effects of a Landslide on Sherman Glacier, Alaska*. Columbus, Ohio: Ohio State University Research Foundation. (Institute of Polar Studies Report No. 30.)

Marchand, P. J. (1987). *Life in the Cold. An Introduction to Winter Ecology*. Hannover: University Press of New England.

Margalef, R. (1968). *Perspectives in Ecological Theory*. Chicago: University of Chicago Press.

Mathews, W. H. (1951). Historic and prehistoric fluctuations of alpine glaciers in the Mount Garibaldi map-area, south-western British Columbia. *Journal of Geology*, **59**, 357–80.

Mathews, W. H. & Mackay, J. R. (1975). Snow creep: its engineering problems and some techniques and results of its investigation. *Canadian Geotechnical Journal*, **12**, 187–98.

Matthews, J. A. (1973). Lichen growth on an active medial moraine, Jotunheimen,

Norway. *Journal of Glaciology*, **12**, 305–13.

Matthews, J. A. (1974). Families of lichenometric dating curves from the Storbreen gletschervorfeld, Jotunheimen, Norway. *Norsk Geografisk Tidsskrift*, **28**, 215–35.

Matthews, J. A. (1975a) Experiments on the reproducibility and reliability of lichenometric dates, Storbreen gletschervorfeld, Jotunheimen, Norway. *Norsk Geografisk Tidsskrift*, **29**, 97–109.

Matthews, J. A. (1975b). The gletschervorfeld: a biogeographical system and microcosm. *University of Edinburgh, Department of Geography, Research Discussion Paper*, **2**, 1–44.

Matthews, J. A. (1976a). 'Little Ice Age' palaeotemperatures from high altitude tree growth in S. Norway. *Nature*, **264**, 243–5.

Matthews, J. A. (1976b). *A Phytogeography of a Gletschervorfeld: Storbreen-i-Leirdalen, Jotunheimen, Norway*. Ph.D. Thesis, University of London (King's College).

Matthews, J. A. (1977a). Glacier and climatic fluctuations inferred from tree-growth variations over the last 250 years, central southern Norway. *Boreas*, **6**, 1–24.

Matthews, J. A. (1977b). A lichenometric test of the 1750 end-moraine hypothesis: Storbreen gletschervorfeld, southern Norway. *Norsk Geografisk Tidsskrift*, **31**, 129–36.

Matthews, J. A. (1978a). Plant colonisation patterns on a gletschervorfeld, southern Norway: a meso-scale geographical approach to vegetation change and phytometric dating. *Boreas*, **7**, 155–78.

Matthews, J. A. (1978b). An application of non-metric multidimensional scaling to the construction of an improved species plexus. *Journal of Ecology*, **66**, 157–73.

Matthews, J. A. (1979a). The vegetation of the Storbreen gletschervorfeld, Jotunheimen, Norway. I. Introduction and approaches involving classification. *Journal of Biogeography*, **6**, 17–47.

Matthews, J. A. (1979b). The vegetation of the Storbreen gletschervorfeld, Jotunheimen, Norway, II. Approaches involving ordination and general conclusions. *Journal of Biogeography*, **6**, 133–67.

Matthews, J. A. (1979c). A study of the variability of some successional and climax assemblage types using multiple discriminant analysis. *Journal of Ecology*, **67**, 255–71.

Matthews, J. A. (1979d). Refutation of convergence in a vegetation succession. *Naturwissenschaften*, **66**, 47–9.

Matthews, J. A. (1981). Vegetation studies. In *Quantitative Geography: A British View*, eds N. Wrigley & R. J. Bennett, pp. 294–308. London: Routledge & Kegan Paul.

Matthews, J. A. (1985). Radiocarbon dating of surface and buried soils: principles, problems and prospects. In *Geomorphology and Soils*, eds. K. S. Richards, R. R. Arnett & S. Ellis, pp. 269–88. London: George Allen & Unwin.

Matthews, J. A. (1987). Regional variation in the composition of Neoglacial end moraines, Jotunheimen, Norway: an altitudinal gradient in clast roundness and its possible palaeoclimatic significance. *Boreas*, **16**, 173–88.

Matthews, J. A. (1989). A distillation of successional wisdom. *Journal of Biogeography*, **16**, 298–9. [Review of *Colonization, Succession and Stability*, eds. A. J. Gray, M. J. Crawley & P. J. Edwards. Oxford: Blackwell Scientific Publications. (1987)]

Matthews, J. A. & Caseldine, C. J. (1987). Arctic-alpine Brown Soils as a source of palaeoenvironmental information: further ^{14}C dating and palynological evidence from Vestre Memurubreen, Jotunheimen, Norway. *Journal of Quaternary Science*, **2**, 59–71.

Matthews, J. A., Dawson, A. G. & Shakesby, R. A. (1986). Lake shoreline development, frost weathering and rock platform erosion in an alpine periglacial environment, Jotunheimen, Norway. *Boreas*, **15**, 33–50.

Matthews, J. A. & Dresser, P. Q. (1983). Intensive ^{14}C dating of a buried palaeosol horizon. *Geologiska Föreningens i Stockholm, Förhandlingar*, **105**, 59–63.

Matthews, J. A., Harris, C. & Ballantyne, C. K. (1986). Studies on a gelifluction lobe, Jotunheimen, Norway: [14]C chronology, stratigraphy and palaeoenvironment.

Geografiska Annaler, **68(A)**, 345–60.

Matthews, J. A., Innes, J. L. & Caseldine, C. J. (1986). [14]C dating and palaeoenvironment of the historic 'Little Ice Age' glacier advance of Nigardsbreen, southwest Norway. *Earth Surface Processes and Landforms*, **11**, 369–75.

Matthews, J. A. & Petch, J. R. (1982). Within-valley asymmetry and related problems of Neoglacial lateral moraine development at certain Jotunheimen glaciers. *Boreas*, **11**, 225–47.

Matthews, J. A. & Shakesby, R. A. (1984). The status of the 'Little Ice Age' in southern Norway: relative-age dating of Neoglacial moraines with Schmidt hammer and lichenometry. *Boreas*, **13**, 333–46.

Matthews, J. A., & Whittaker, R. J. (1987). Vegetation succession on the Storbreen glacier foreland, Jotunheimen, Norway: a review. *Arctic and Alpine Research*, **19**, 385–95.

McCarroll, D. (1989a). Potential and limitations of the Schmidt hammer for relative-age dating: field tests on Neoglacial moraines, Jotunheimen, southern Norway. *Arctic and Alpine Research*, **21**, 268–75.

McCarroll, D. (1989b). Schmidt hammer relative-age evaluation of a possible pre-'Little Ice Age' Neoglacial moraine, Leirbreen, southern Norway. *Norsk Geologisk Tidsskrift*, **69**, 125–30.

McCarroll, D. (1990). Differential weathering of feldspar and pyroxene in an arctic-alpine environment. *Earth Surface Processes and Landforms*, **15**, 641–51.

McCarroll, D. & Ware, M. (1989). The variability of soil development on Preboreal moraine ridge crests, Breiseterdalen, southern Norway. *Norsk Geografisk Tidsskrift*, **43**, 31–6.

McCormick, J. (1968). Succession. *Via*, **1**, 22–35 & 131–2.

McCune, B. & Allen, T. F. H. (1985). Will similar forests develop on similar sites? *Canadian Journal of Botany*, **63**, 367–76.

McGreevy, J. P. (1981). Perspectives on frost shattering. *Progress in Physical Geography*, **5**, 56–75.

McIntosh, R. P. (1967). The continuum concept of vegetation. *Botanical Review*, **33**, 130–87.

McIntosh, R. P. (1980). The relationship between succession and the recovery process in ecosystems. In *The Recovery Process in Damaged Ecosystems*, ed. J. Cairns Jnr, pp. 11–62. Ann Arbor, Michigan: Ann Arbor Science.

McIntosh, R. P. (1981). Succession and ecological theory. In *Forest Succession: Concepts and Application*, eds. D. C. West, H. H. Shugart & D. B. Botkin, pp. 10–23. New York: Springer Verlag.

McIntosh, R. P. (1985). *The Background of Ecology: Concept and Theory*. Cambridge: Cambridge University Press.

McKenna-Neumann, C. & Gilbert, R. (1986). Aeolian processes and landforms in glaciofluvial environments of southeastern Baffin Island, N.W.T., Canada. In *Aeolian Geomorphology*, ed. W. G. Nickling, pp. 213–35. Boston: Allen & Unwin.

McKenzie, G. D. & Goldthwait, R. P. (1971). Glacial history of the last eleven thousand years in Adams Inlet, southeastern Alaska. *Geological Society of America Bulletin*, **82**, 1767–82.

McKenzie, G. D. & Goodwin, R. G. (1987). Development of collapsed glacial topography in the Adams Inlet area, Alaska, U.S.A. *Journal of Glaciology*, **33**, 55–9.

Meeker, D. O. Jnr & Merkel, D. L. (1984). Climax theories and a recommendation for vegetation classification. *Journal of Range Management*, **37**, 427–30.

Meier, M. F. (1965). Glaciers and climate. In *The Quaternary of the United States*, ed. H. E. Wright Jnr & D. G. Frey, pp. 795–805. Princeton, New Jersey: Princeton University Press.

Meier, M. F. & Post, A. (1969). What are glacier surges? *Canadian Journal of Earth Sciences*, **6**, 807–17.

Mellor, A. (1984). *An Investigation of Pedogenesis on Selected Neoglacial Moraine Ridge Sequences, Jostedalsbreen and Jotunheimen, Southern Norway*. Ph.D. Thesis. University of Hull.

Mellor, A. (1985). Soil chronosequences on Neoglacial moraine ridges, Jostedalsbreen and Jotunheimen, southern Norway: a quantitative pedogenic approach. In *Geomorphology and Soils*, eds. K. S. Richards, R. R. Arnett & S. Ellis, pp. 289–308. London: George Allen & Unwin.

Mellor, A. (1986a). A micromorphological examination of two alpine soil chronosequences, southern Norway. *Geoderma*, **39**, 41–57.

Mellor, A. (1986b). Textural and scanning-electron microscope observations of some arctic-alpine soils developed in Weichselian and Neoglacial till deposits in southern Norway. *Arctic and Alpine Research*, **18**, 327–36.

Mellor, A. (1986c). Hydrobiotite formation in some Norwegian arctic-alpine soils developing in Neoglacial till. *Norsk Geologisk Tidsskrift*, **66**, 183–5.

Mellor, A. (1987). A pedogenic investigation of some soil chronosequences on Neoglacial moraine ridges, southern Norway: examination of soil chemical data using principal components analysis. *Catena*, **14**, 369–81.

Menzies, J. (1979). A review of the literature on the formation and location of drumlins. *Earth Science Reviews*, **14**, 315–59.

Messer, A. C. (1984). *A Geographical Investigation of Soil Development on Glacier Forelands in South-central Norway*. Ph.D. Thesis, University of Wales (University College Cardiff).

Messer, A. C. (1988). Regional variations in rates of pedogenesis and the influence of climatic factors on moraine chronosequences, southern Norway. *Arctic and Alpine Research*, **20**, 31–9.

Messer, A. C. (1989). An alternative approach to the study of pedogenic chronosequences. *Norsk Geografisk Tidsskrift*, **43**, 221–9.

Miall, A. D. (1983). Glaciofluvial transport and deposition. In *Glacial Geology*, ed. N. Eyles, pp. 168–83. Oxford: Pergamon Press.

Miles, J. (1979). *Vegetation Dynamics*. London: Chapman & Hall.

Miles, J. (1987). Vegetation succession: past and present perceptions. In *Colonization, Succession and Stability*, eds. A. J. Gray, M. J. Crawley & P. J. Edwards, pp. 1–29. Oxford: Blackwell Scientific Publications.

Miller, G. H. (1973). Variations in lichen growth from direct measurements: preliminary curves for *Alectoria miniscula* from Eastern Baffin Island, N.W.T., Canada. *Arctic and Alpine Research*, **13**, 333–9.

Miller, P. C. (1982). Environment and vegetation variation across a snow accumulation area in montane tundra in central Alaska. *Holarctic Ecology*, **5**, 85–98.

Milne, G. (1935). Composite units for the mapping of complex soil associations. *Transactions 3rd International Congress of Soil Science*, **1**, 345–7.

Milner, A. M. (1987). Colonization and ecological development of new streams in Glacier Bay National Park, Alaska. *Freshwater Biology*, **18**, 53–70.

Milner, A. M. & Bailey, R. G. (1989). Salmonid colonization of new streams in Glacier Bay National Park, Alaska. *Aquaculture and Fisheries Management*, **20**, 179–92.

Moen, A. (1987). The regional vegetation of Norway; that of Central Norway in particular. *Norsk Geografisk Tidsskrift*, **41**, 179–226.

Moiroud, A. & Gonnet, J. F. (1977). *Jardins de Glaciers*. Grenoble: Editions Allier.

Monte, J. A. (1973). The successional convergence of vegetation from grassland and bare soil on the Piedmont of New Jersey. *William L. Hutcheson Memorial Forest Bulletin*, **3**, 3–13.

Moravec, J. (1969). Succession of plant communities and soil development. *Folia Geobotanica et Phytotaxonomica*, **4**, 133–64.

Mottershead, D. N. (1980). Lichenometry – some recent applications. In *Timescales in Geomorphology*, eds. R. A. Cullingford, D. A. Davidson & J. Lewin, pp. 95–108.

352   *References*

Chichester: Wiley.

Mueller-Dombois, D. & Ellenberg, H. (1974). *Aims and Methods of Vegetation Ecology.* New York: Wiley.

Muhs, D. R. (1984). Intrinsic thresholds in soil systems. *Physical Geography*, **5**, 99–110.

Muller, C. H. (1940). Plant succession in the Larrea-Flourensia climax. *Ecology*, **21**, 206–12.

Muller, C. H. (1952). Plant succession in Arctic heath and tundra in northern Scandinavia. *Bulletin of the Torrey Botanical Club*, **79**, 296–309.

Muller, C. H. (1966). The role of chemical inhibition (allelopathy) in vegetational composition. *Bulletin of the Torrey Botanical Club*, **93**, 332–51.

Muller, C. H. (1969). Allelopathy as a factor in ecological process. *Vegetatio*, **18**, 348–57.

Naveh, Z. & Leberman, A. S. (1984). *Landscape Ecology: Theory and Applications.* New York: Springer Verlag.

Negri, G. (1934). La vegetazione delle morene del Ghiacciaio del Lys (Monte Rosa). *Bollettino Comitato Glaciologico Italiano* (Roma), **14**, 105–72.

Negri, G. (1936). Osservazioni di U. Monterin su alcuni casi di invasione della morene galleggianti dei Ghiacciai del Monte Rosa da parte della vegetazione (contav. XVII e XVIII). *Nuovo Giornale Botanico Italiano* (Firenze), **42**, 699–712.

Nichols, R. L. & Miller, M. M. (1952). The Moreno Glacier, Lago Argentino, Patagonia. Advancing glaciers and nearby simultaneously retreating glaciers. *Journal of Glaciology*, **2**, 41–50.

Nicholson, F. H. (1976). Patterned ground formation and description as suggested by low Arctic and subarctic examples. *Arctic and Alpine Research*, **8**, 329–42.

Nickling, W. G. & Brazel, A. J. (1985). Surface wind characteristics along the Icefield Ranges, Yukon Territory, Canada. *Arctic and Alpine Research*, **17**, 125–34.

Nicolet, A. C. (1841). Catalogue des plantes de la moraine médiane du glacier inférieur de l'Aar recueillies en août 1840. *Bibliothèque Univ. de Genève n.s.*, **32**, 394–5. (Paris 1841)

Nitychoruk, J. & Dzierzek, J. (1988). Morphological features of talus cones in northwestern Wedel Jarlsberg Land, Spitzbergen. *Polish Polar Research*, **9**, 73–85.

Noble, I. R. (1981). Predicting successional change. *Proceedings of the Conference on Fire Regimes and Ecosystem Properties* (Honolulu, Hawaii), pp. 278–300. (United States Department of Agriculture Forest Service, General Technical Report WO–26)

Noble, I. R. & Slatyer, R. O. (1977). Post-fire succession of plants in mediterranean ecosystems. In *Proceedings of the Symposium on the Environmental Consequences of Fire and Fire Management in Mediterranean Ecosystems* (Palo Alto, California), eds. H. A. Mooney & C. E. Conrad, pp. 27–36. (United States Department of Agriculture Forest Service, General Technical Report WO–3)

Noble, I. R. & Slatyer, R. O. (1980). The use of vital attributes to predict successional changes in plant communities subject to recurrent disturbances. *Vegetatio*, **43**, 5–21.

Noble, M. G. (1978). Sherman Glacier vegetation. *Northwest Science*, **52**, 351–7.

Noble, M. G., Lawrence, D. B. & Streveler, G. P. (1984). *Sphagnum* invasion beneath an evergreen forest canopy in southeastern Alaska. *The Bryologist*, **87**, 119–27.

Odum, E. P. (1969). The strategy of ecosystem development. *Science*, **164**, 262–70.

Oechslin, M. (1935). Beitrag zur Kenntis der pflanzlichen Besiedelung der durch Gletscher freigegebenen Grundmoränenböden. Gebiet des Griesgletschers, Klausen, Kanton Uri. *Berichte der Naturforschenden Gesellschaft Uri*, **4**, 27–48.

Oke, T. R. (1987). *Boundary Layer Climates.* London: Methuen.

Oliver, C. D. (1981). Forest development in North America following major disturbances. *Forest Ecology and Management*, **3**, 153–68.

Oliver, C. D. & Adams, A. B. (1979). Vegetation dynamics. In *Nooksack Cirque Natural History*, eds. C. D. Oliver & R. J. Zazoski (principal investigators), pp. 42–96. United States National Park Service Report CX–9000–6–0418.

Oliver, C. D., Adams, A. B. & Zazoski, R. J. (1985). Disturbance patterns and forest development in a recently deglaciated valley in the northwestern Cascade Range of Washington, U.S.A. *Canadian Journal of Forest Research*, **15**, 221–32.

Olson, J. S. (1958). Rates of succession and soil changes on southern Lake Michigan sand dunes. *Botanical Gazette*, **119**, 125–70.

O'Neill, R. V., DeAngelis, D. L., Waide, J. B. & Allen, T. F. H. (1986). *A Hierarchical Concept of Ecosystems*. Princeton, New Jersey: Princeton University Press.

Orwin, J. (1970). Lichen succession on recently deposited rock surfaces. *New Zealand Journal of Botany*, **8**, 452–77.

Orwin, J. (1972). The effect of environment on assemblages of lichens growing on rock surfaces. *New Zealand Journal of Botany*, **10**, 37–47.

Østrem, G. (1974). Present alpine ice cover. In *Arctic and Alpine Environments*, eds. J. D. Ives & R. G. Barry, pp. 225–50. London: Methuen.

Østrem, J. (1975). Sediment transport in glacial meltwater streams. In *Glaciofluvial and Glaciolacustrine Sedimentation*, eds. A. V. Joplin & B. C. McDonald, pp. 101–22. Tulsa, Oklahoma: Society of Economic Palaeontologists and Mineralogists. (SEPM Special Publication No. 23)

Østrem, G., Haakensen, N. & Eriksson, T. (1981). The glaciation level in southern Alaska. *Geografiska Annaler*, **63(A)**, 251–60.

Østrem, G., Liestøl, O. & Wold, B. (1976). Glaciological investigations at Nigardsbreen, Norway. *Norsk Geografisk Tidsskrift*, **30**, 187–209.

Østrem, G. & Olsen, H. C. (1987). Sedimentation in a glacier lake. *Geografiska Annaler*, **69(A)**, 123–38.

Palmer, W. H. & Miller, A. K. (1961). Botanical evidence for the recession of a glacier. *Oikos*, **12**, 75–86.

Parkinson, R. J. & Gellatly, A. F. (in prep.). Soil formation on Holocene moraines: Cirque de Troumouse, French Pyrenees.

Parkinson, R. J. & Roberts, A. H. (1985). Microtopographical control of soil development in the Okstindan Region, North Norway. *Okstindan Research Project Preliminary Report for 1983*, eds. J. Rose & C. A. Whiteman, pp. 1–9. London: University of London, Birkbeck College.

Parrish, J. A. D. & Bazzaz, F. A. (1982). Competitive interactions in plant communities of different ages. *Ecology*, **63**, 314–20.

Pastor, J. & Bockheim, J. G. (1980). Soil development on moraines of Taylor Glacier, Lower Taylor Valley, Antarctica. *Soil Science Society of America Journal*, **44**, 341–8.

Paton, T. R. (1978). *The Formation of Soil Material*. London: George Allen & Unwin.

Patterson, W. S. B. (1981). *The Physics of Glaciers*. Oxford: Pergamon Press.

Patzelt, G. (1974). Holocene variations of glaciers in the Alps. *Colloques Internationaux du Centre National de la Recherche Scientifique*, **219**, 51–9.

Patzelt, G. & Bortenschlager, S. (1973). Die postglazialen Gletscher- und Klimaschwankungen in der Venedigergruppe (Hohe Tauern, Ostalpen). *Zeitschrift für Geomorphologie N.F. Supplement Band*, **16**, 25–72.

Paul, M. A. (1983). The supraglacial landsystem. In *Glacial Geology*, ed. N. Eyles, pp. 71–90. Oxford: Pergamon Press.

Paul, M. A. & Eyles, N. (1990). Constraints on the preservation of diamict facies (melt-out tills) at the margins of stagnant glaciers. *Quaternary Science Reviews*, **9**, 51–69.

Payette, S., Gauthier, L. & Grenier, I. (1986). Dating ice-wedge growth in subarctic peatlands following deforestation. *Nature*, **322**, 724–7.

Pérez, F. C. (1987). Downslope stone transport by needle ice in a high Andean area (Venezuela). *Revue de Géomorphologie Dynamique*, **36**, 33–51.

Perfect, E., Miller, R. D. & Burton, B. (1987). Root morphology and vigour effects on winter heaving of established Alfalfa. *Agronomy Journal*, **79**, 1061–7.

Perfect, E., Miller, R. D. & Burton, B. (1988). Frost upheaval of overwintering plants: a quantitative field study of the displacement process. *Arctic and Alpine Research*, **20**, 70–5.

Perring, F. (1958). A theoretical approach to the study of chalk grassland. *Journal of Ecology*, **46**, 665–79.

Persson, Å. (1964). The vegetation at the margin of the receding glacier Skaftafellsjökull, southeastern Iceland. *Botaniska Notiser*, **117**, 323–54.

Pickett, S. T. A. (1976). Succession: an evolutionary interpretation. *The American Naturalist*, **110**, 107–19.

Pickett, S. T. A. (1988). Space-for-time substitution as an alternative to long-term studies. In *Long-term Studies in Ecology: Approaches and Alternatives*, ed. G. E. Likens, pp. 110–35. New York: Springer Verlag.

Pickett, S. T. A., Collins, S. L. & Armesto, J. J. (1987a). Models, mechanisms and pathways of succession. *Botanical Review*, **53**, 336–71.

Pickett, S. T. A., Collins, S. L. & Armesto, J. J. (1987b). A hierarchical consideration of causes and mechanisms of succession. *Vegetatio*, **69**, 109–14.

Pickett, S. T. A. & McDonnell, M. J. (1989). Changing perspectives in community dynamics: a theory of successional forces. *Trends in Ecology and Evolution*, **4**, 241–5.

Pickett, S. T. A. & White, P. S. (1985). *The Ecology of Natural Disturbance and Patch Dynamics*. London: Academic Press.

Pielou, E. C. (1975). *Ecological Diversity*. New York: Wiley.

Pignatelli, O. & Bleuler, M. (1988). Anni caratteristici come indicatori di attacchi della tortice grigia del larice (*Zeiraphera diniana* Gn.). *Dendrochronologia*, **6**, 163–70.

Pineda, F. D., Nocolas, J. P., Ruiz, M., Peco, B. & Beraldez, F. G. (1981). Ecological succession in oligotrophic pastures in central Spain. *Vegetatio*, **44**, 165–76.

Porter, S. C. (1981a). Glaciological evidence of Holocene climatic change. In *Climate and History*, eds. T. M. L. Wigley, M. J. Ingram & G. Farmer, pp. 82–110. Cambridge: Cambridge University Press.

Porter, S. C. (1981b). Recent glacier variations and volcanic eruptions. *Nature*, **291**, 139–42.

Porter, S. C. (1981c). Lichenometric studies in the Cascade Range of Washington: Establishment of *Rhizocarpon geographicum* growth curves at Mount Rainier. *Arctic and Alpine Research*, **13**, 11–23.

Porter, S. C. (1981d). The use of tephrochronology in the Quaternary Geology of the United States. In *Tephra Studies*, eds. S. Self & R.S.J. Sparks, pp. 135–60. Dordrecht: Reidel.

Porter, S. C. (1986). Pattern and forcing of northern hemisphere glacier variations during the last millenium. *Quaternary Research*, **26**, 27–48.

Porter, S. C. (1989). Late Holocene fluctuations of the fiord glacier system in Icy Bay, Alaska, U.S.A. *Arctic and Alpine Research*, **21**, 364–79.

Porter, S. C. & Denton, G. H. (1967). Chronology of Neoglaciation in the North American Cordillera. *American Journal of Science*, **265**, 177–210.

Post, A. & Streveler, G. (1976). The tilted forests: glaciological-geological implications of vegetated Neoglacial ice at Lituya Bay, Alaska. *Quaternary Research*, **6**, 111–7.

Prentice , I. C. (1977). Non-metric ordination methods in ecology. *Journal of Ecology*, **65**, 85–94.

Price, L. W. (1981). *Mountains and Man: A Study of Process and Environment*. Berkeley: University of California Press.

Price, R. J. (1971). The development and destruction of a sandur, Breidamerkurjökull, Iceland. *Arctic and Alpine Research*, **3**, 225–37.

Price, R. J. (1973). *Glacial and Fluvioglacial Landforms*. Edinburgh: Oliver & Boyd.

Price, R. J. (1980). Rates of geomorphic changes in proglacial areas. In *Timescales in Geomorphology*, ed. R. A. Cullingford, D. A. Davidson & J. Lewin, pp. 79–93. Chichester: Wiley.

Proctor, M. C. F. (1983). Sizes and growth-rates of thalli of the lichen *Rhizocarpon geographicum* on moraines of the Glacier de Valsorey, Valais, Switzerland. *The Lichenologist*, **15**, 249–61.

Rabassa, J., Rubulis, S. & Suarez, J. (1981). Moraine in-transit as parent material for soil development and the growth of Valdivian rain forest on moving ice: Casa Pangue Glacier, Mount Tronador (Lat. 41° 10'S) Chili. *Annals of Glaciology*, **2**, 97–102.

Rampton, V. (1970). Neoglacial fluctuations of the Natazhat and Klutlan Glaciers, Yukon Territory, Canada. *Canadian Journal of Earth Sciences*, **7**, 1236–63.

Rannie, W. F. (1977). A note on the effect of a glacier on the summer thermal climate of an ice-marginal area. *Arctic and Alpine Research*, **9**, 301–4.

Rapp, A. (1960). Recent development of mountain slopes in Kärkevagge and surroundings, northern Scandinavia. *Geografiska Annaler*, **42**, 65–200.

Raup, H. M. (1951). Vegetation and cryoplanation. *Ohio Journal of Science*, **51**, 105–16.

Raup, H. M. (1957). Vegetation adjustment to the instability of the site. *Proceedings and Papers of the 6th Technical Meeting of the International Union for the Conservation of Nature and Natural Resources (Edinburgh, June 1956)*, 36–48.

Raup, H. M. (1971). The vegetational relations of weathering, frost action and patterned ground processes. *Meddelelser øm Grønland*, **194**, 1–92.

Raup, H. M. (1981). Physical disturbance in the life of plants. In *Biotic Crises in Ecological and Evolutionary Time*, ed. M. H. Nitecki, pp. 39–52. New York: Academic Press.

Reheis, M. J. (1975). Source, transport and deposition of debris on Arapaho Glacier, Front Range, Colorado, U.S.A. *Journal of Glaciology*, **14**, 407–20.

Reid, J. R. (1969). Effects of a debris slide on "Sioux Glacier", south Alaska. *Journal of Glaciology*, **8**, 353–67.

Reiners, W. A., Worley, I. A. & Lawrence, D. B. (1971). Plant diversity in a chronosequence at Glacier Bay, Alaska. *Ecology*, **52**, 55–69.

Remmert, H. (1980). *Arctic Animal Ecology*. Berlin: Springer Verlag.

Retzer, J. T. (1974). Alpine soils. In *Arctic and Alpine Environments*, eds. J. D. Ives & R. G. Barry, pp. 749–802. London: Methuen.

Rice, E. L. (1974). *Allelopathy*. New York: Academic Press.

Rice, E. L. (1979). Allelopathy – an update. *Botanical Review*, **45**, 17–109.

Richard, J. L. (1968). Les groupements végétaux de la Réserve D'Aletsch (Valais, Suisse) avec une carte en couleurs. *Matériaux pour le levé Géobotanique de la Suisse*, **51**, 1–30. (Berne: Editions Hans Huber)

Richard, J. L. (1973). Dynamique de la végétation au bord du grand glacier d'Aletsch (Alpes suisses). *Bericht der Schweizerischen Botanischen Gesellschaft* (Bern), **83**, 159–74.

Richard, J. L. (1975). Dynamique de la végétation au bord du grand glacier d'Aletsch (Alpes suisses). In *Sukzessionsforschung*, ed. W. Schmidt, pp. 189–209. Vaduz: J. Cramer.

Richards, K. (1984). Some observations on suspended sediment dynamics in Storbregrova, Jotunheimen. *Earth Surface Processes and Landforms*, **9**, 101–12.

Rieger, S. (1983). *The Genesis and Classification of Cold Soils*. New York: Academic Press.

Riezebos, P. A., Boulton, G. S., van der Meer, J. J. M., Ruegg, G. H. J., Beets, D. J., Castel, I. I. Y., Hart, J., Quinn, I., Thornton, M. & van der Wateren, F. M. (1986). Products and effects of modern eolian activity on a nineteenth-century glacier-pushed ridge in West Spitzbergen, Svalbard. *Arctic and Alpine Research*, **18**, 389–96.

Risser, P. G., Karr, J. R. & Forman, R. T. T. (1984). *Landscape Ecology: Directions and Approaches*. Champaign, Illinois: Illinois Natural History Society. (Illinois Natural History Survey Special Publication No. 2)

Roberts, D. W. (1987). A dynamical systems perspective on vegetation theory. *Vegetatio*, **69**, 27–33.

Rodbell, D. T. (1990). Soil-age relationships on Late Quaternary moraines, Arrowsmith Range, Southern Alps, New Zealand. *Arctic and Alpine Research*, **22**, 355–65.

Romans, J. C. C., Robertson, L. & Dent, D. L. (1980). The micromorphology of young

soils from south-east Iceland. *Geografiska Annaler*, **62(A)**, 93–103.
Rosswall, T., Veum, A. K. & Karenlampi, L. (1975). Plant litter decomposition at Fennoscandian tundra sites. In *Fennoscandian Tundra Ecosystems, Part I, Plants and Microorganisms*, pp. 268–78. Berlin: Springer Verlag.
Röthlisberger, F. (1976). Gletscher- und Klimaschwankungen im Raum Zermatt, Ferpècle und Arolla. *Die Alpen: Zeitschrift der Schweizer Alpen Club*, **52(3–4)**, 59–152.
Röthlisberger, F. (1986). *1000 Jahre Gletschergeschichte der Erde*. Aarau: Verlag Sauerländer.
Röthlisberger, F. & Geyh, M. A. (1985). Glacier variations in Himalayas and Karakorum. *Zeitschrift für Gletscherkunde und Glazialgeologie*, **21**, 237–49.
Röthlisberger, F., Haas, P., Holzhauser, H., Keller, W., Bircher, W. & Renner, F. (1980). Holocene climatic fluctuations – radiocarbon dating of fossil soils (fAh) and woods from moraines and glaciers in the Alps. *Geographica Helvetica*, **35(5)**, 21–52.
Röthlisberger, F. & Schneebeli, W. (1979). Genesis of lateral moraine complexes, demonstrated by fossil soils and trunks; indicators of postglacial climatic fluctuations. In *Moraines and Varves: Origin, Genesis, Classification*, ed. Ch. Schlüchter, pp. 387–419. Rotterdam: Balkema.
Röthlisberger, H. & Iken, A. (1981). Plucking as an effect of water-pressure variations at the glacier bed. *Annals of Glaciology*, **2**, 57–62.
Röthlisberger, H. & Lang, H. (1987). Glacial hydrology. In *Glacio-fluvial Sediment Transfer: An Alpine Perspective*, eds. A. M. Gurnell & M. J. Clark, pp. 207–84. Chichester: Wiley.
Rudolph, E. D. (1966). Ecological succession: a summary. In *Soil Development and Ecological Succession in a Deglaciated Area of Muir Inlet, Southeast Alaska*, ed. A. Mirsky, pp. 163–7. Columbus, Ohio: Ohio State University Research Foundation. (Institute of Polar Studies, Report No. 20)
Runge, E. C. A. (1973). Soil development sequences and energy models. *Soil Science*, **115**, 183–93.
Ryder, J. M. & Thomson, B. (1986). Neoglaciation in the southern Coast Mountains of British Columbia: chronology prior to the late Neoglacial maximum. *Canadian Journal of Earth Science*, **23**, 273–87.
Rydin, H. & Borgegård, S. O. (1988). Primary succession over sixty years on hundred-year old islets in Lake Hjälmaren, Sweden. *Vegetatio*, **77**, 159–68.
Ryvarden, L. (1971). Studies in seed dispersal I. Trapping of diaspores in the alpine zone at Finse, Norway. *Norwegian Journal of Botany*, **18**, 215–26.
Ryvarden, L. (1975). Studies in seed dispersal II. Winter-dispersed species at Finse, Norway. *Norwegian Journal of Botany*, **22**, 21–4.
Salzberg, K., Fredriksson, S. & Webber, P. J., eds. (1987). Restoration and Vegetation Succession in Circumpolar Lands. *Arctic and Alpine Research*, **19**, 337–577. (Proceedings 7th Conference Comité Arctique Internationale, Reykjavik, 1986).
Sandgren, C. D. & Noble, M. G. (1978). A floristic survey of a subalpine meadow on Mt Wright, Glacier Bay National Monument, Alaska. *Northwest Science*, **52**, 329–36.
Schoener, T. W. (1983). Field experiments on interspecific competition. *American Naturalist*, **122**, 240–85.
Schoenike, R. E. (1958). Influence of mountain avens (*Dryas drummondii*) on the growth of young cottonwoods (*Populus trichocarpa*) at Glacier Bay, Alaska. *Proceedings of the Minnesota Academy of Science*, pp. 25–6, 55–8.
Schoenike, R. E. (1984). The role of *Dryas drummondii* in plant succession on newly deglaciated terrain at Glacier Bay, Alaska. In *Proceedings 1st Glacier Bay Science Symposium*, eds. J. D. Wood, M. Gladziszewski, I. A. Worley & G. Vequist, pp. 27–8. Atlanta, Georgia: United States Department of the Interior, National Parks Service.
Schreckenthal-Schimitschek, G. (1933). Klima, Boden und Holzarten an der Wald- und

Baumgrenze in einzelnen Gebieten Tirols. *Veröffentlichungen des Museum Ferdinandeum* (Innsbruck), **13**, 115–252.

Schreiber, K. F. (1990). The history of landscape ecology in Europe. In *Changing Landscapes: An Ecological Perspective*, eds. I. S. Zonneveld & R. T. T. Forman, pp. 21–33. New York: Springer Verlag.

Schroeder-Lanz, H. (1983). Establishing lichen growth curves by repeated size (diameter) measurements of lichen individua in a test area – a mathematical approach. In *Late- and Postglacial Oscillations of Glaciers: Glacial and Periglacial Forms*, ed. H. Schroeder-Lanz, pp. 393–409. Rotterdam: Balkema.

Schubiger-Bossard, C. M. (1988). Die Vegetation des Rhonegletschervorfeldes, Ihre Sukzession und Naturräumliche Gliederung. *Beiträge zur Geobotanischen Landesaufnahme der Schweiz*, **64**, 1–228. (Teufen AR, Switzerland: F. Flück-Wirth)

Schweingruber, F. H. (1988). *Tree Rings: Basics and Applications of Dendrochronology*. Dordrecht: Kluwer.

Scott, A. J. (1990). Changes in vegetation on Heard Island 1947–48. In *Antarctic Ecosystems : Ecological Change and Conservation*, eds. K. R. Kerry & G. Hempel, pp. 61–76. Berlin: Springer Verlag.

Scott, R. W. (1974a). The effect of snow duration on alpine plant community composition and distribution. In *Icefield Ranges Research Project, Scientific Results: Volume 4*, eds. V. C. Bushnell & M. G. Marcus, pp. 307–18. New York & Montreal: American Geographical Society & Arctic Institute of North America.

Scott, R. W. (1974b). Successional patterns on moraines and outwash of the Frederika Glacier, Alaska. In *Icefield Ranges Research Project, Scientific Results: volume 4*, eds. V. C. Bushnell & M. G. Marcus, pp. 319–29. New York & Montreal: American Geographical Society & Arctic Institute of North America.

Selleck, G. W. (1960). The climax concept. *Botanical Review*, **26**, 534–45.

Seppelt, R. D., Broady, P. A., Pickard, J. & Adamson, D. A. (1988). Plants and landscape in the Vestfold Hills, Antarctica. *Hydrobiologia*, **165**, 185–96.

Shakesby, R. A. (1985). Geomorphological effects of jökulhlaups and ice-dammed lakes, Jotunheimen, Norway. *Norsk Geografisk Tidsskrift*, **39**, 1–16.

Shakesby, R. A. & Matthews, J. A. (1987). Frost weathering and rock platform erosion on periglacial lake shorelines: a test of a hypothesis. *Norsk Geologisk Tidsskrift*, **67**, 197–203.

Sharp, M. (1984). Annual moraine ridges at Skálafellsjökull, south-east Iceland. *Journal of Glaciology*, **30**, 82–93.

Sharp, M. (1985a). Sedimentation and stratigraphy at Eyjabakkajökull – an Icelandic surging glacier. *Quaternary Research*, **24**, 268–84.

Sharp, M. (1985b). "Crevasse-fill" ridges – a landform type characteristic of surging glaciers? *Geografiska Annaler*, **67(A)**, 213–20.

Sharp, M. & Dugmore, A. (1985). Holocene glacier fluctuations in eastern Iceland. *Zeitschrift für Gletscherkunde und Glazialgeologie*, **21**, 341–9.

Sharp, M., Gemmell, J. C. & Tison, J.-L. (1989). Structure and stability of the former subglacial drainage system of the Glacier de Tsanfleuron, Switzerland. *Earth Surface Processes and Landforms*, **14**, 119–34.

Sharp, R. P. (1958). The latest major advance of Malaspina Glacier, Alaska. *Geographical Review*, **48**, 16–26.

Shaw, J. (1977). Till deposited in arid polar environments. *Canadian Journal of Earth Science*, **14**, 1239–45.

Shaw, J. (1985). Subglacial and ice marginal environments. In *Glacial Sedimentary Environments*, eds. G. M. Ashley, J. Shaw & N. D. Smith, pp. 7–84. Tulsa, Oklahoma: Society of Economic Palaeontologists and Mineralogists. (SEPM Short Course No. 16)

Shimwell, D. W. (1971). *The Description and Classification of Vegetation*. London:

Sidgwick & Jackson.

Shroder, J. F. Jnr (1980). Dendrogeomorphology: review and new techniques of tree-ring dating. *Progress in Physical Geography*, **4**, 161–88.

Shugart, H. H. (1984). *A Theory of Forest Dynamics: The Ecological Implications of Forest Succession Models*. New York: Springer Verlag.

Shugart, H. H., West, D. C. & Emanuel, W. R. (1981). Patterns and dynamics of forests: an application of simulation models. In *Forest Succession: Concepts and Application*, eds. D. C. West, H. H. Shugart & D. B. Botkin, pp. 74–94. New York: Springer Verlag.

Shure, D. J. & Ragsdale, H. L. (1977). Patterns of primary succession on granite outcrop surfaces. *Ecology*, **58**, 993–1006.

Sidle, R. C. & Milner, A. M. (1989). Stream development in Glacier Bay National Park, Alaska, U.S.A. *Arctic and Alpine Research*, **21**, 350–63.

Sigafoos, R. S. (1951). Soil instability in tundra vegetation. *Ohio Journal of Science*, **51**, 281–98.

Sigafoos, R. S. & Hendricks, E. L. (1961). Botanical evidence for the modern history of Nisqually Glacier, Washington. *United States Geological Survey, Professional Paper*, 387-A, 1–20.

Sigafoos, R. S. & Hendricks, E. L. (1969). The time interval between stabilization of alpine glacial deposits and establishment of tree seedlings. *United States Geological Survey, Professional Paper*, 650-B, 89–93.

Sigafoos, R. S. & Hendricks, E. L. (1972). Recent activity of glaciers of Mount Rainier, Washington. *United States Geological Survey, Professional Paper*, 387-B, 1–24.

Simonson, R. W. (1959). Outline of a generalized theory of soil genesis. *Soil Science Society of America, Proceedings*, **23**, 152–6.

Simonson, R. W. (1978). A multi-process model of soil genesis. In *Quaternary Soils*, ed. W. C. Mahaney, pp. 1–25. Norwich: Geo Abstracts Ltd.

Small, R. J. (1983). Lateral moraines of Glacier de Tsidjiore Nouve: form, development, and implications. *Journal of Glaciology*, **29**, 250–9.

Small, R. J. (1987a). Englacial and supraglacial sediment: transport and deposition. In *Glacio-fluvial Sediment Transfer: An Alpine Perspective*, eds. A. M. Gurnell & M. J. Clark, pp. 111–45. Chichester: Wiley.

Small, R. J. (1987b). The glacial sediment system: an alpine perspective. In *Glacio-fluvial Sediment Transfer: An Alpine Perspective*, eds. A. M. Gurnell & M. J. Clark, pp. 199–203. Chichester: Wiley.

Smeck, N. E., Runge, E. C. A. & Mackintosh, E. E. (1983). Dynamics and genetic modelling of soil systems. In *Pedogenesis and Soil Taxonomy. I. Concepts and Interactions*, eds. L. P. Wilding, N. E. Smeck & G. F. Hall, pp. 51–81. Amsterdam: Elsevier.

Smith, D. G. (1976). Effect of vegetation on lateral migration of anastomosed channels of a glacial meltwater river. *Geological Society of America Bulletin*, **87**, 857–60.

Smith, N. D. (1985). Proglacial fluvial environment. In *Glacial Sedimentary Environments*, eds. G. M. Ashley, J. Shaw & N. D. Smith, pp. 85–134. Tulsa, Oklahoma: Society of Economic Palaeontologists and Mineralogists. (SEPM Short Course No. 16)

Smith, N. D. & Ashley, G. (1985). Proglacial lacustrine environment. In *Glacial Sedimentary Environments*, eds. G. M. Ashley, J. Shaw & N. D. Smith, pp. 135–216. Tulsa, Oklahoma: Society of Economic Palaeontologists and Mineralogists. (SEPM Short Course No. 16)

Smith, R. I. Lewis (1972). Vegetation of the South Orkney Islands with particular reference to Signy Island. *British Antarctic Survey Scientific Reports*, **68**, 1–124.

Smith, R. I. Lewis (1982). Plant succession and re-exposed moss banks on a deglaciated headland in Arthur Harbour, Anvers Island. *British Antarctic Survey Bulletin*, **51**, 193–9.

Smith, R. I. Lewis (1984). Terrestrial plant biology of the sub-Antarctic and Antarctic. In

Antarctic Ecology, Volume I, ed. R. M. Laws, pp. 61–162. London: Academic Press.

Smith, R. I. Lewis (1985a). A unique community of pioneer mosses dominated by *Pterygoneurum* cf. *ovatum* in the Antarctic. *Journal of Bryology*, **13**, 509–14.

Smith, R. I. Lewis (1985b). Studies on plant colonization and community development in Antarctic fellfields. *British Antarctic Survey Bulletin*, **68**, 109–13.

Smith, R. I. Lewis (1987). The bryophyte propagule bank of Antarctic fellfield soils. *Symposia Biologica Hungarica*, **35**, 233–45.

Smith, R. I. Lewis (1990). Signy Island as a paradigm of biological and environmental change in Antarctic terrestrial ecosystems. In *Antarctic Ecology: Ecological Change and Conservation*, eds K. R. Kerry & G. Hempel, pp. 32–50. Berlin: Springer Verlag.

Smith, R. I. Lewis (1991). Bryophyte propagule banks: a case study of an Antarctic fellfield soil. In *Primary Succession on Land*, eds. J. Miles & D. H. Walton, Oxford: Blackwell Scientific Publications. (in press)

Smith, R. I. Lewis & Coupar, A. M. (1987). The colonization potential of bryophyte propagules in Antarctic fellfield soils. *Comité National Francais des Recherches Antarctiques*, **58**, 189–204.

Solomina, O. N. (1989). Phytocoenic method. In *Moraines as a Source of Glaciological Information*, eds. L. R. Serebryanny, A. V. Orlov & O. N. Solomina, pp. 139–146. Moscow: Nauka. [in Russian]

Sommerville, P., Mark, A. F. & Wilson, J. B. (1982). Plant succession on moraines of the upper Dart Valley, southern South Island, New Zealand. *New Zealand Journal of Botany*, **20**, 227–44.

Sondheim, M. W. & Standish, J. T. (1983). Numerical analysis of a chronosequence, including assessment of variability. *Canadian Journal of Soil Science*, **63**, 501–17.

Souchez, R. A. & Lemmens, M. M. (1987). Solutes. In *Glacio-fluvial Sediment Transfer: An Alpine Perspective*, eds. A. M. Gurnell & M. J. Clark, pp. 285–303. Chichester: Wiley.

Souchez, R. A. & Lorrain, R. D. (1987). The subglacial sediment system. In *Glacio-fluvial Sediment Transfer: An Alpine Perspective*, eds. A. M. Gurnell & M. J. Clark, pp. 147–64. Chichester: Wiley.

Sousa, W. P. (1984). The role of disturbance in natural communities. *Annual Review of Ecology and Systematics*, **15**, 353–91.

Southwood, T. R. E. (1977). Habitat, the templet for ecological strategies. *Journal of Animal Ecology*, **46**, 337–65.

Southwood, T. R. E. (1988). Tactics, strategies and templets. *Oikos*, **53**, 3–18.

Spaltenstein, H. & Ugolini, F. C. (1988). Soil dynamics in a short-term chronosequence in spruce-hemlock forest in SE Alaska. *Agronomy Abstracts*, 232.

Spence, J. R. (1981). Comments on the cryptogam vegetation in front of glaciers in the Teton Range. *The Bryologist*, **84**, 564–8.

Spence, J. R. (1985). A floristic analysis of Neoglacial deposits in the Teton Range, Wyoming, U.S.A. *Arctic and Alpine Research*, **17**, 19–30.

Spence, J. R. (1989). Plant succession on glacial deposits of Mount Kenya, East Africa. In *Quaternary and Environmental Research on East African Mountains*, ed. W. C. Mahaney, pp. 279–90. Rotterdam: Balkema.

Spence, J. R. & Shaw, R. J. (1983). Observations on alpine vegetation near Schoolroom Glacier, Teton Range, Wyoming. *Great Basin Naturalist*, **43**, 483–91.

Sprague, R. & Lawrence, D. B. (1959a). The fungi of deglaciated Alaskan terrain of known age (Part I). *Washington State University, Research Studies*, **27**, 110–28.

Sprague, R. & Lawrence, D. B. (1959b). The fungi of deglaciated Alaskan terrain of known age (Part II). *Washington State University, Research Studies*, **27**, 214–29.

Sprague, R. & Lawrence, D. B. (1960). The fungi of deglaciated Alaskan terrain of known age (Part III). *Washington State University, Research Studies*, **28**, 1–20.

Staschel, R. (1989). Die Kartierung von Weiserjahren als Methode der

Dendroökologischen Differenzierung von Baumstandorten. *Dendrochronologia*, **7**, 83–95.

Stephens, C. G. (1947). Functional synthesis in pedogenesis. *Transactions of the Royal Society of South Australia*, **71**, 168–81.

Stephens, F. R. (1969). A forested ecosystem on a glacier in Alaska. *Arctic*, **22**, 441–4.

Stephens, F. R. (unpublished) Primary ecosystems developing below receding glaciers in southeastern Alaska. [F. R. Stephens is deceased; he wrote this paper *ca* 1969. A version of the manuscript, edited by E. A. Alexander, will be published in Watershed '91: Proceedings of the Watershed '91 conference, April 16–17, 1991. Juneau, Alaska: U.S. Dept. of Agriculture Forest Service.]

Stevens, P. R. (1963). *A chronosequence of soils and vegetation near the Franz Josef Glacier.* M.Agr.Sci. thesis, University of Canterbury, New Zealand (Lincoln College).

Stevens, P. R. (1968). *A chronosequence of soils near the Franz Josef Glacier.* Ph.D. Thesis, University of Canterbury, New Zealand (Lincoln College).

Stevens, P. R. & Walker, T. W. (1970). The chronosequence concept and soil formation. *Quarterly Review of Biology*, **45**, 333–50.

Stork, A. (1963). Plant immigration in front of retreating glaciers, with examples from the Kebnekajse area, northern Sweden. *Geografiska Annaler*, **45**, 1–22.

Streten, N. A., Ishikawa, N. & Wendler, G. (1974). Some observations of the local wind regime on an Alaskan Arctic glacier. *Archiv für Meteorologie, Geophysik und Bioklimatologie, Ser B*, **22**, 337–50.

Streten, N. A. & Wendler, G. (1968). Some observations of Alaskan glacier winds in midsummer. *Arctic*, **21**, 98–102.

Stuiver, M. (1978). Radiocarbon timescale tested against magnetic and other dating methods. *Nature*, **273**, 271–4.

Stuiver, M. (1982). A high precision calibration of the A. D. radiocarbon time scale. *Radiocarbon*, **24**, 1–26.

Sugden, D. (1982). *Arctic and Antarctic.* Oxford; Basil Blackwell.

Sugden, D. & John, B. S. (1976). *Glaciers and Landscape.* London: Edward Arnold.

Sukachev, V. & Dylis, N. (1964). *Fundamentals of Forest Biogeocoenology.* Edinburgh: Oliver & Boyd.

Sutherland, D. G. (1984). Modern glacial characteristics as a basis for inferring former climates with particular reference to the Loch Lomond Stadial. *Quaternary Science Reviews*, **3**, 291–309.

Svoboda, J. & Henry, G. H. R. (1987). Succession in marginal Arctic environments. *Arctic and Alpine Research*, **19**, 373–84.

Tansley, A. G. (1920). The classification of vegetation and the concept of development. *Journal of Ecology*, **8**, 118–49.

Tansley, A. G. (1929). Succession, the concept and its value. *Proceedings of the International Congress of Plant Science, Ithaca*, 677–86.

Tansley, A. G. (1935). The use and abuse of vegetational concepts and terms. *Ecology*, **16**, 284–307.

Tansley, A. G. (1949). *The British Islands and their Vegetation.* Cambridge: Cambridge University Press.

Tarr, R. S. & Martin, L. (1914). *Alaskan Glacier Studies.* Washington D.C.: National Geographic Society.

Taylor, D. R., Aarssen, L. W. & Loehle, C. (1990). On the relationship between r/k selection and environmental carrying capacity: a new habitat templet for plant life history strategies. *Oikos*, **58**, 239–50.

Tedrow, J. C. F. & Ugolini, F. C. (1966). Antarctic soils. In *Antarctic Soils and Soil Forming Processes*, ed. J. C. F. Tedrow, pp. 161–77. Washington D.C.: American Geophysical Union. (Antarctic Research Series No. 8)

Ter Braak, C. J. F. (1986). Canonical correspondence analysis: a new eigenvector technique for multivariate direct gradient analysis. *Ecology*, **67**, 1167–79.

Ter Braak, C. J. F. (1987). The analysis of vegetation-environment relationships by canonical correspondence analysis. *Vegetatio*, **69**, 69–77.

Tessier, L., Coûteaux, M. & Guiot, J. (1986). An attempt at an absolute dating of a sediment from the last glacial recurrence through correlations between pollenanalytical and tree-ring data. *Pollen et Spores*, **28**, 61–76.

Theakstone, W. H. (1982). Sediment fans and sediment flows generated by snowmelt: observations at Austerdalsisen, Norway. *Journal of Geology*, **90**, 583–8.

Theakstone, W. H. & Knighton, D. (1979). The moss *Aongstroemia longipes*, an environmentally sensitive colonizer of sediments at Austerdalsisen, Norway. *Arctic and Alpine Research*, **11**, 353–6.

Thompson, A. & Jones, A. (1986). Rates and causes of proglacial river terrace formation in southeast Iceland: an application of lichenometric dating techniques. *Boreas*, **15**, 231–46.

Thompson, K. (1987). The resource ratio hypothesis and the meaning of competition. *Functional Ecology*, **1**, 297–315.

Thompson, R. (1984). A global view of palaeomagnetic results from wet lake sediments. In *Lake Sediments and Environmental History*, eds. E. Y. Haworth & J. W. G. Lund, pp. 145–64. Leicester: Leicester University Press.

Thompson, R. & Oldfield, F. (1986). *Environmental Magnetism*. London: Allen & Unwin.

Thorarinsson, S. (1956). On the variations of Svinafellsjökull, Skaftafellsjökull and Kviárjökull in Öraefi. *Jökull*, **6**, 1–15.

Thorarinsson, S. (1964). On the age of the terminal moraines of Brúarjökull and Hálsajökull: a tephrochronological study. *Jökull*, **14**, 67–75.

Thorarinsson, S. (1966). The age of the maximum postglacial advance of Hagafellsjökul eystri: a tephrochronological study. *Jökull*, **16**, 207–10.

Thorarinsson, S. (1981). The application of tephrochronology in Iceland. In *Tephra Studies*, eds. S. Self & R. S. J. Sparks, pp. 109–34. Dordrecht: Reidel.

Thorn, C. E. (1976). Quantitative evaluation of nivation in the Colorado Front Range. *Geological Society of America Bulletin*, **87**, 1169–78.

Thorn, C. E. (1988). Nivation: a geomorphic chimera. In *Advances in Periglacial Geomorphology*, ed. M. J. Clark, pp. 3–31. Chichester: Wiley.

Thorn, C. E. & Hall, K. J. (1980). Nivation: an arctic-alpine comparison and reappraisal. *Journal of Glaciology*, **25**, 109–24.

Thorn, C. E. & Loewenherz, D. S. (1987). Spatial and temporal trends in alpine periglacial studies: implications for paleo reconstruction. In *Periglacial Processes and Landforms in Britain and Ireland*, ed. J. Boardman, pp. 57–65. Cambridge: Cambridge University Press.

Tilman, D. (1985). The resource-ratio hypothesis of plant succession. *American Naturalist*, **125**, 827–52.

Tilman, D. (1986). Resources, competition and the dynamics of plant communities. In *Plant Ecology*, ed. M. J. Crawley, pp. 51–75. Oxford: Blackwell Scientific Publications.

Tilman, D. (1987). On the meaning of competition and the mechanisms of competitive superiority. *Functional Ecology*, **1**, 304–15.

Tilman, D. (1988). *Plant Strategies and the Dynamics and Structure of Plant Communities*. Princeton, New Jersey: Princeton University Press.

Tipping, R. M. (1988). The recognition of glacial retreat from palynological data: a review of recent work in the British Isles. *Journal of Quaternary Science*, **3**, 171–82.

Tisdale, E. W., Fosberg, M. A. & Poulton, C. E. (1966). Vegetation and soil development on a recently glaciated area near Mount Robson, British Columbia. *Ecology*, **47**, 517–23.

Tollner, H. (1931). Gletscherwinde in den Ostalpen. *Meteorologische Zeitschrift* (Braunschweig), **48**, 414–21.

Tollner, H. (1934). Gletscherwinde und ihr Einfluß auf die Pflanzenwelt. *Österreichische*

Botanische Zeitschrift, **81**, 64–5.

Tollner, H. (1935). Gletscherwinde auf der Pasterze. *Jahresbericht des Sonnblick-Vereines*, **44**, 38–54.

Torrent, J. & Nettleton, W. D. (1978). Feedback processes in soil genesis. *Geoderma*, **20**, 281–7.

Trautman, M. B. (1966). Birds. In *Soil Development and Ecological Succession in a Deglaciated Area of Muir Inlet, Southeast Alaska*, ed. A. Mirsky, pp. 121–43. Columbus, Ohio: Ohio State University Research Foundation. (Institute of Polar Studies, Report No. 20)

Tricart, J. & Cailleux, A. (1972). *Introduction to Climatic Geomorphology*. London: Longman.

Troll, C. (1939a). Luftbildplan und ökologische Bodenforschung. *Zeitschrift der Gesellschaft für Erdkunde zu Berlin*, 1939, Nr 7/8, 241–98.

Troll, C. (1939b). Das Pflanzenkleid des Nanga Parbat. Begleitworte zer Vegetationskarte der Nanga Parbat-Gruppe (Nordwest-Himalaya) 1: 50,000. *Wissenschaftliche Veröffentlichungen. Museum für Länderkunde zu Leipzig, N. F.*, **7**, 149–93. [not seen]

Troll, C. (1958). Structure soils, solifluction, and frost climates of the earth. *United States Army, Snow, Ice and Permafrost Establishment, Translation*, **43**, 1–121.

Troll, C. (1963). Über Landschafts-Sukzession. *Arbeiten zur Rheinschen Landeskunde Bonn*, **19**, 5–12.

Troll, C. (1971). Landscape ecology (geoecology) and biogeocenology – a terminological study. *Geoforum*, **8**, 43–6.

Troll, C. (1972). Geoecology and the world-wide differentiation of high-mountain ecosystems. In *Geoecology of the High-mountain Regions of Eurasia*, ed. C. Troll, 1–16. Wiesbaden: Franz Steiner Verlag.

Turmanina, V. I. & Volodina, E. R. (1978). Dynamics of the vegetation in the Mount Elbrus Region, U.S.S.R. *Arctic and Alpine Research*, **10**, 325–34.

Turmanina, V. I. & Volodina, E. R. (1979). Seasonal and long-term dynamics of vegetation near the Dzhankuat Glacier, Caucasus. *Bulletin Moscow O. for the Exploration of Nature. Biological Division*, **84**, 64–75. [in Russian]

Tüxen, R. (1933). Klimaxprobleme des nordwesteuropaischen Festlandes. *Nederlandsch Kruidkundig Archief*, **43**, 293–309.

Ugolini, F. C. (1966). Soils. In *Soil Development and Ecological Succession in a Deglaciated Area of Muir Inlet, Southeast Alaska*, ed. A. Mirsky, pp. 29–72. Columbus, Ohio: Ohio State University Research Foundation. (Institute of Polar Studies, Report No. 20)

Ugolini, F. C. (1968). Soil development and alder invasion in a recently deglaciated area of Glacier Bay, Alaska. In *The Biology of Alder*, eds. J. M. Trappe, J. F. Franklin, R. F. Tarrant & G. M. Hansen, pp. 115–40. Portland, Oregon: United States Forestry Service.

Ugolini, F. C., Bormann, B. T. & Bowers, F. H. (in preparation). The effect of treefall on forest soil development in southeast Alaska.

Ugolini, F. C. & Edmonds, R. L. (1983). Soil biology. In *Pedogenesis and Soil Taxonomy, I. Concepts and Interactions*, eds. L. P. Wilding, N. E. Smeck & G. F. Hall, pp. 193–231. Amsterdam: Elsevier.

Ugolini, F. C. & Mann, D. H. (1979). Biopedological origin of peatlands in south east Alaska. *Nature*, **281**, 366–8.

Uhl, C. & Jordan, C. F. (1984). Succession and nutrient dynamics following forest cutting and burning in Amazonia. *Ecology*, **65**, 1476–90.

Underwood, A. J. (1986). The analysis of competition by field experiments. In *Community Ecology: Pattern and Process*, eds. J. Kikkawa & D. J. Anderson, pp. 240–68. Oxford: Blackwell Scientific Publications.

Urban, D. L., O'Neill, R. V. & Shugart, H. H. Jr. (1987). Landscape ecology. *Bioscience*, **37**, 119–27.

van der Maarel, E. (1988). Vegetation dynamics: patterns in time and space. *Vegetatio*, **77**, 7–19.

van der Maarel, E., de Cock, N. & de Wildt, E. (1985). Population dynamics of some major woody species in relation to long-term succession in the dunes of Voorne. *Vegetatio*, **61**, 209–19.

van der Valk, A. G. (1981). Succession in wetlands: a Gleasonian approach. *Ecology*, **62**, 688–96.

Van Vliet-Lanoë, B. (1985). Frost effects in soils. In *Soils and Quaternary Landscape Evolution*, ed. J. Boardman, pp. 117–58. Chichester: Wiley.

Van Vliet-Lanoë, B., Coutard, J. P. & Pissart, A. (1984). Structures caused by repeated freezing and thawing in various loamy sediments. A comparison of active, fossil and experimental data. *Earth Surface Processes and Landforms*, **12**, 467–73.

Veblen, T. T. & Ashton, S. R. (1978). Catastrophic influences on the vegetation of the Valdivian Andes, Chili. *Vegetatio*, **36**, 149–67.

Veblen, T. T., Ashton, S. R., Rubulis, S., Lorenz, D. C. & Cortes, M. (1989). *Nothofagus* stand development on in-transit moraines, Casa Pangue Glacier, Chili. *Arctic and Alpine Research*, **21**, 144–55.

Veblen, T. T. Schlegel, F. M. & Escobar, R. (1980). Structure and dynamics of old-growth *Nothofagus* forests in the Valdivian Andes. *Journal of Ecology*, **68**, 1–31.

Vere, D. M. & Benn, D. I. (1989). Structure and debris characteristics of medial moraines in Jotunheimen, Norway: implications for moraine classification. *Journal of Glaciology*, **35**, 276–80.

Vere, D. M. & Matthews, J. A. (1985). Rock glacier formation from a lateral moraine at Bukkeholsbreen, Jotunheimen, Norway: a sedimentological approach. *Zeitschrift für Geomorphologie, N.F.*, **29**, 397–415.

Vetaas, O. R. (1986). *Økologiske Faktorer i en Primærsuksesjon på Daterte Endmorener i Bødalen, Stryn*. Ph.D. Thesis, University of Bergen. (Botanical Institute)

Viereck, L. A. (1966). Plant succession and soil development on gravel outwash of the Muldrow Glacier, Alaska. *Ecological Monographs*, **36**, 181–99.

Viereck, L. A. (1968). Botanical dating of recent glacier activity in Western North America. In *Arctic and Alpine Environments*, eds. H. E. Wright Jnr & W. H. Osburn, pp. 189–204. Bloomington: Indiana University Press.

Villalba, R., Leiva, J. C., Rubulis, S., Suarez, J. & Lenzano, L. (1990). Climate, tree-ring, and glacial fluctuations in the Rio Frias Valley, Rio Negro, Argentina. *Arctic and Alpine Research*, **22**, 215–32.

Virtanen, A. I. (1957). Investigations on nitrogen fixation by alder. II. Associated culture of spruce and inoculated alder without combined nitrogen. *Physiologia Plantarum*, **10**, 164–9.

Virtanen, A. I. & Meittinen, J. K. (1952). Free amino acids in the leaves, roots and root nodules of the alder (*Alnus*). *Nature*, **170**, 283–4.

Virtanen, A. I., Moisio, T., Allison, R. M. & Burris, R. H. (1955). Fixation of nitrogen by excised nodules of the alder. *Acta Chemica Scandinavica*, **9**, 184–6.

Vreeken, W. J. (1975). Principal kinds of chronosequences and their significance in soil history. *Journal of Soil Science*, **26**, 378–94.

Walker, L. R. (1989). Soil nitrogen changes during primary succession on a floodplain in Alaska, U.S.A. *Arctic and Alpine Research*, **21**, 341–9.

Walker, L. R. (1991). Nitrogen fixers and species replacements in primary succession. In *Primary Succession on Land*, eds. J. Miles & D. H. Walton, Oxford: Blackwell Scientific Publications. (in press).

Walker, L. R. & Chapin III, F. S. (1986). Physiological controls over seedling growth in primary succession on an Alaskan floodplain. *Ecology*, **67**, 1508–23.

Walker, L. R. & Chapin III, F. S. (1987). Interactions amongst processes controlling successional change. *Oikos*, **50**, 131–5.

Walker, L. R., Zasada, J. C. & Chapin III, F. S. (1986). The role of life history processes in primary succession on an Alaskan floodplain. *Ecology*, **67**, 1243–53.

Walker, T. W. & Syers, J. K. (1976). The fate of phosphorus during pedogenesis. *Geoderma*, **15**, 1–19.

Ward, J. H. (1963). Hierarchical grouping to optimize an objective function. *Journal of the American Statistical Association*, **58**, 236–44.

Wardle, P. (1977). Plant communities of Westland National Park (New Zealand) and neighbouring lowland and coastal areas. *New Zealand Journal of Botany*, **15**, 323–98.

Wardle, P. (1980). Primary succession in Westland National Park and its vicinity. *New Zealand Journal of Botany*, **18**, 221–32.

Washburn, A. L. (1956). Classification of patterned ground and review of suggested origins. *Geological Society of America Bulletin*, **67**, 823–65.

Washburn, A. L. (1979). *Geocryology: A Survey of Periglacial Processes and Environments*. London: Edward Arnold.

Watanabe, T., Shiraiwa, T. & Ono, Y. (1989). Distribution of periglacial landforms in the Langtang Valley, Nepal Himalaya. *Bulletin of Glacier Research* (Japan), **7**, 209–20.

Watson, R. A. (1980). Landform development on moraines of the Klutlan Glacier, Yukon Territory, Canada. *Quaternary Research*, **14**, 50–9.

Watt, A. S. (1947). Pattern and process in the plant community. *Journal of Ecology*, **35**, 1–22.

Webber, P. J. (1974). Tundra primary productivity. In *Arctic and Alpine Environments*, eds. J. D. Ives & R. G. Barry, pp. 445–73. London: Methuen.

Webber, P. J. & Andrews, J. T. (1973). Lichenometry: a commentary. *Arctic and Alpine Research*, **5**, 295–302.

Webster, R. & Burrough, P. A. (1974). Multiple discriminant analysis in soil survey. *Journal of Soil Science*, **25**, 120–34.

Weidick, A. (1985). Review of glacier changes in West Greenland. *Zeitschrift für Gletscherkunde und Glazialgeologie*, **21**, 301–9.

Weisbrod, A. R. & Dragavon, J. (1979). Mammals and birds. In *Nooksack Cirque Natural History, Final Report*, eds. C. D. Oliver & R. J. Zasoski (principal investigators), pp. 96–129. U.S. National Park Service Report CX–9000–6–0418.

Welch, D. M. (1970). Substitution of space for time in a study of slope development. *Journal of Geology*, **78**, 234–9.

West, D. C., Shugart, H. H. & Botkin, D. B. (1981). Introduction. In *Forest Succession: Concepts and Application*, eds. D. C. West, H. H. Shugart & D. B. Botkin, pp. 1–9. New York: Springer Verlag.

Whalley, W. B. (1974). Rock glaciers and their formation as part of a glacier debris-transport system. *Reading Geographical Papers*, **24**, 1–60. (Department of Geography, University of Reading)

Whalley, W. B. (1982). A preliminary scanning electron microscope study of quartz grains from a dirt band in the Tuto ice tunnel, northwest Greenland. *Arctic and Alpine Research*, **14**, 355–60.

Whalley, W. B., Gordon, J. E. & Thompson, D. L. (1981). Periglacial features on the margins of a receding plateau ice cap, Lyngen, north Norway. *Journal of Glaciology*, **27**, 492–6.

Whalley, W. B. & Krinsley, D. H. (1974). A scanning electron microscope study of surface textures of quartz grains from glacial environments. *Sedimentology*, **21**, 87–105.

White, P. S. (1979). Pattern, process, and natural disturbance in vegetation. *Botanical Review*, **45**, 229–99.

White, S. E. (1972). Alpine subnival boulder pavements in the Colorado Front Range.

Geological Society of America Bulletin, **83**, 195–200.

Whitehouse, I. E., McSaveney, M. J., Kneupfer, P. L. K. & Chinn, T. J. H. (1986). Growth of weathering rinds on Torlesse sandstone, Southern Alps, New Zealand. In *Rates of Chemical Weathering of Rocks and Minerals*, eds. S. M. Colman & D. P. Dethier, pp. 419–35. Orlando, Florida: Academic Press.

Whitmore, T. C. (1982). On pattern and process in forests. In *The Plant Community as a Working Mechanism*, ed. E. I. Newman, pp. 45–59. Oxford: Blackwell Scientific Publications. (British Ecological Society, Special Publication No. 1.)

Whittaker, R. H. (1951). A criticism of the plant association and climatic climax concepts. *Northwest Science*, **25**, 17–31.

Whittaker, R. H. (1953). A consideration of climax theory: the climax as a population and pattern. *Ecological Monographs*, **23**, 41–78.

Whittaker, R. H. (1967). Gradient analysis of vegetation. *Biological Reviews of the Cambridge Philosophical Society*, **42**, 207–64.

Whittaker, R. H. (1974). Climax concepts and recognition. In *Vegetation Dynamics*, ed. R. Knapp, pp. 139–54. The Hague: Junk.

Whittaker, R. H. & Levin, S. A. (1977). The role of mosaic phenomena in natural communities. *Theoretical Population Biology*, **12**, 117–39.

Whittaker, R. J. (1985). *Plant Community and Plant Population Studies of a Successional Sequence: Storbreen Glacier Foreland, Jotunheimen, Norway*. Ph.D. Thesis, University of Wales. (University College Cardiff)

Whittaker, R. J. (1987). An application of detrended correspondence analysis and non-metric multidimensional scaling to the identification and analysis of environmental factor complexes and vegetation structures. *Journal of Ecology*, **75**, 363–76.

Whittaker, R. J. (1989). The vegetation of the Storbreen gletschervorfeld, Jotunheimen, Norway. III. Vegetation-environment relationships. *Journal of Biogeography*, **16**, 413–33.

Whittaker, R. J. (1991). The vegetation of the Storbreen gletschervorfeld, Jotunheimen, Norway. IV. Short-term vegetation change. *Journal of Biogeography*, **18**, 41–52.

Whittaker, R. J. (unpublished). Small-scale frost disturbance on the Storbreen glacier foreland, Norway.

Wijk, S. (1986a). Performance of *Salix herbacea* in an alpine snow-bed gradient. *Journal of Ecology*, **74**, 675–84.

Wijk, S. (1986b). Influence of climate and age on shoot increment in *Salix herbacea*. *Journal of Ecology*, **74**, 685–92.

Wilde, S. A. (1946). *Forest Soils and Forest Growth*. Waltham, Massachusetts: Chronica Botanica.

Williams, L. D. & Wigley, T. M. L. (1983). A comparison of evidence for late Holocene summer temperature variations in the northern hemisphere. *Quaternary Research*, **20**, 286–307.

Williams, P. F. & Rust, B. R. (1969). The sedimentology of a braided river. *Journal of Sedimentary Petrology*, **39**, 649–79.

Williams, R. S. Jnr (1986). Glacier inventories of Iceland: evaluation and use of sources of data. *Annals of Glaciology*, **8**, 184–91.

Wilson, M. J. (1975). Chemical weathering of some primary rock-forming minerals. *Soil Science*, **119**, 349–55.

Wilson, M. J. & McHardy, W. J. (1980). Experimental etching of a microcline perthite and implications regarding natural weathering. *Journal of Microscopy*, **120**, 291–302.

Wise, S. M. (1980). Caesium-137 and Lead-210: a review of the techniques and some applications in geomorphology. In *Timescales in Geomorphology*, eds. R. A. Cullingford, D. A. Davidson & J. Lewin, pp. 109–27. London: Wiley.

Wishart, D. (1969). An algorithm for hierarchical classification. *Biometrics*, **25**, 165–70.

Wójcik, G. (1973). The results of the meteorological investigations on the foreland of Skeidarárjökull. *Geographia Polonica*, **26**, 157–83.

Wood, D. M. & del Moral, R. (1987). Mechanisms of early primary succession in subalpine habitats on Mount St. Helens. *Ecology*, **68**, 780–90.

Woodward, F. I. (1987). *Climate and Plant Distribution*. Cambridge: Cambridge University Press.

Worley, I. A. (1973). The "black crust" phenomenon in upper Glacier Bay, Alaska. *Northwest Science*, **47**, 20–9.

Worsley, P. (1974a). On the significance of the age of a buried tree stump by Engabreen, Svartisen. *Norsk Polarinstitutt Årbok for 1972*, 111–7.

Worsley, P. (1974b). Recent "annual" moraine ridges at Austre Okstindbreen, Okstindan, North Norway. *Journal of Glaciology*, **13**, 265–77.

Worsley, P. (1981). Lichenometry. In *Geomorphological Techniques*, ed. A. Goudie, pp. 302–5. London: George Allen & Unwin.

Worsley, P. & Alexander, M. J. (1976a). Neoglacial palaeoenvironmental change at Engabrevatn, Svartisen Holandsfjord, North Norway. *Norges Geologiske Undersøkelse*, **321**, 37–66.

Worsley, P. & Alexander, M. J. (1976b). Glacier and environmental changes: Neoglacial data from the outermost moraine ridges at Engabreen, northern Norway. *Geografiska Annaler*, **58(A)**, 55–69.

Worsley, P. & Ward, M. R. (1974). Plant colonization of recent "annual" moraine ridges at Austre Okstindbreen, North Norway. *Arctic and Alpine Research*, **6**, 217–230.

Wright, H. E. Jnr (1980). Surge moraines of the Klutlan Glacier, Yukon Territory, Canada: origin, wastage, vegetation succession, lake development, and application to the Late-Glacial of Minnesota. *Quaternary Research*, **14**, 2–18.

Wright, H. E. Jnr & Osburn, W. H. (1968). *Arctic and Alpine Environments*. Bloomington, Indiana: Indiana University Press.

Wynn-Williams, D. D. (1985). The biota of a lateral moraine and hinterland of the Blue Glacier, South Victoria Land, Antarctica. *British Antarctic Survey Bulletin*, **66**, 1–5.

Wynn-Williams, D. D. (1986). Microbial colonization of Antarctic fellfield soils. In *Perspectives in Microbial Ecology*, eds. F. Megusar & M. Cantar, pp. 191–200. Ljubljana: Slovene Society for Microbiology. (Proceedings of the 4th International Symposium on Microbial Ecology, Ljubljana)

Wynn-Williams, D. D. (1988). Television image analysis of microbial communities in Antarctic fellfields. *Polarforschung*, **58**, 239–49.

Wynn-Williams, D. D. (1991). Microbial processes and initial stabilization of fellfield. In *Primary Succession on Land*, eds. J. Miles & D. H. Walton, Oxford: Blackwell Scientific Publications. (in press)

Yaalon, D. H. (1971). Soil forming processes in time and space. In *Paleopedology: Origin, Nature and Dating of Paleosols*, ed. D. H. Yaalon, pp. 29–39. Jerusalem: International Society of Soil Science & Israel Universities Press.

Yaalon, D. H. (1975). Conceptual models in pedogenesis: can soil-forming functions be solved? *Geoderma*, **14**, 189–205.

Yaalon, D. H. (1976). Conceptual models in pedogenesis – a reply. *Geoderma*, **16**, 263–4.

Yaalon, D. H. (1983). Climate, time and soil development. In *Pedogenesis and Soil Taxonomy. I. Concepts and Interactions*, eds. L. P. Wilding, N. E. Smeck & G. F. Hall, 233–51. Amsterdam: Elsevier.

Young, K. R. (1989), The tropical Andes as a morphoclimatic zone. *Progress in Physical Geography*, **13**, 13–22.

Zasoski, R. J. & Bardo, K. (1979). Soils and surficial geology. In *Nooksack Cirque Natural History, Final Report*, eds. C. D. Oliver & R. J. Zasoski (principal investigators), pp. 24–42. U.S. National Park Service Report CX–9000–6–0418.

Zollitsch, B. (1969). Die Vegetationsentwicklung im Pasterzenvorfeld. *Wissenschaftliche Alpenvereinsheft* (München), **21**, 267–90.

Zumbühl, H. J. (1980). *Die Schwankungen der Grindelwaldgletscher, in den Historischen Bild- und Schriftquellen des 12. bis 19. Jahrhunderts.* Basel: Birkhäuser Verlag.

Zumbühl, H., Budmiger, G. & Haeberli, W. (1981). Historical documents. In *Switzerland and Her Glaciers*, eds. P. Kasser & W. Haeberli. Berne: Kümmerly & Frey.

Index